COMMUNITY, ENVIRONMENT AND HEALTH: Geographic Perspectives

COMMUNITY, ENVIRONMENT AND HEALTH: Geographic Perspectives

edited by

MICHAEL V. HAYES
LESLIE T. FOSTER
HAROLD D. FOSTER

Western Geographical Series Volume 27

Department of Geography, University of Victoria
Victoria, British Columbia
Canada

1992 University of Victoria

Western Geographical Series, Volume 27

editorial address

Harold D. Foster, Ph.D.
Department of Geography
University of Victoria
Victoria, British Columbia
Canada

Publication of the Western Geographical Series has been generously supported bythe Leon and Thea Koerner Foundation, the Social Science Federation of Canada, the National Centre for Atmospheric Research, the International Geographical Union Congress, the University of Victoria, the Natural Sciences Engineering Research Council of Canada, the Institute of the North American West, and the British Columbia Ministry of Health and Ministry Responsible for Seniors.

COMMUNITY, ENVIRONMENT AND HEALTH: GEOGRAPHIC PERSPECTIVES

(Western geographical series; ISSN 0315-2022; v. 27)
Edited papers from a conference held at Simon Fraser University, Feb. 21-22, 1991.
Includes bibliographical references.
ISBN 0-919838-17-0

1. Medical geography--Canada--Congresses. 2. Environment health--Canada--Congresses. 3. Public health--Canada--Congresses. 4. Medical care--Canada--Congresses. I. Hayes, Michael V. II. Foster, Leslie T., 1947- III. Foster, Harold D., 1943- IV. University of Victoria (B.C.). Dept. of Geography. V. Series.
RA809.C64 1992 614.4'271 C92-091478-0

ACKNOWLEDGEMENTS

Several members of the Department of Geography, University of Victoria co-operated to ensure the successful publication of this volume of the Western Geographical Series. Special thanks are due to members of the technical services division under the direction of Ian Norie. Diane Macdonald undertook the very demanding task of typesetting, while cartography was in the expert hands of Ken Josephson. Their dedication and hard work is greatly appreciated.

The generous financial support of the British Columbia Ministry of Health and Ministry Responsible for Seniors, for both the symposium and this associated volume, is acknowledged with thanks.

University of Victoria
Victoria, British Columbia
June 25, 1992

Harold D. Foster
Series Editor

PREFACE

The papers in this collection were initially presented at a symposium entitled *Community, Environment and Health in British Columbia: Geographic Perspectives*, held at Simon Fraser University's Harbour Centre in Vancouver, February 21 and 22, 1991. The purpose of the symposium was to bring together persons from diverse backgrounds - academics, service providers, policy makers and community representatives - to explore themes and issues in health and health care with particular reference to their geographic aspects. Dr. Harold Foster of the University of Victoria, co-host of the symposium, invited half of those who presented papers. These were concerned primarily with the influence of the physical environment upon human health. I invited the others, which emphasized the influence of the social environment upon human health. Papers were written by both geographers and non-geographers. Although we suggested topics to authors (based on our knowledge of individual interest and experience), they were free to write about any subject they chose in keeping with the objective of the symposium. The various topics addressed at the symposium related to three basic themes; (i) the spatial distribution of disease, (ii) geophysical correlates of cancer and other diseases, and (iii) the interplay between geographic and political, economic, cultural and demographic dimensions of health care policy, planning and service delivery.

Contemporary frameworks for understanding the determinants of health[1-3] emphasize multidimensional, interconnected and dynamic relationships between the individual (or self) and the environment (or not-self) that impinge directly or indirectly upon individual well-being. One implication of such models is that interdisciplinary and intersectoral communication and collaboration between persons with diverse interests and backgrounds will be required to enrich our understanding of these relationships. But communication and understanding require at least a common language of discourse, if not some general consensus regarding conceptions of science and related ideologies, theories and methodologies. As these papers illustrate, no such consensus existed between the invited speakers. The papers reflect different analytical approaches, differences in interpretive practices, and differences (implicitly and explicitly) in ideology. Notions of "the geographic" also differ considerably between papers.

This volume represents an attempt to foster a better understanding of the determinants of health, and a step in the direction of creating a common discourse among persons interested in a richer appreciation of human health and its influences. Future symposia will provide opportunities to explore other themes of general interest to an interdisciplinary community of researchers and practitioners.

Neither the symposium nor this volume would have been possible without the generous support of the British Columbia Ministry of Health and Ministry Responsible for Seniors.

Michael Hayes, Ph.D.
Department of Geography
Simon Fraser University

---

[1] White, N.F. 1981. Modern Health Concepts, in White, N.F. (ed.) *The Health Conundrum*. Toronto: TVOntario.

[2] Health and Welfare Canada. 1986. *Achieving Health for All: A Framework for Health Promotion*. Ottawa: Ministry of Supply and Services.

[3] Evans, R.G. and Stoddart, G.L. 1990. Producing Health, Consuming Health Care, *Social Science and Medicine* 31(12), pp. 1347-1363.

TABLE OF CONTENTS

LIST OF TABLES

LIST OF FIGURES

LIST OF PLATES

All photos by Ministry of Health and Ministry Responsible for Seniors

1

DEATH IN PARADISE: Considerations and Caveats in Mapping Mortality in British Columbia (1985-1989)

L.T. Foster
S.H. Uh
M.A. Collison

Province of British Columbia,
Ministry of Health and Ministry Responsible for Seniors,
Division of Vital Statistics

INTRODUCTION

The mapping of mortality and morbidity has a long tradition that goes back nearly 200 years. Howe[1] makes reference to a map published in 1798 showing the spatial distribution of yellow fever. Kemp and Boyle[2] refer to the 1844 work of Perry who portrayed the distribution of what is now believed to be an influenza outbreak in Glasgow. Perhaps the most important work is the now classic 1854 study by Snow.[3] Indeed, it has been suggested that this study, which mapped the location of cholera victims in London, leading to the identification of the exact location and elimination of the cause of the outbreak, was the origin of the scientific approach to epidemiology. It also clearly demonstrated the potential value of mortality mapping. Other notable pioneering works are Haviland's[4] 1855 work, the first known cancer atlas, and the innovative cancer mapping work of Stocks,[5] in the 1920s and 1930s, which really brought cancer mortality mapping into prominence. One of the first comprehensive national mortality atlases is the work of Howe,[6] who mapped United Kingdom death data for the period 1959-63.

Since these early works, there has been a steady growth in mortality mapping. A recent review of mortality atlases by Walter and Birnie[7] shows that there have been no less than 39 published works between 1976 and 1989, in nearly 25 different countries. Of these atlases, no less than 33 have been devoted to cancer, while the remainder were broader. Three of the reviewed atlases are Canadian national atlases[8] and one is provincial[9] in scope. In addition to the atlases noted by Walter and Birnie, two recent Canadian contributions are the cancer atlases for British Columbia and Quebec.[10] Mortality mapping has clearly become a well established and growing field in health geography, although it is not without its critics.[11]

The growing activity in mortality mapping over the last decade is undoubtedly related to the increasing affordability and ease of use of computer cartographic technology. This chapter presents a selection of some of the work being undertaken within British Columbia's Division of Vital Statistics. A description of the materials, measures and methods being used to develop mortality maps is provided, as well as a selection of maps which demonstrate various aspects of mapping mortality data and the various patterns that emerge. Next, some of the more common problems associated with mortality mapping, in general, are discussed briefly. Other uses for mortality maps and possible future applications are presented in the conclusion.

Mapping Death or Mapping Health?

In the 1980s, some major reviews of the Canadian health system were undertaken. There have been no less than seven Royal Commissions[12] appointed by Provincial Governments to find how the current system can be improved. British Columbia is no exception to this trend, with the appointment of the Seaton Royal Commission, in February 1989.[13] Such challenges to the current health systems have indicated the need for change, including focussing on health status and service outcome, while emphasizing health promotion initiatives. This involves, among other things, empowering communities and individuals to address their own specific health problems and priorities. Providing appropriate health status information to communities can help them to determine these priorities. Increasingly, the British Columbia Ministry of Health is providing that information on a local geographical basis.

The 1988 Annual Report[14] of the Division of Vital Statistics, which contained selected birth, death and marriage information, included one map showing preventable, externally caused deaths by the 21 Provincial Health Units. The Ministry of Health Annual Report for 1988/89[15] contained three maps illustrating crude rates for all births, all deaths and deaths related to external causes (Figure 1,1). The 1989 Vital Statistics

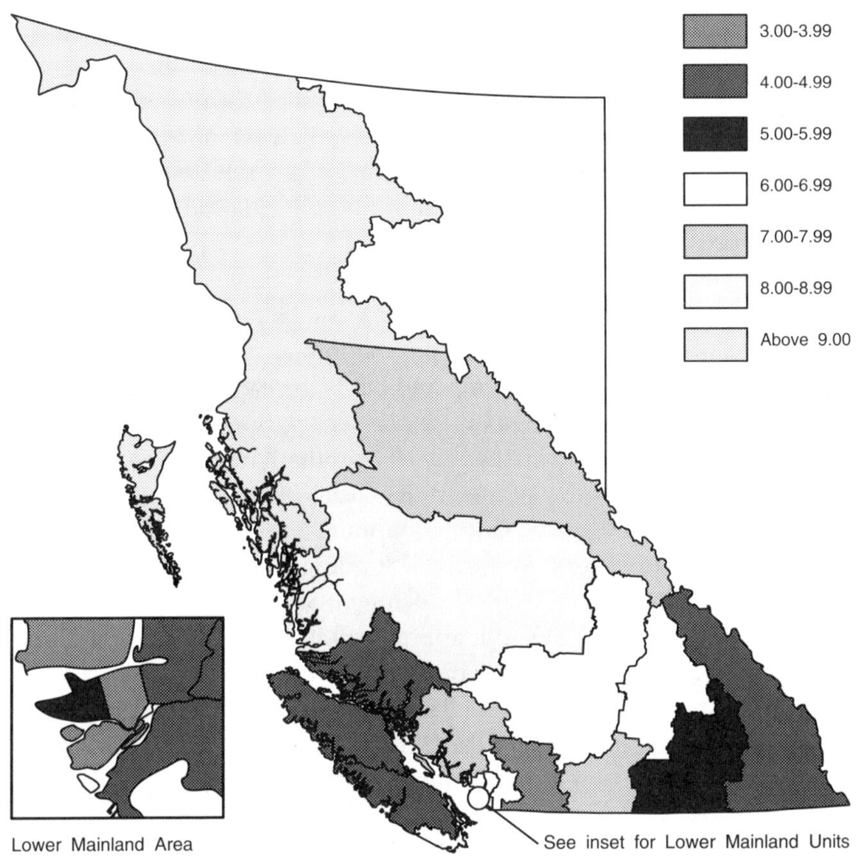

Figure 1,1 Deaths Related to External Causes, Ministry of Health Annual Report for 1988/89.

Annual Report[16] moved up a level in sophistication by presenting Standardized Mortality Ratios (SMR) and Potential Years of Life Lost (PYLL) data by Health Unit, and also indicated levels of statistical significance to help focus in on "high" and "low" rate areas.

Medical Health Officers, who administer Health Units, have also been provided with more detailed information about their own Health Unit, relating SMR and PYLL data to the leading causes of death aggregated over several years. The next step is to supply them with a compendium of mortality maps by 80 Local Health Areas. Providing information in map form "is the simplest, most elegant and informative way of presenting data which vary across a surface".[17] The provision of such maps increases the capability of professionals and citizens to make informed decisions regarding local health care priorities. Furthermore, the provision of relevant health care information at the "community level" allows for a focus on local concerns in a manner that highlights issues pertaining to the local community and lessens the biases of the data being viewed from a provincial perspective.

However, while mapping death can hardly be viewed as mapping health status,[18] it is a convenient, readily available, unambiguous surrogate for health status for the total population, as all deaths have to be recorded by statute. Indeed, Kemp and Boyle have gone so far as to state that mortality maps have helped in "the recognition of health services research as a legitimate area of study".[19] Although most emphasis in the British Columbia Ministry of Health has been placed on mapping vital events data such as births and deaths, mapping is taking on greater importance for a variety of reasons: hospital funding is increasingly based on the demographics of geographical catchment areas; analysis of response times for ambulances looks to mapping to identify problem areas; and mapping the distribution of health professionals and other services throughout the province points to potential service "surpluses" and "gaps", especially with respect to urban/rural differences, a point that has not gone unnoticed by the Seaton Royal Commission[20].

CAUTIONS AND CAVEATS

Mortality maps have been developed to highlight spatial variations in mortality measures. It is often easier to see patterns on a color-coded map than to sift through numerous tables of numbers.[21] As Charlton has noted, "Mapping is not intended to produce definitive results, but to raise questions and hypotheses which may be answered by further investigations"[22], an observation echoed by numerous other authors.[23] Various problems and cautions must be taken into consideration when dealing with data reflecting mortality measures and, in particular, spatial variations. Being forewarned, however, helps lessen the pitfalls associated with the use of data and their interpretation. Many of the issues are closely interrelated, but for ease of discussion in this chapter they are grouped into four major categories: data integrity; areal units; time period; and interpretation.

Data Integrity

Although mortality data are readily available, their accuracy and consistency are often open to question. Charlton has noted that there are many problems inherent in the mortality data themselves, so much so that he has concluded, rather gloomily, "inaccuracies which limit the usefulness of mortality data are difficult, if not impossible to quantify".[24]

One problem of underlying significance is that diagnosis is not always easy, especially in older patients who may be suffering from several major concurrent illnesses.[25] Furthermore, physicians have varying levels of diagnostic skill[26] and there may be differences in diagnostic practices and preferences among physicians, not to mention differences in death certification and coding practices.[27] In some instances, the attending physician or coroner may have limited knowledge of the decedent's medical history, having seen the decedent only briefly (eg. emergency admission to hospital), or never having seen the decedent prior to death. While autopsies may give more accurate diagnoses, these are usually only performed where there is a reasonable degree of uncertainty regarding the cause of death. In British Columbia, autopsies are performed in only 24 percent of registered deaths.

Areal Units of Analysis

While geographers are inclined to search for natural or ecological regions (eg. watershed) as a basis for mapping, in reality, most areal mapping units chosen are artificial and arbitrary. As Kemp and Boyle have aptly put it "death occurs in people, not in regions",[28] and areal units are usually based on administrative units which have enough cases to do appropriate analyses of one kind or another.

Choosing appropriate areal units is a trade-off between obtaining population and sample sizes of sufficient magnitude for the intended statistical purposes, and yet small enough not to obscure important anomalies and to increase the likelihood of homogeneity of environmental and socio-economic factors.[29] Pickle and Mason have demonstrated how using smaller areal units picks up patterns which had been previously missed because the analytic region chosen was too large.[30] This point is demonstrated later in this chapter.

The residential mailing address of a decedent is usually captured on death registrations. In analyzing the data, determining residential location from the records may be difficult, particulary in rural areas, as a postal address using a box number may be relatively distant from the actual residence. Very different patterns may emerge, depending on the areal units selected for the analysis.

Time Period of Analysis

The choice of an appropriate time period over which to aggregate mortality data for analysis of spatial variances also involves trade-offs. Mapping data aggregates have ranged from time periods of a single year to as long as 30 years.[31] Short time periods limit the number of observations, making good statistical measures difficult to obtain. Alternatively, long time periods are fraught with problems, as they may obscure real trends over time.[32] In addition, diagnoses may improve over time. This improvement may be differential by areal unit and, though changes may be small, it would be difficult to determine whether such differential changes have had a significant effect on the results of the analysis.

Pickle et al., in a study of 30 years of data relating to cancer among U.S. Whites, noted a decreasing geographical variability over time. They felt this was due, in part, to increased standardization of diagnostic measures and death certification practices, and also to the wider distribution of improved medical care and better survival rates, rather than to changes in cancer incidence.[33] Gardner et al. also note that amalgamating data over long time periods may create problems related to changes in age/gender distributions arising from migration patterns.[34]

Interpretation Issues

It is important to recognize that the absence of statistical significance, where such measures are employed, cannot in itself be interpreted to mean that there are not health problems of concern. Conversely, areas recording significantly high or low rates require further investigation to ensure that these rates are indeed "true" and actually reflect underlying natural characteristics of the areas in which they occur. For example, Gardner et al. noted that high cancer death rates in one area were related, not so much to the local environment, but rather to the location of a cancer treatment facility which was "home" for terminally ill patients.[35]

For externally caused deaths, such as homicides or motor vehicle accidents, mapping the data by location of death may be much more appropriate than by residence of the decedent, although this also has its limitations, as will be illustrated later in this chapter.

Interpretations of geographical variations must take into account other factors, particularly those related to migration. Deaths may be related to causal factors in earlier life which are masked or exaggerated by migration patterns. Furthermore, selective migration may occur for reasons of health: families with a chronically ill member may change their residence to be closer to a specific treatment facility; and seniors may move to a long term care facility or from rural to urban areas to enjoy the relative security of knowing that a variety of health services are close at hand. In British Columbia these factors may be of particular concern, as certain areas are known to attract retirees both from within the province and from across the country. This is especially so for southern and eastern Vancouver Island, the lower mainland and the Okanagan Valley, because of relatively mild climates.

These are all factors that warn against hasty interpretations and conclusions. The mapping of mortality data is a precursory step to further, detailed epidemiological investigations, and should not be used in isolation as an indicator of health status. Clearly, user education is an important component in the development and distribution of such information.

MATERIALS, MEASURES, METHODS

Data Sources

Most health mapping initiatives undertaken by the British Columbia Ministry of Health have utilized mortality data. As Charlton[36] points out, mortality data, like most data, have limitations in terms of availability, economy and completeness. However, no other routine system of health related data collection rivals death records.

Death information collected by the Division of Vital Statistics has several characteristics that make it particularly attractive as a data source. The *Vital Statistics Act* requires that all deaths occurring in the province be reported within a prescribed period of time, and that the personal particulars of the deceased person be recorded. Among the items recorded on death records are: the cause of death, as well as other antecedent causes or conditions; residence, postal code and standard geocode of the decedent; age and gender; and other particulars related to the death. In order to ensure accuracy in the coding of cause of death information, the Division employs trained medical coders to record such information according to a format prescribed by the World Health Organization (International Classification of Diseases). Once coded, this information is stored on computer readable media. Aside from the relative completeness and accuracy of these data, other benefits include the availability of several years of data to act as a baseline, in time series analyses, and the ability to access original documents via microfilm, for verification purposes, in the case of questionable data.

Population and Sample Sizes

As with other statistical undertakings, when mapping mortality data, one must consider population size as well as the number of cases being used in the calculation of various statistical measures. Here, the choices of the areal unit of analysis and time period of analysis become important interrelated factors.

Areal Units of Analysis

The first major consideration in selecting areal units is to ensure that they provide a basis for the determination of reliable population estimates. This suggests that the areal units should be based on some areal unit employed for census purposes. While not a census area, the postal code is collected as part of census data. It is, therefore, possible to derive census population data based on aggregates of postal codes, as well as census tracts, census divisions, etc. It may also be desirable to relate to census based socioeconomic data. The Division of Vital Statistics currently employs the postal code as its basic unit of geographic coding, with the standard geocode employed as an edit check. With few exceptions, postal code areas may be readily aggregated to LHAs (Figure 2,1), Health Units (Figure 3,1) and other administrative areas.

The second major consideration is that there be sufficient population sizes and number of cases (deaths), in order to ensure the desired degree of reliability of any statistical analysis. This, of course, depends on the nature of the specific analysis. In some instances, the unit chosen may be based on administrative considerations. This may be the case when preparing data for Medical Health Officers or others who are responsible for dealing with such prescribed areas as Health Units, or Regional Hospital Districts. It may sometimes be appropriate to use a finer areal unit such as Local Health Area, Census Tract, or even a small postal area, for statistical/analytical purposes, where there is a need to be more precise in terms of attributing certain characteristics to an area.

Time Period of Analysis

In terms of the number of cases, the time period of analysis must be considered in conjunction with the choice of areal units of analysis. For example, there may be a sufficient number of cases (deaths) for statistical

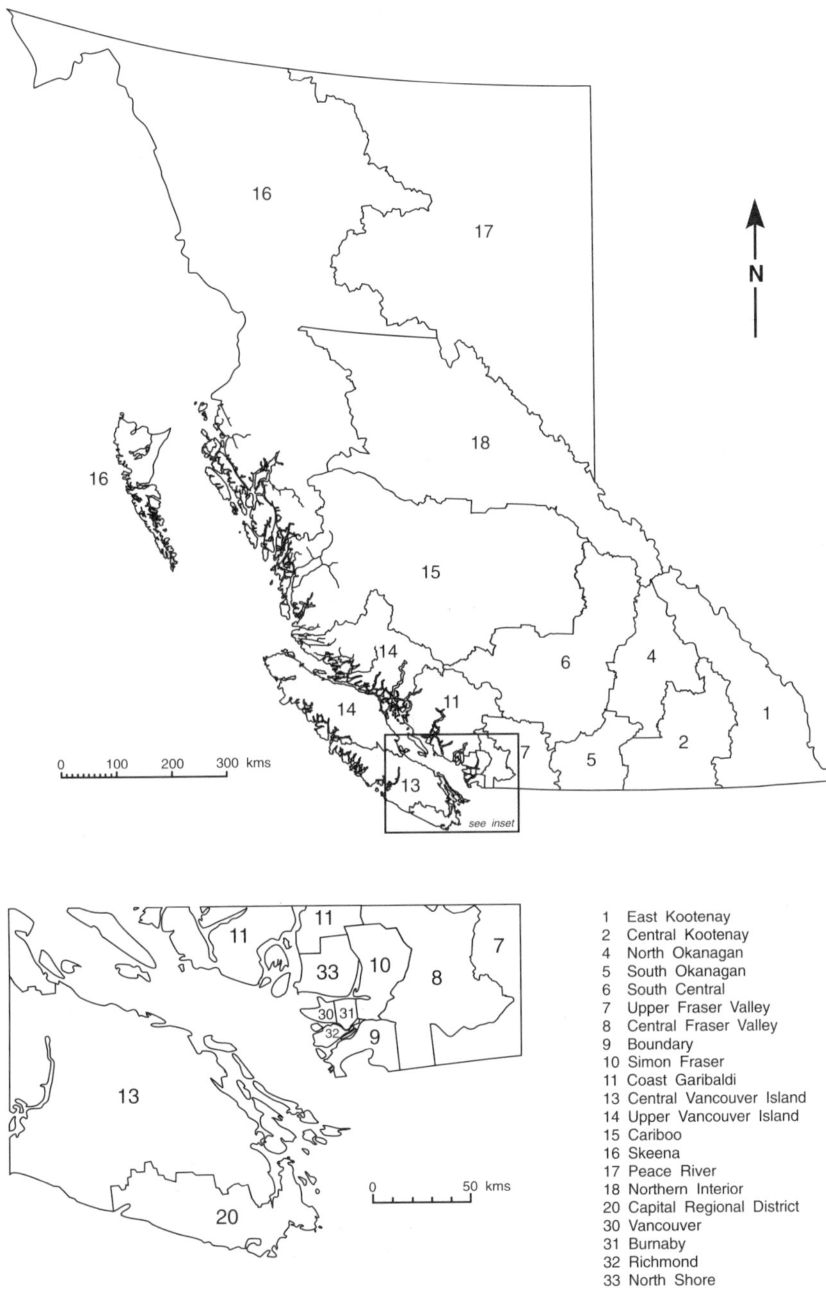

Figure 2,1 British Columbia's 21 Health Units.

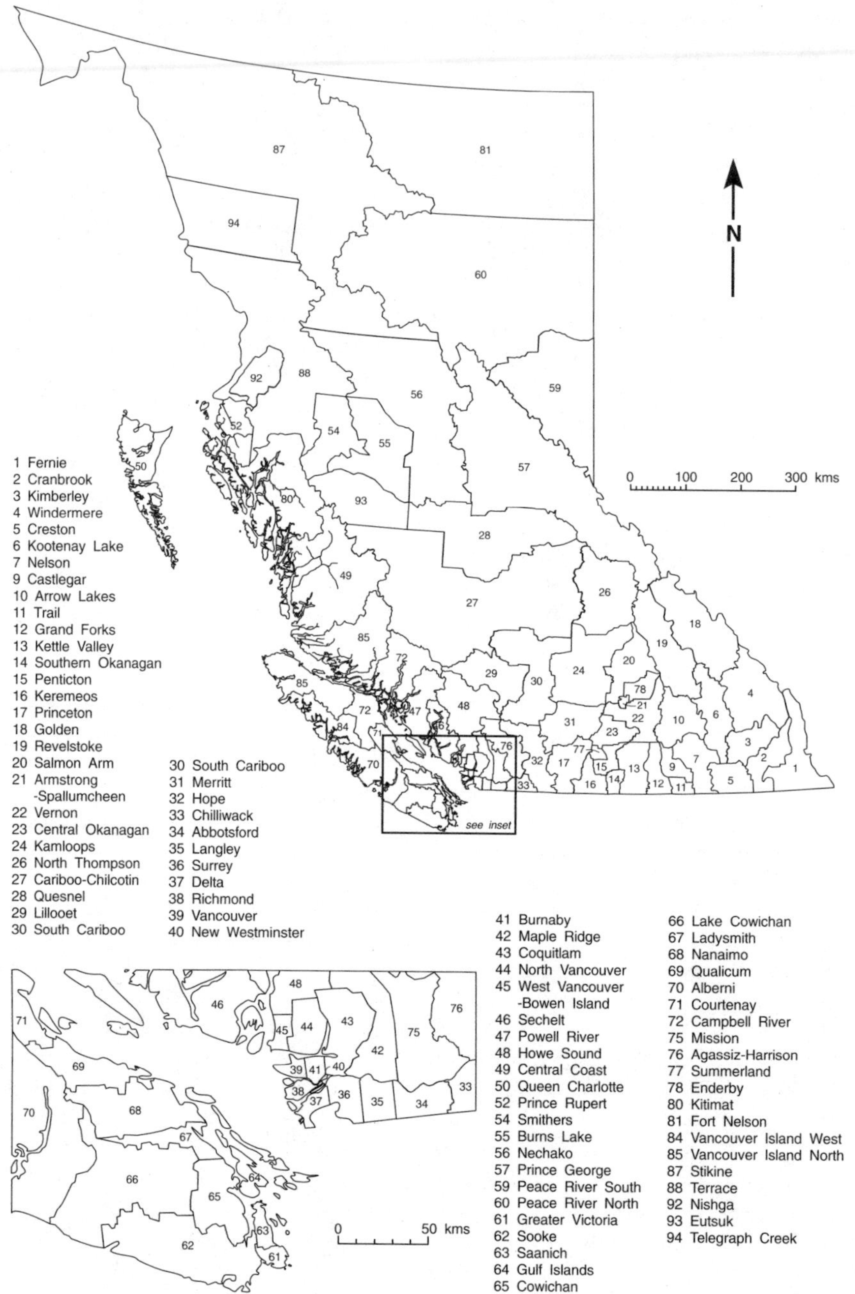

Figure 3,1 British Columbia's 80 Local Health Areas.

analysis using one year of data for the 21 Health Units, whereas it may be necessary to combine several years of data if the 80 Local Health Areas are being used. As a general rule of thumb, the Division of Vital Statistics recognizes five cases as a minimum number to include in the calculation of any statistical measure. This both ensures a level of protection on the confidentiality of personally identifiable data, and eliminates the production of highly unreliable statistical measures based on very small case numbers. Additionally, the Division of Vital Statistics combines several years of data when looking at any areal unit smaller than a Health Unit, in order to ensure some reliability in statistical analyses, and to avoid anomalous "peaks" and "troughs" (although these are interesting in themselves).

Maps included in this chapter were developed based on postal codes and employ population data, categorized by gender and five-year age groups, and data relating to death events occurring in British Columbia over the five year period 1985-1989.[37] Any deaths occurring in the province to non-residents were excluded. After eliminating cases that lacked important data elements, a total of 108,888 cases were compiled of which 59,665 were male and 49,223 were female.

Statistical Measures

A number of statistical measures have been employed by the Ministry of Health in its mortality mapping. There has been a progression in terms of the sophistication of statistical measures being used in these undertakings, starting with crude rates and progressing to standardized rates, and normalized indices, as well as moving toward measures that better reflect the impact of mortality. Adjusted measures, in order to account for population compositional differences, use various weighting factors derived from a "standard population". This allows more meaningful mortality measure comparisons to be made between different areas or time periods. Comparisons of such adjusted measures may be made directly without further manipulation of the data (e.g. measures calculated by different provinces for areas within their own jurisdictions may be compared if they employ the same standard population).

The choice of the standard population is generally governed by the purpose of the analysis, the ease with which standard population factors may be obtained, and the "degree of relevancy" to the populations being studied. In Canada, the 1971 census population, national or provincial, has often been employed as the standard for such calculations. For international comparisons, the World Standard Population may be used.

The adjusted measures used in this chapter are limited to adjustment for age and gender, the most significant factors. Adjustment may be done separately for males and females to account for population gender compositional differences, or the data may be both age and gender adjusted.

It is important to bear in mind that all measures derived using standard population weighting factors are surrogate measures, not real measures for the areas. It is the relative values of the adjusted measures that are important for comparative analysis of different time periods or areas, rather than the absolute values of the measures. In addition, all such measures are subject to some degree of bias, as they cannot totally compensate for population compositional differences. However, they lessen the potentially significant biases of unadjusted measures.

The statistical measures used in this chapter are summarized below.

CRUDE RATES (CR): The crude annual death rate per *k* population for an area represents the total number of deaths of area residents in a given year divided by the midyear population of the area multiplied by *k*. The general formula for the crude rate per *k* population is as follows:

$$CR = \frac{D}{P} \times k$$

where: x = multiply
P = total area population
D = total deaths in area
k = a rate conversion factor, most commonly 1,000 or 10,000

While crude rates are useful in terms of a cursory examination of data for patterns or trends, their analytical utility is limited. Comparisons of the crude mortality rates between different areas or time periods are apt to be misleading, as they do not take into account differences in population compositional factors.

AGE STANDARDIZED MORTALITY RATE (ASMR): The ASMR utilizes the area's mortality rate and weighting factors based on the proportion of the standard population in various age-groups. This is referred to as the "direct method" of adjustment. The general formula for calculating the ASMR per *k* population is as follows:

$$ASMR = \frac{\sum_{i=1}^{n} m_i \times \pi_i}{\Pi} \times k$$

where: x = multiply
Σ = sum of
p_i = area population in age group i
π_i = standard population in age group i
Π = $\Sigma\pi_i$ = total standard population
d_i = deaths in area population in age group i
m_i = d_i/p_i = mortality rate in area age group i
n = number of age groups
k = a rate conversion factor, most commonly 1,000 or 10,000

All things being equal, the ASMR is the preferred measure for examining time trends in mortality and making comparisons between relatively disparate geographical areas.[38] *The Atlas of U.S. Cancer Mortality Among Whites (1950-1980)*[39] and the *Quebec Atlas of Cancer*[40] are examples of the use of this measure in mapping.

STANDARDIZED MORTALITY RATIO (SMR): The SMR utilizes the area's population and weighting factors based on age-group mortality rates of the standard population. This method of adjustment, also referred to as the "indirect method", derives a ratio of the total observed deaths and the total expected deaths for an area, based upon the age-specific mortality rate of the standard population. The general formula is as follows:

$$\text{SMR} = \frac{D}{\sum_{i=1}^{n} \mu_i \times p_i}$$

where: x = multiply
Σ = sum of
p_i = area population in age group i
π_i = standard population in age group i
D = total deaths in area
δ_i = deaths in standard population in age group i
μ_i = δ_i / π_i = mortality rate in standard population age group i
n = number of age groups

Like the ASMR, the SMR is, to a large degree, able to compensate for the differential impact of age compositional variations, although in a somewhat different manner. The advantage of the SMR is that, as a normalized index (all values are relative to 1, the value of the standard population), it is much easier to compare relative values. Also, the SMR is the preferred measure when making geographical comparisons involving a small number of cases (e.g. areas with relatively small populations).[41] The *Atlas of Cancer Mortality in British Columbia (1956-1983)*[42] and the *Atlas of Cancer in Scotland (1975-1980)*[43] are examples of the use of the SMR in mapping.

POTENTIAL YEARS OF LIFE LOST STANDARD RATE (PYLLSR): This differs from other measures in that the impact of mortality is measured in terms of the number of years of life lost. The assumption made is that an individual's "most valuable" years of life fall between the "beginning of years of life" and a selected cutoff year, and that only those years within the "most valuable" years are of consequence. A number of age cutoffs have been proposed in the calculation of PYLL,[44] as well as the inclusion or exclusion of infants (<1 year). This has resulted in three different age categories generally being used: 0-75, 1-70, and 0-70.

The PYLL Standard Rate uses the area's age-group deaths multiplied by the "average" number of years remaining to the cut off age and weighting factors based on the proportion of the standard population in various age groups. The PYLLSR is similar to the ASMR and the general formula (for a cut off age of 75) per *k* population is as follows:

$$\text{PYLLSR} = \frac{\sum_{i=1}^{n} m_i \times \pi_i \ (75 - Y_i)}{\Pi} \times k$$

where: x = multiply
Σ = sum of
p_i = area population in age group i
π_i = standard population in age group i
Π = $\Sigma\pi_i$ = total standard population
d_i = deaths in area population in age group i
m_i = d_i/p_i = mortality rate in area age group i
n = number of age groups
k = a rate conversion factor, most commonly 1,000 or 10,000
Y_i = mid-point age of age group i

The PYLLSR is a useful measure to emphasize the "impact" of mortality, which is weighted more heavily towards younger age groups, by expressing deaths in terms of years of life lost rather than number of deaths.

Although use of PYLLSR in mapping is quite common, no atlas has been published, to date, using this measure.

POTENTIAL YEARS OF LIFE LOST INDEX (PYLLI): This measure, being very similar to the SMR, is a ratio of the observed potential years of life lost to the expected years of life lost. The general formula (for a cut off age of 75) is as follows:

$$\text{PYLLI} = \frac{\sum_{i=1}^{n} d_i \ (75 - Y_i)}{\sum_{i=1}^{n} \mu_i \times p_i \ (75 - Y_i)}$$

where: x = multiply
Σ = sum of
p_i = area population in age group i
π_i = standard population in age group i
d_i = deaths in area population in age group i
δ_i = deaths in standard population in age group i
μ_i = δ_i/π_i = mortality rate in standard population age group i
n = number of age groups
Y_i = mid-point age of age group i

As in the case of the PYLLSR, no PYLLI atlas has been published to date.

Statistical Method

In the case of SMR calculations for each geographic region and selected cause of death, a Chi-Square (x^2) test is employed to determine whether the observed number of deaths show a statistically significant difference from the expected number of deaths. In using this test of significance ($p<0.05$), a Poisson distribution is assumed and a 95 percent confidence interval is employed.

In the case of PYLL indices, due to the intrinsic characteristics of the measure (i.e. the use of number of years of life lost rather than number of deaths), statistical significance is determined utilizing a 95 percent confidence interval when determining if a significant difference exists between a specific geographic region and the province as a whole.

As it was felt that mortality mapping based on significance levels alone does not give any indication of the actual magnitude of the various measures, an alternative plotting method is used in this chapter. This method incorporates both significance and size of the measure. While many of the Division's maps are now being produced in color, the examples contained in this chapter use a four shading scheme as follows: darkest shade indicates that the actual or observed deaths of a region are statistically significantly higher than the expected deaths relative to the province, and that the region has a SMR or PYLLI that falls within the top decile; less dark shade indicates a region that is statistically significantly higher than the province, but does not have a SMR or PYLLI that falls within the top decile; lightest shade indicates a region that is not statistically significant relative to the province; and no shade indicates a region that is statistically significantly lower than the province.

DEATH IN PARADISE: SOME SELECTED PATTERNS

This section provides a series of figures to show selected trends and patterns of mortality measures in British Columbia over the period 1985 to 1989. Figures 4,1 to 6,1 provide five year mortality trend graphs for the province. Figures 7,1 and 8,1 show how different patterns emerge depending on what areal unit is chosen. In this case, PYLLI All Causes maps were used to compare Health Units to Local Health Areas. Figures 7,1 and 9,1 compare different measures by illustrating maps of PYLLIs and SMRs by Health Units. Figures 10,1 and 11,1 show the different patterns that emerge for the same disease (Ischaemic Heart Disease) when plotted for males and females individually. Lastly, Figure 12,1 and Tables 6,1 to 8,1 and accompanying text demonstrate the need to fully understand the data that are being used. Here, motor vehicle traffic accident deaths are compared using place of residence of decedent versus place of death.

FIVE YEAR TRENDS

Although the total number of annual deaths increased from 21,127 in 1985 to 22,770 in 1989, the ASMR for all causes of death showed a steady decline from 54.57 to 51.47 per 10,000 population during the five year period. This reduction was apparent for both males and females and occurred at approximately the same rate (Figure 4,1). Much of the decline can be attributed to a reduction in death rates from circulatory diseases, and to a lesser extent, death rates from external causes. Of the other major

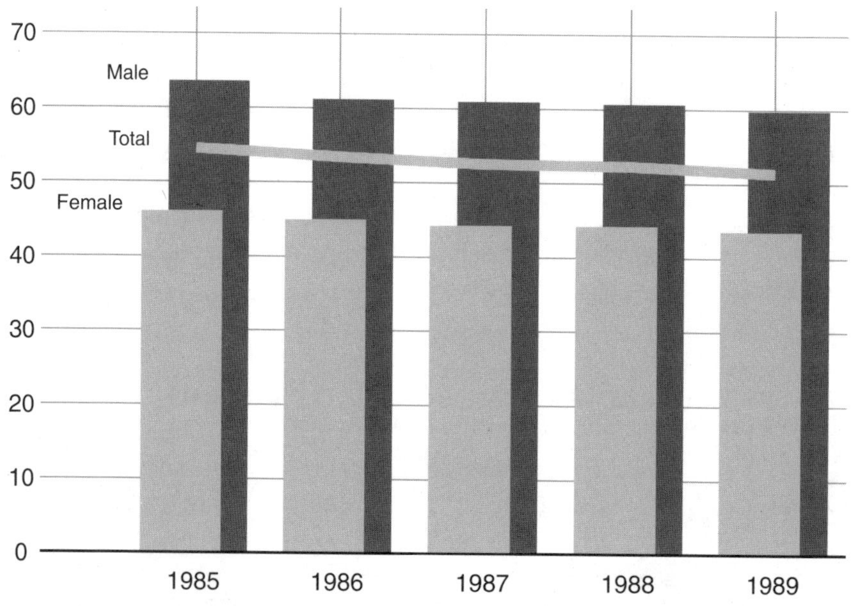

Figure 4,1 Trends of ASMRs* Due to All Causes of Death (British Columbia, 1985-1989)

causes of death, such as cancer and diseases of the respiratory and digestive systems, no overall change in death rates was observed, whereas motor vehicle traffic accidents, a sub-component of external causes of death, showed a slight overall increase (Figure 5,1).

Total deaths are higher for males than for females, as is the case for PYLL. The difference, however, is much more pronounced for the PYLL, showing the higher preponderance of earlier death for males. For the ASMR, the leading causes of death, in order of magnitude are: total diseases of the circulatory system, cancer, and external causes. Conversely, when looking at the PYLLSR, the order of leading causes of death is reversed to assume the following order: external causes, cancer, and total

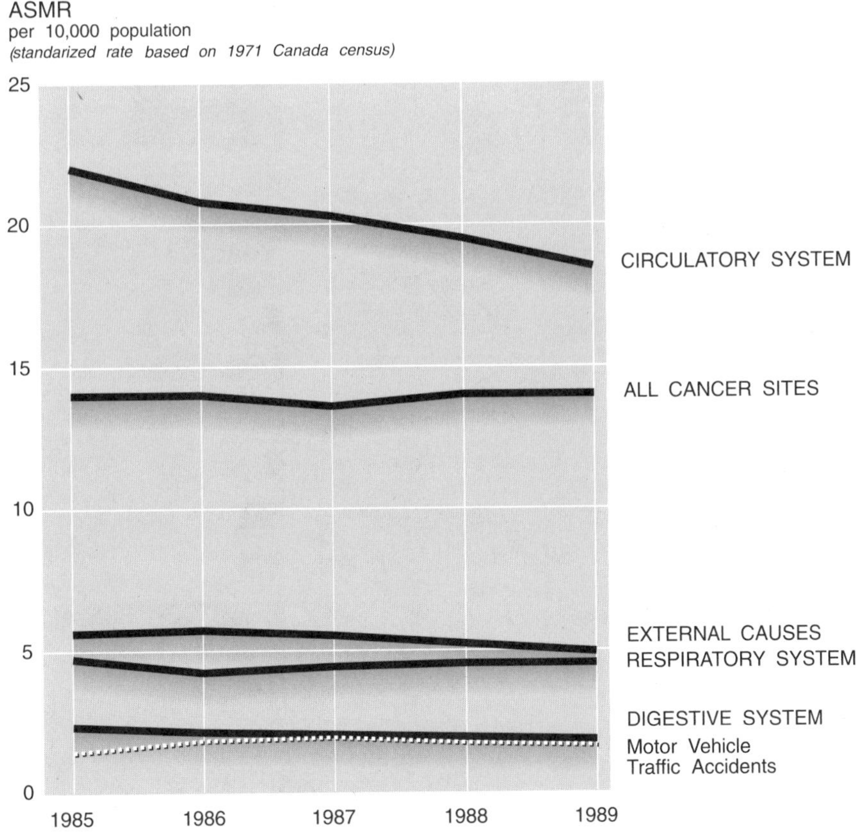

Figure 5,1 Trends of Major Causes of Death (British Columbia, 1985-1989)

diseases of the circulatory system (Figure 6,1). This shift can be attributed to the characteristic of PYLLSR which gives greater weight to deaths at an earlier age. Use of the ASMR ranks diseases of the circulatory system (e.g. ischaemic heart disease and stroke) as the leading cause of death, reflecting deaths of the elderly. Alternately, use of the PYLLSR ranks deaths due to external causes (e.g. motor vehicle traffic accidents and suicides) as the leading cause of years of lost life, many of which are preventable and reflect cause of death in younger people.

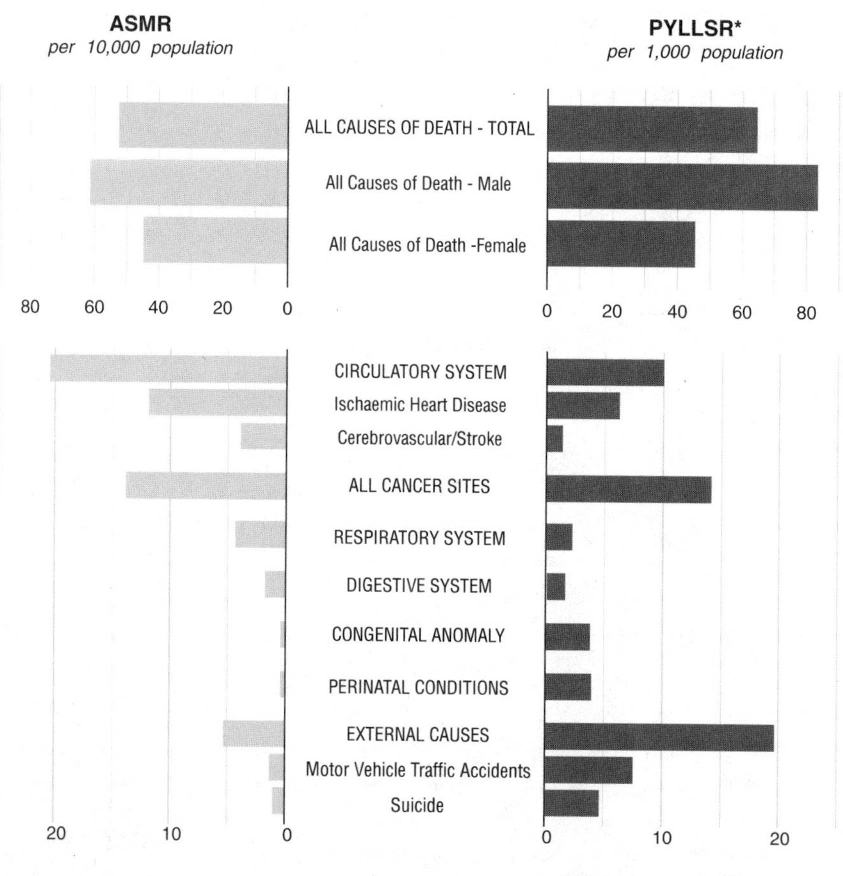

Figure 6,1 Age Standardized Mortality Rates and Potential Years of Life Lost: British Columbia, 1985-1989
(all rates are averaged over five year period and standardized rates are based on 1971 Canada census)

MAPPING ASPECTS AND PATTERNS

Health Units versus Local Health Areas

Figure 7,1 and Table 1,1 show the variation in PYLLI for all causes of death for both genders, by Health Unit. Significantly high PYLLIs, within the top 10 percent, occur in the Cariboo and Skeena units. Significantly high rates are also evident in the adjoining Northern Interior, South Central, Coast Garibaldi, Central Vancouver Island and Peace River units. Significantly low rates occur in the urban areas in parts of the lower mainland and southern Vancouver Island. The City of Vancouver, however, has a significantly high rate, and is anomalous in the lower mainland. Other significantly low rates occur in the southeast of the province and in the South Okanagan unit. Using a finer areal unit for mapping (i.e. 80 Local Health Areas) demonstrates a pattern with some important differences (Figure 8,1 and Table 2,1). Substantially more of the lower mainland emerges as being significantly low, while the Qualicum area on Vancouver Island changes from being significantly high to significantly low. The area in the southeast of the province that had been significantly low shrinks substantially, a pattern that is also reflected in the South Okanagan. New areas of significantly high values emerge in the northern part of Vancouver Island and facing mainland area, as well as in the Hope and Enderby areas.

In general, the pattern for Local Health Areas is quite similar to the Health Unit pattern. However, if there are both significantly high and low Local Health Areas within one Health Unit, these geographic variations can be "masked" at the Health Unit level. For example, the North Okanagan Health Unit is neither significantly low nor high, but contains the smaller Local Health Areas of Salmon Arm and Enderby which are, respectively, significantly low and significantly high. A comparison of these two maps clearly shows how the geographical unit chosen influences the mortality pattern that emerges in terms of statistical significance.

PYLLI versus SMRs

While a comparison of Figure 7,1/Table 1,1 (PYLLI - All Causes) and Figure 9,1/Table 3,1 (SMR - All Causes) shows a largely similar pattern, some interesting differences emerge. Two Health Units, Simon Fraser and Upper Vancouver Island, emerge as having significantly high SMRs, with only the former having a significantly low PYLLI. While the exact reason for these differences needs further examination, Simon Fraser Health Unit also has a significantly high Ischaemic Heart Disease SMR for males and

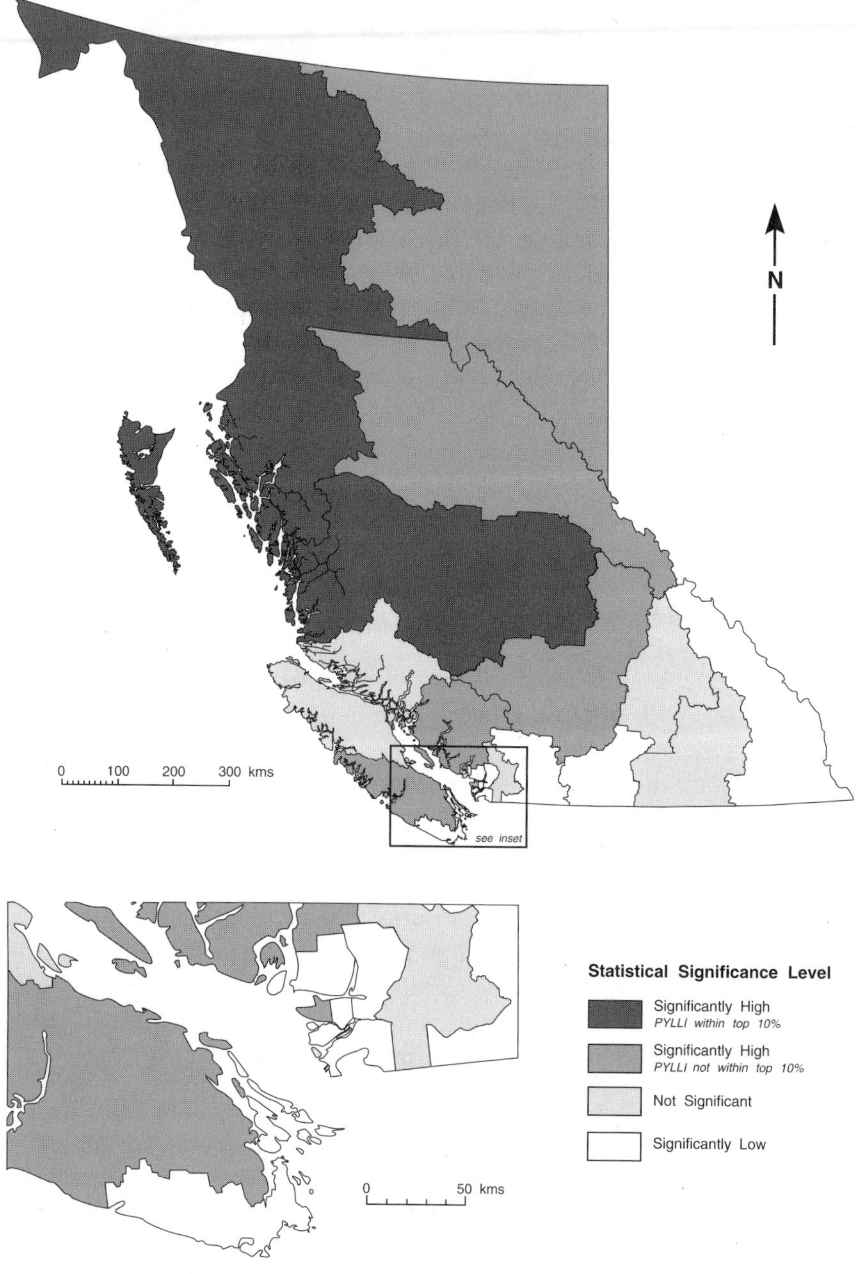

Figure 7,1 All Causes of Death: PYLL by Health Unit - Male and Female (British Columbia, 1985-1989)

Table 1,1 Potential Years of Life Lost by Health Unit: All Causes of Death (Age 0 - 75 Years) (British Columbia, 1985 to 1989)

	Health Unit	*Observed Deaths*	*Total PYLL*	*Expected PYLL*	*PYLL Index*	*95% C.I. Lower*	*Upper*
15	Cariboo	1178	29668	20521.84	1.45	1.34 -	1.55
16	Skeena	1225	35561	24891.13	1.43	1.33 -	1.53
18	Northern Interior	1673	43334	35059.68	1.24	1.16 -	1.31
06	South Central	1862	38909	32972.71	1.18	1.11 -	1.25
30	Vancouver	9565	164861	145674.94	1.13	1.10 -	1.16
11	Coast Garibaldi	1011	19015	16998.48	1.12	1.02 -	1.22
17	Peace River	799	19680	17794.03	1.11	1.01 -	1.21
13	Central Vancouver Is.	3661	62663	56257.58	1.11	1.06 -	1.17
07	Upper Fraser Valley	2301	40390	42971.87	0.94	0.88 -	1.00
10	Simon Fraser	2694	49537	52536.38	0.94	0.89 -	0.99
05	South Okanagan	3142	47689	51391.70	0.93	0.88 -	0.98
20	Capital Regional Dist.	5235	78144	89063.69	0.88	0.84 -	0.91
09	Boundary	4504	79374	93769.81	0.85	0.81 -	0.88
01	East Kootenay	1083	18655	22079.11	0.84	0.77 -	0.92
31	Burnaby	2705	40842	49536.20	0.82	0.78 -	0.87
33	North Shore	2394	38629	49761.45	0.78	0.73 -	0.82
32	Richmond	1599	27524	36680.79	0.75	0.70 -	0.80

Statistically Significant High & PYLLI within top 10%
Statistically Significant High & PYLLI NOT within top 10%
Statistically Significant Low

females (Figures 10,1 and 11,1) - a disease which is more prevalent in older people than in the younger population.

Significantly low PYLLIs occur in the East Kootenay and Burnaby Health Units, but this is not the case for SMRs. On the other hand, significantly low SMRs occur in the Upper Fraser Valley and North Okanagan Health Units, but these same areas do not have significantly low PYLLIs. While the reasons for these differing patterns are not immediately clear, this comparison clearly demonstrates that using differing measures results in differing patterns of significance, even though substantially the same data are used. Understanding the differences in the measures, and the patterns that result, is important if the targeting of health promotion/ prevention programs or the provision of care services are to be based on these results.

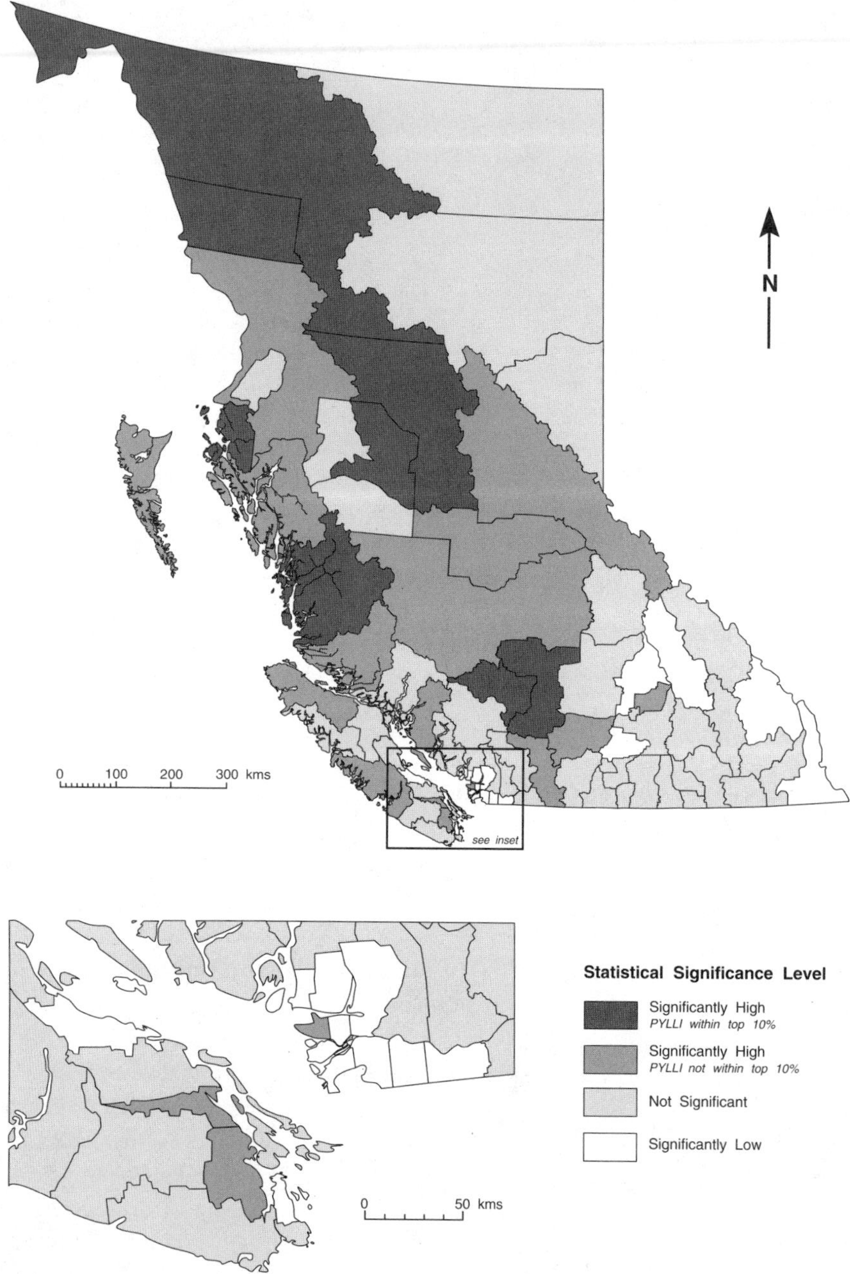

Figure 8,1 All Causes of Death: PYLL by Local Health Area - Male and Female (British Columbia, 1985-1989)

Table 2,1 Potential Years of Life Lost by Local Health Area: All Causes of Death (Age Under 75 Years) (1985 to 1989)

	Local Health Area	*Observed Deaths*	*Total PYLL*	*Expected PYLL*	*PYLL Index*	*95% C.I. Lower*	*95% C.I. Upper*
94	Telegraph Creek	15	648	226.06	2.86	1.30	4.43
49	Central Coast	89	2815	1047.21	2.69	1.99	3.38
29	Lillooet	112	3047	1492.51	2.04	1.55	2.53
87	Stikine	32	1068	580.26	1.84	1.08	2.60
30	South Cariboo	173	3884	2342.10	1.66	1.32	1.99
56	Nechako	247	7704	4827.69	1.60	1.34	1.85
55	Burns Lake	142	3907	2542.58	1.54	1.21	1.86
52	Prince Rupert	329	8594	5677.27	1.51	1.30	1.73
50	Queen Charlotte	90	2445	1692.83	1.44	1.06	1.82
27	Cariboo-Chilcotin	667	16753	11910.11	1.41	1.27	1.55
54	Smithers	209	6319	4490.61	1.41	1.16	1.65
85	Vancouver Island North	216	6519	4669.69	1.40	1.17	1.62
88	Terrace	361	10721	7829.69	1.37	1.20	1.54
31	Merritt	204	4299	3149.61	1.36	1.12	1.61
78	Enderby	124	2329	1720.06	1.35	1.02	1.69
28	Quesnel	422	10100	7564.45	1.34	1.16	1.51
32	Hope	169	3321	2502.58	1.33	1.05	1.61
67	Ladysmith	330	5583	4191.89	1.33	1.12	1.55
70	Alberni	605	12736	9857.99	1.29	1.15	1.44
80	Kitimat	164	5062	3916.52	1.29	1.05	1.53
40	New Westminster	1009	16930	13487.68	1.26	1.14	1.37
47	Powell River	373	7008	5850.96	1.20	1.03	1.37
65	Cowichan	769	14145	11779.27	1.20	1.08	1.33
57	Prince George	1284	31724	27689.38	1.15	1.06	1.23
39	Vancouver	9565	164861	145674.94	1.13	1.10	1.16
61	Greater Victoria	3621	52375	57730.12	0.91	0.86	0.95
35	Langley	1152	21286	23366.67	0.91	0.83	0.99
23	Central Okanagan	1693	27444	30878.44	0.89	0.83	0.95
36	Surrey	3493	61110	68530.19	0.89	0.85	0.93
34	Abbotsford	1074	19943	22828.09	0.87	0.80	0.95
69	Qualicum	551	6978	8239.39	0.85	0.74	0.96
43	Coquitlam	1685	32607	39048.61	0.84	0.78	0.89
44	North Vancouver	1700	29354	35556.79	0.83	0.77	0.88
20	Salmon Arm	479	7049	8453.02	0.83	0.73	0.94
41	Burnaby	2705	40842	49536.20	0.82	0.78	0.87
04	Windermere	96	1697	2210.17	0.77	0.56	0.97
38	Richmond	1599	27524	36680.79	0.75	0.70	0.80
37	Delta	1011	18265	25240.02	0.72	0.66	0.79
18	Golden	69	1624	2255.95	0.72	0.49	0.94
63	Saanich	758	10554	14608.19	0.72	0.65	0.80
01	Fernie	192	3511	4965.37	0.71	0.56	0.85
45	W. Vancouver-Bowen I.	694	9275	14204.54	0.65	0.58	0.73

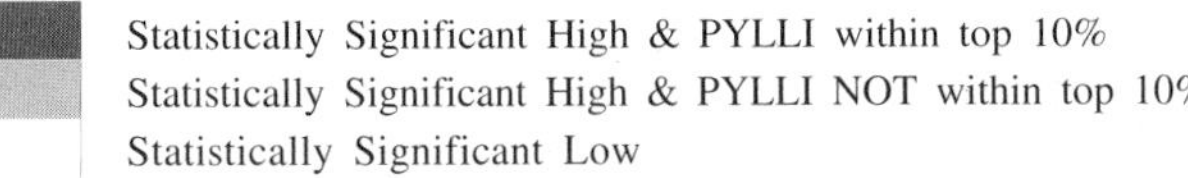

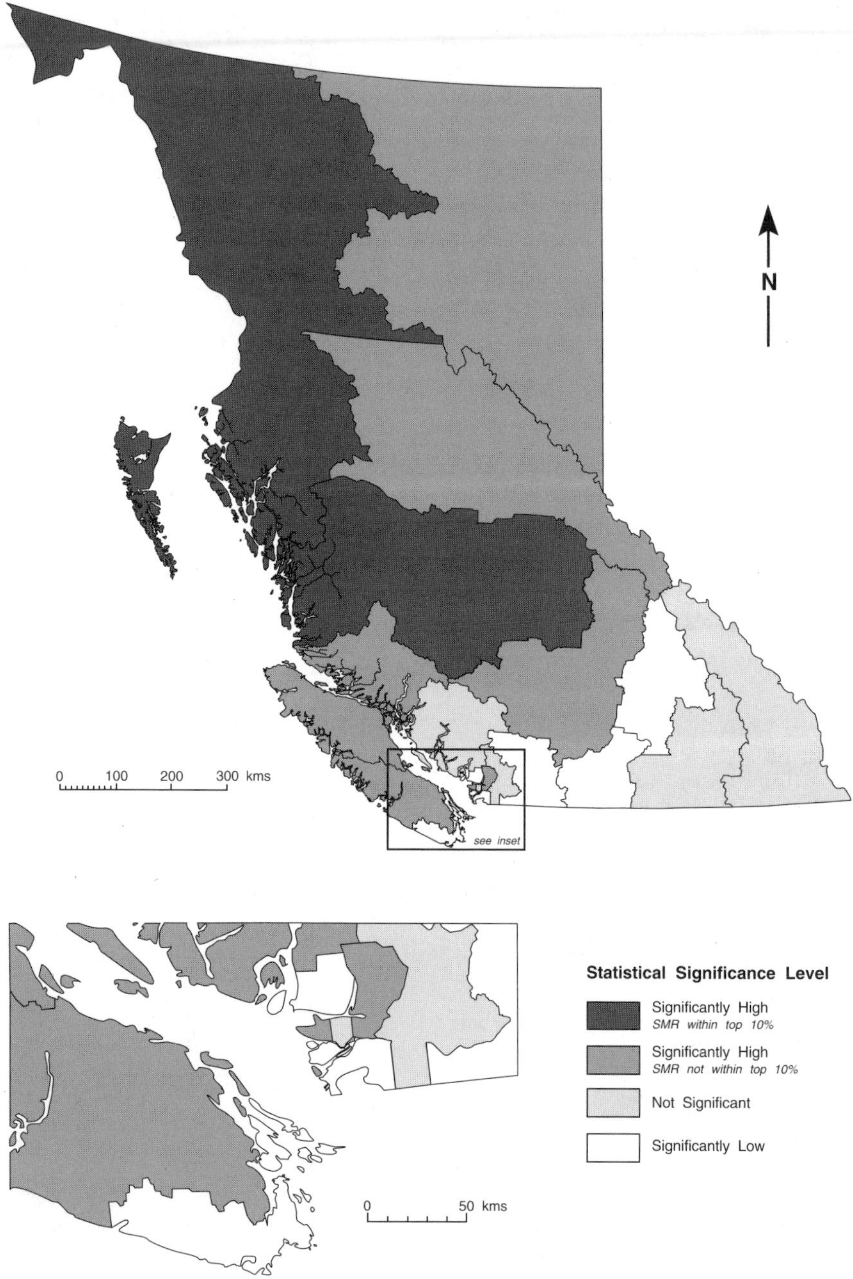

Figure 9,1 All Causes of Death: SMR by Health Unit - Male and Female (British Columbia, 1985-1989)

Table 3,1 Standardized Mortality Ratio by Health Unit: All Causes of Death - Both Genders (British Columbia, 1985 to 1989)

	Health Unit	*Observed Deaths*	*Expected Deaths*	*SMR*	*95% C.I. Lower*		*Upper*
15	Cariboo	1680	1368.45	1.23	1.17	-	1.29
16	Skeena	1759	1435.97	1.22	1.17	-	1.28
18	Northern Interior	2365	1955.27	1.21	1.16	-	1.26
06	South Central	3106	2768.12	1.12	1.08	-	1.16
17	Peace River	1220	1128.17	1.08	1.02	-	1.14
10	Simon Fraser	5329	5022.65	1.06	1.03	-	1.09
14	Upper Vancouver Island	2490	2351.24	1.06	1.02	-	1.10
13	Central Vancouver Island	6827	6478.46	1.05	1.03	-	1.08
30	Vancouver	21111	20022.14	1.05	1.04	-	1.07
33	North Shore	5375	5626.95	0.96	0.93	-	0.98
09	Boundary	8689	9119.07	0.95	0.93	-	0.97
07	Upper Fraser Valley	4831	5118.05	0.94	0.92	-	0.97
04	North Okanagan	3374	3624.49	0.93	0.9	-	0.96
20	Capital Regional District	13232	14373.64	0.92	0.9	-	0.94
32	Richmond	2956	3248.65	0.91	0.88	-	0.94
05	South Okanagan	7015	7842.88	0.89	0.87	-	0.92

Statistically Significant High & SMR within top 10%
Statistically Significant High & SMR NOT within top 10%
Statistically Significant Low

Males versus Females

Significantly high SMR values for ischaemic heart disease are found for both males and females in the Simon Fraser, Boundary and Burnaby Health Units. Both males and females have significantly low values in the North and South Okanagan Health Units and the Capital Regional District.

Significantly high SMRs for males (Table 4,1) occur in both Peace River and Cariboo (top 10 percent), but these areas do not have significantly high SMRs for females (Table 5,1). Conversely, a significantly high SMR for females occurs in the Central Kootenay Health Unit, but not for males. The high SMR for males in the Peace River appears to reflect a high

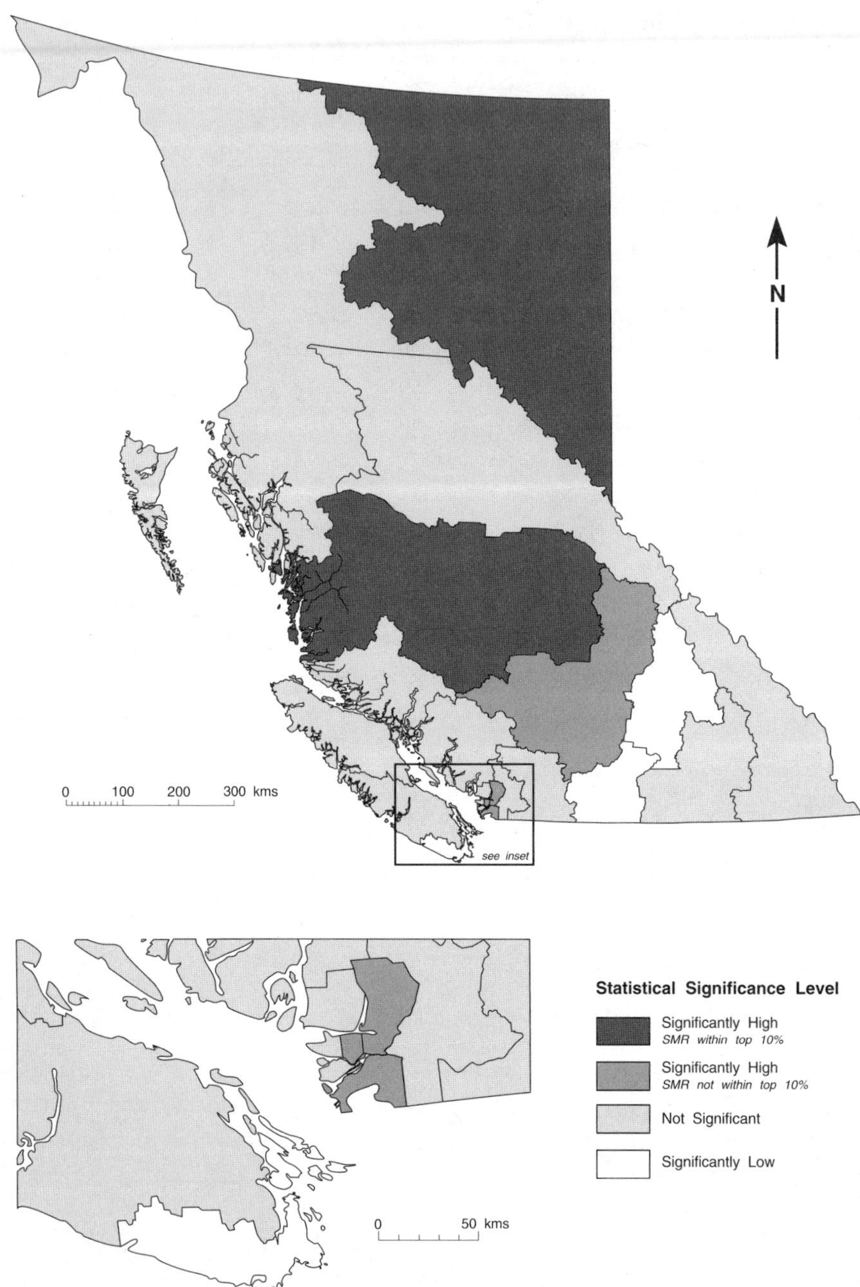

Figure 10,1 Ischaemic Heart Disease: SMR by Health Unit - Male (British Columbia, 1985-1989)

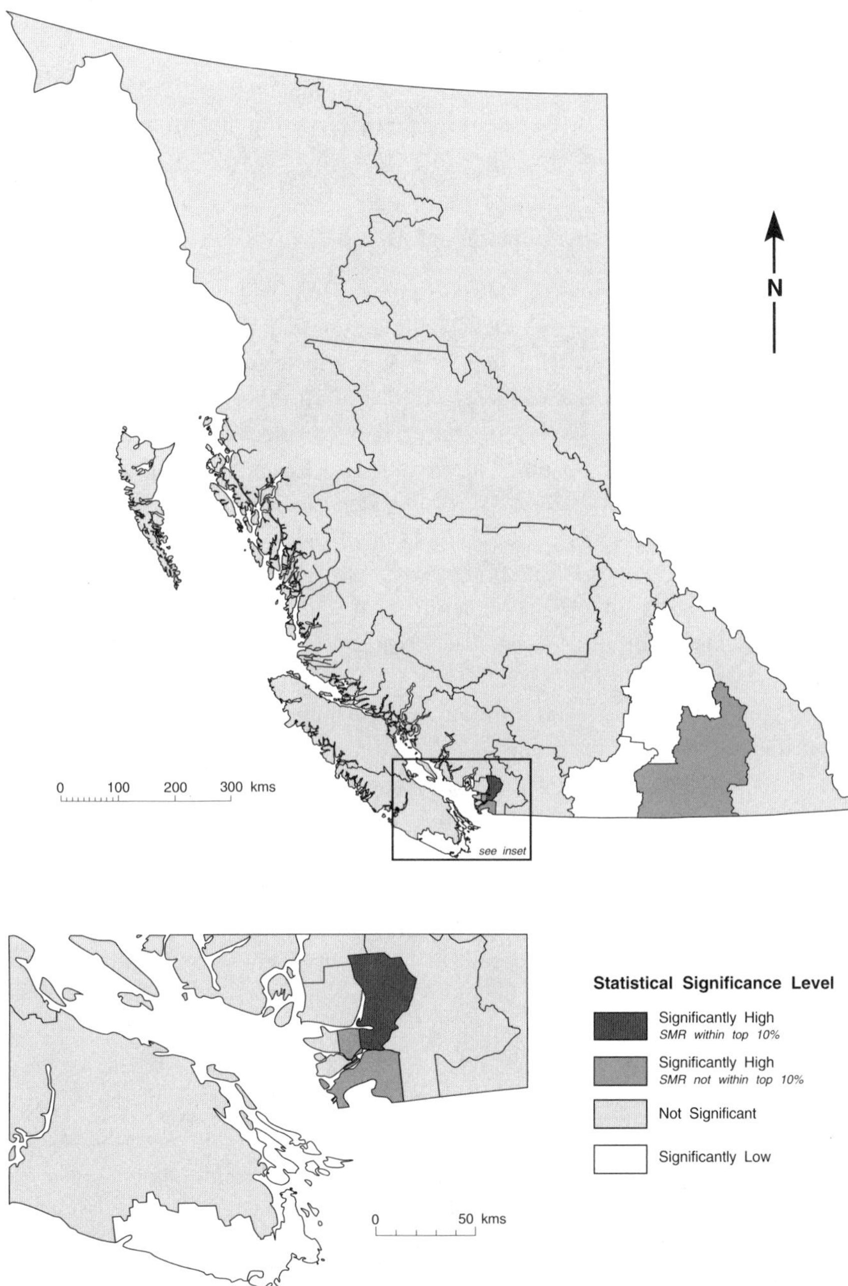

Figure 11,1 Ischaemic Heart Disease: SMR by Health Unit - Female (British Columbia, 1985-1989)

value in one area of that Health Unit, the Peace River South Local Health Area.

Generally, differentiating on the basis of gender does lead to differing patterns which need to be recognized in targeting health programming efforts.

Usual Residence versus Location of Death

The last map (Figure 12,1) and accompanying tables (Tables 6,1 to 8,1) illustrate the importance of clearly understanding the nature and limitations of the data. Figure 12,1 depicts motor vehicle traffic accident (MVTA) deaths for males by usual residence, by Local Health Areas. Motor vehicle traffic accident death data were analyzed by residency information, which is routinely collected on death registrations. From the results, it can be noted that the Enderby, South Cariboo, Keremeos, Princeton, Kettle Valley, Nechako and Cariboo-Chilcotin Local Health Areas have the significantly highest values (top 10 percent), while the more urban areas of Burnaby, Vancouver, North Vancouver, Coquitlam, Saanich, Richmond, West Vancouver-Bowen Island and Greater Victoria have the significantly lowest values (Table 6,1).

Prior to attempting interpretation, it is important to fully understand that this map depicts residence of the victim, *not* place of death. A number of MVTA deaths also occur in Local Health Areas where the victim is not a resident. This is demonstrated in Table 7,1, which draws a comparison between the percentages of male MVTA deaths that occur within the victim's LHA of usual residence and those that take place elsewhere. For example, of the total 19 male MVTA deaths of residents of Sooke, only one occurred within the Sooke Local Health Area. This suggests that educating drivers in Sooke would be more successful in reducing MVTAs than redesigning roads in the Sooke area.

Another consideration that must be taken into account when mapping MVTA deaths by usual residence is that one is *not* looking at the location where the death occurred. This analysis can, however, be carried out utilizing the data element *place of death*, which is captured on the death registration form. An example of this is provided in Table 8,1, which, in the case of Hope, reveals that of the total 38 male MVTA deaths that occurred in Hope, only two involved residents. Educating Hope residents on MVTA would not likely result in fewer deaths in the Hope area, but doing something to, or on, the road system, such as signage indicating frequent accident spots, might result in improvements.

Table 4,1 Standardized Mortality Ratio by Health Unit: Ischaemic Heart Disease - Male (British Columbia, 1985 to 1989)

	Health Unit	Observed Deaths	Expected Deaths	SMR	95% C.I. Lower		Upper
17	Peace River	188	159.52	1.18	1.02	-	1.36
15	Cariboo	232	201.26	1.15	1.01	-	1.31
06	South Central	457	400.77	1.14	1.04	-	1.25
10	Simon Fraser	731	640.72	1.14	1.06	-	1.23
31	Burnaby	894	787.81	1.13	1.06	-	1.21
09	Boundary	1377	1257.00	1.10	1.04	-	1.15
04	North Okanagan	488	553.43	0.88	0.81	-	0.96
05	South Okanagan	1007	1206.54	0.83	0.78	-	0.89
20	Capital Reg. Dist.	1556	1896.83	0.82	0.78	-	0.86

Statistically Significant High & SMR within top 10%
Statistically Significant High & SMR NOT within top 10%
Statistically Significant Low

Table 5,1 Standardized Mortality Ratio by Health Unit: Ischaemic Heart Disease - Female (British Columbia, 1985 to 1989)

	Health Unit	Observed Deaths	Expected Deaths	SMR	95% C.I. Lower		Upper
10	Simon Fraser	658	514.43	1.28	1.18	-	1.38
02	Central Kootenay	318	267.03	1.19	1.06	-	1.33
31	Burnaby	753	669.01	1.13	1.05	-	1.21
09	Boundary	922	850.1	1.08	1.02	-	1.16
04	North Okanagan	289	327.22	0.88	0.78	-	0.99
20	Capital Reg. Dist.	1437	1706.19	0.84	0.8	-	0.89
05	South Okanagan	617	750.66	0.82	0.76	-	0.89

Statistically Significant High & SMR within top 10%
Statistically Significant High & SMR NOT within top 10%
Statistically Significant Low

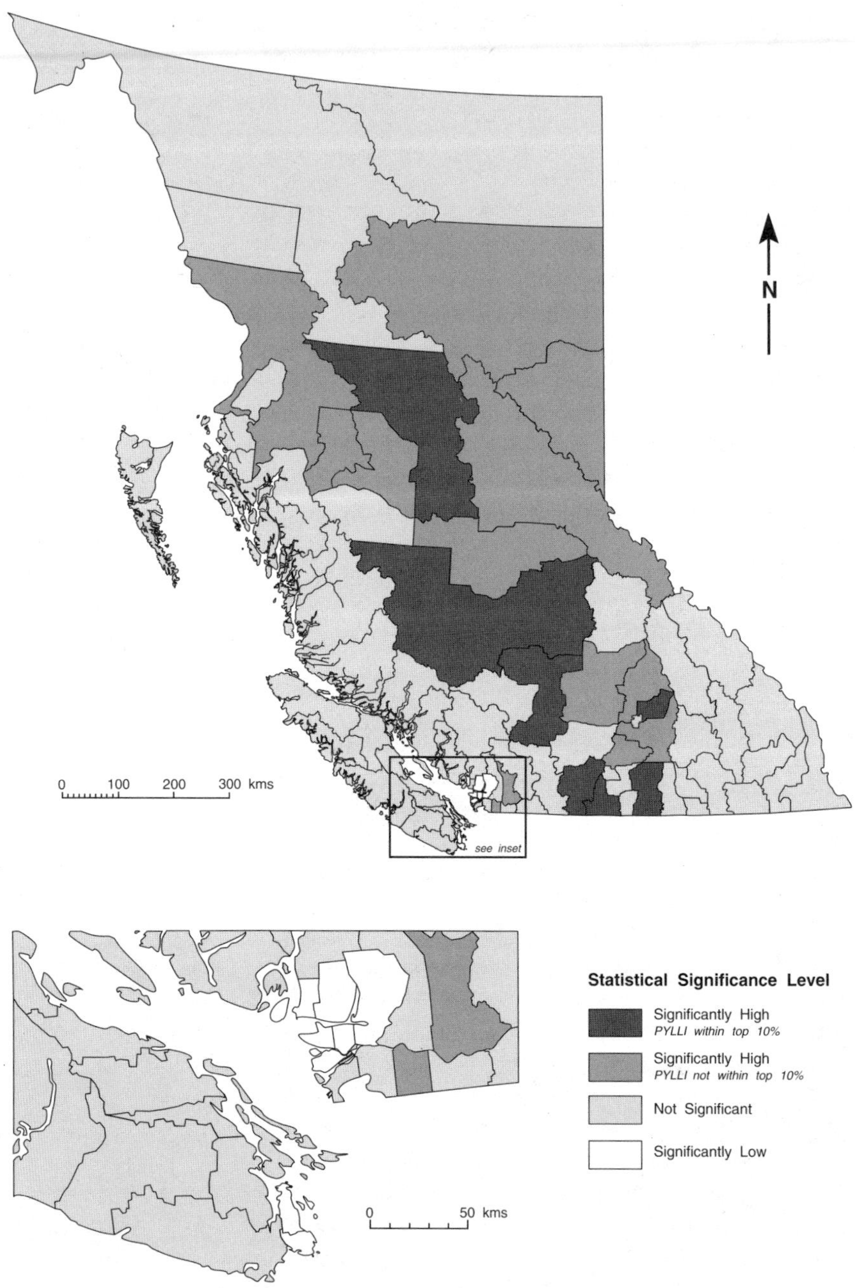

Figure 12,1 Motor Vehicle Traffic Accidents: SMR by Local Health Area for Usual Residence - Male (British Columbia, 1985-1989)

Table 6,1 Standardized Mortality Ratio by Local Health Area: Motor Vehicle Traffic Accident - Male (British Columbia, 1985 to 1989)

	Local Health Area	*Observed Deaths*	*Expected Deaths*	*SMR*	*95% C.I. Lower*	*95% C.I. Upper*
78	Enderby	16	3.08	5.20	2.97	8.45
30	South Cariboo	20	4.45	4.49	2.74	6.94
16	Keremeos	9	2.20	4.09	1.87	7.76
17	Princeton	10	2.86	3.49	1.67	6.43
13	Kettle Valley	6	1.87	3.22	1.17	7.00
56	Nechako	24	9.33	2.57	1.65	3.83
27	Cariboo-Chilcotin	55	21.84	2.52	1.90	3.28
54	Smithers	22	8.87	2.48	1.55	3.76
55	Burns Lake	11	4.75	2.32	1.15	4.14
20	Salmon Arm	27	14.10	1.91	1.26	2.79
88	Terrace	28	15.29	1.83	1.22	2.65
59	Peace River South	30	16.74	1.79	1.21	2.56
60	Peace River North	26	14.73	1.77	1.15	2.59
57	Prince George	92	53.37	1.72	1.39	2.11
75	Mission	26	15.66	1.66	1.08	2.43
28	Quesnel	23	14.06	1.64	1.04	2.45
22	Vernon	38	25.33	1.50	1.06	2.06
24	Kamloops	65	45.85	1.42	1.09	1.81
23	Central Okanagan	75	54.14	1.39	1.09	1.74
35	Langley	60	43.14	1.39	1.06	1.79
41	Burnaby	59	92.72	0.64	0.48	0.82
39	Vancouver	177	284.05	0.62	0.53	0.72
44	North Vancouver	36	64.79	0.56	0.39	0.77
43	Coquitlam	40	74.05	0.54	0.39	0.74
63	Saanich	11	23.81	0.46	0.23	0.83
38	Richmond	30	66.63	0.45	0.30	0.64
45	West Vancouver-Bowen Island	10	23.85	0.42	0.20	0.77
61	Greater Victoria	39	111.22	0.35	0.25	0.48

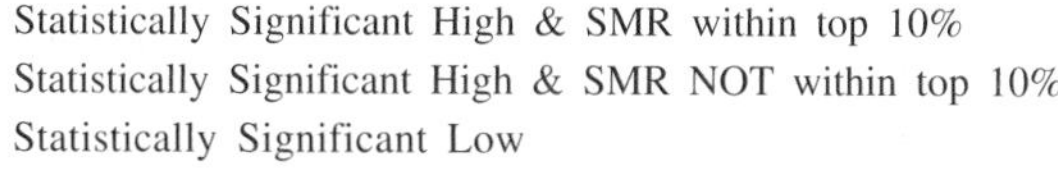

Table 7,1 Motor Vehicle Traffic Accidents by Local Health Area for Usual Residence - Male (British Columbia, 1985 to 1989)

LHA for Usual Residence		*Relationship to Place of Death* *Same LHA* Number	%	*Different LHA* Number	%	*Total* Number
45	West Vancouver-Bowen I.	0	0.0	10	100.0	10
62	Sooke	1	5.3	18	94.7	19
69	Qualicum	1	7.1	13	92.9	14
63	Saanich	1	9.1	10	90.9	11
43	Coquitlam	5	12.5	35	87.5	40
41	Burnaby	9	15.3	50	84.7	59
37	Delta	7	20.6	27	79.4	34
19	Revelstoke	2	22.2	7	77.8	9

Note: Prepared by Vital Statistics, Health - February 21, 1992.

Table 8,1 Motor Vehicle Traffic Accidents by Local Health Area for Place of Death - Male (British Columbia, 1985 to 1989)

LHA for Usual Residence		*Relationship to Place of Death* *Same LHA* Number	%	*Different LHA* Number	%	*Total* Number
32	Hope	2	5.3	36	94.7	38
40	New Westminster	9	6.9	121	93.1	130
69	Qualicum	1	16.7	5	83.3	6
19	Revelstoke	2	18.2	9	81.8	11
62	Sooke	1	20.0	4	80.0	5
31	Merritt	3	21.4	11	78.6	14
13	Kettle Valley	4	28.6	10	71.4	14
09	Castlegar	2	28.6	5	71.4	7

Note: Prepared by Vital Statistics, Health - February 21, 1992.

This is an interesting finding, as it points out the external nature of much of the motor vehicle traffic in this community, which is bisected by major highway routes. This finding applies to the other Local Health Areas listed in Table 8,1, with the exception of New Westminster. Here, the non-resident death phenomenon would more likely be attributed to the location of the Royal Columbian Hospital, which acts as a main trauma centre for MVTAs occurring in the lower mainland/Fraser Valley area. From this viewpoint, place of death has its limitations in mapping, as it may just indicate the location where the pronouncement of death was made. Furthermore, due to the transitory nature of the population at risk, the ascertainment of a population denominator to be used in the calculation of mortality rates is not possible.

In summary, when mapping events such as MVTA deaths one should be fully aware of the intricacies and limitations of the data being mapped. In the case of examining various indicators to be used in preventative efforts aimed at reducing MVTA deaths, place of death might be preferable to place of residence, while "place of occurrence" might be the optimum indicator.

CONCLUSION: THE FUTURE OF MORTALITY MAPPING

The recent progress made in mapping mortality and other correlates of health has been substantial. However, the full potential of such techniques and their incorporation as viable tools in our health planning and prevention efforts have yet to be fully realized in British Columbia and Canada. The Chinese have been mapping health correlates for several decades, and have taken the exercise a step further in overlaying environmental and demographic variables in an attempt to shed light on etiological or epidemiological questions. This pursuit has resulted in a number of measures that have helped to reduce the incidence of several diseases and generally promote human longevity.[45]

One of the principal goals in British Columbia's mortality mapping initiatives is the dissemination of relevant community level information to local health care practitioners and planners to assist them in addressing their specific health care challenges. Such data may also help identify pertinent questions about the health of residents in an area that in turn lead to the generation of hypotheses which may be tested through further detailed investigations. It is partially through efforts to provide individual communities with information that is of high utility and quality that local health care priorities can be identified and appropriate decisions made. This is consistent with the Ministry's initiative to empower communities to take more responsibility for their residents' health by allowing them to examine local data, which may help them to identify local problems they can address. The problems of Atlin in the extreme north are substantially different from the problems of Victoria in the south. Locally based data can assist in the development of locally based, targeted programming.

Furthermore, communities can be provided with comparative data in order to measure their progress against neighboring communities or other similar areas. A "mentor community" is selected on the basis of having achieved a desirable health status in regard to a specific issue (e.g. PYLL) or condition (e.g. coronary heart disease). This community then acts as the benchmark for other communities to strive toward attaining or even surpassing. This concept is presently being utilized at the national level.[46]

It is important to note that measures that act to enhance the health of a population can only be made truly effective when there is recognition by the communities that many health-debilitating conditions are "lifestyle related". These conditions can be closely associated with diet and behavior, although genetics is being viewed increasingly as a predisposing agent. A certain degree of responsibility must be placed with the individual for his or her own well-being.

Notwithstanding the benefits of providing communities with health information such as mortality maps, there is an inherent danger in doing so, as indicated earlier. In the absence of proper explanation and analyses, hasty and inappropriate conclusions may be made that misrepresent the data. The trade-off, however, is probably worth it, and the danger can be reduced if the release of such information is accompanied by caveats and user education.[47] This point is an important issue that will need to be addressed by a partnership of public officials; health providers, planners and researchers; the media; and last but not least, instructors in our education system.

REFERENCES

[1] Howe, G.M. 1989. Historical Evolution of Disease Mapping in General and Specifically Cancer Mapping, in Boyle, P., Muir, G.S. and Grundmann, E. (eds.), *Cancer Mapping*. Paris: Springer-Verlag, pp. 1-21.

[2] Kemp, I. and Boyle, P. (eds.) 1985. *Atlas of Cancer in Scotland (1975-1980): Incidence and Epidemiological Perspective*. Lyon: IARC Scientific Publications, p.11.

[3] *Ibid.*

[4] Haviland, A. 1855. The Geographical Distribution of Diseases in Great Britain. London: Smith Elder.

[5] Stocks, P. 1936, 1937 and 1989. Distribution in England and Wales of Cancer of Various Sites, *Annual Reports of the British Empire Cancer Campaign*.

[6] Howe, G.M. 1970. *National Atlas of Disease Mortality in the United Kingdom*. London: Nelson.

[7] Walter, S.D. and Birnie, S.E. 1991. Mapping Mortality and Morbidity Patterns: An International Comparison. *International Journal of Epidemiology* 20(3).

[8] Health and Welfare Canada. 1980. *Mortality Atlas of Canada, Volume 1: Cancer; Volume 2: General Mortality; Volume 3: Urban Mortality*. Hull: Canadian Government Publishing Centre.

[9] Saskatchewan Cancer Foundation. 1988. *Saskatchewan Cancer Atlas 1970-1987*. Regina: Cancer Registry Report.

[10] Band, P.R., Spinelli, J.J., Gallagher, R.P., Threfall, W.J., Ng, V.T.Y., McBride, M.L., Hislop, T.G., and Coldman, A.J. 1989. *Atlas of Cancer Mortality in British Columbia: 1956-1983*. Ottawa: Statistics Canada; Ghadirian, P., Thouez, J.P., Petitclerc, C., Rannou, A., and Beaudoin, Y. 1989. *Cancer Incidence: Atlas of the Province of Quebec 1982-1983*, Montreal, Hotel-Dieu Hospital, Epidemiology Research Unit.

[11] Alderson, M.R. 1988. *Mortality, Morbidity and Health Statistics*. Basingstoke: MacMillan Press Ltd., pp. 96-98

[12] Along with British Columbia, other Provinces with major reviews of health systems include: Alberta, New Brunswick, Nova Scotia, Ontario, Quebec and Saskatchewan.

[13] British Columbia. 1990. *Royal Health Commissionaires Named*, News Release. Victoria: Ministry of Health.

[14] Danderfer, R. and Foster, L. (eds.) 1989. *Vital Statistics of the Province of British Columbia, 1988 - 117th Annual Report*. Victoria: Division of Vital Statistics, Ministry of Health.

[15] British Columbia. 1990. *Ministry of Health - 1988/89 Annual Report.* Victoria: Ministry of Health.

[16] Danderfer, R. and Foster, L. (eds.) 1990. *Vital Statistics of the Province of British Columbia, 1989 - 118th Annual Report.* Victoria: Division of Vital Statistics, Ministry of Health.

[17] Bliss, G., Foster, L. and Ho, K. 1990. Mapping Community Health Information, in Coward, J.H. (ed.), *Proceedings of the Third National and the First International Conference on Information Technology in Community Health.* Victoria: University of Victoria Conference Services.

[18] Charlton, J.R.H. 1987. Avoidable Deaths and Diseases as Monitors of Health Promotion, in Ablin, T., Brzezniski, Z.J. and Carstairs, V. (eds.), *Measurement in Health Promotion and Protection.* Copenhagen: World Health Organization, Regional Office for Europe, p. 470.

[19] Kemp, I. and Boyle, P., *op. cit.*, p. 9.

[20] "Rural Hospitals Voice Complaints to Judge", *Vancouver Sun*, Saturday, August 25, 1990, p. A6.

[21] Gardner, M.J., Winter, P.D., Taylor, C.P., and Acheson, E.D. 1979. *Atlas of Cancer Mortality in England and Wales (1968-1978).* Chichester: John Wiley & Sons.

[22] Charlton, J.R.H., *op. cit.*, p. 469.

[23] Kemp, I. and Boyle, P., *op. cit.*, p. 41; Gardner, M.J. et al., *op. cit.*, p. 1.

[24] Charlton, J.R.H., *op. cit.*, p. 468.

[25] Gardner, M.J. et al., *op. cit.*, p. 3.

[26] Baird, P. 1990. Underlying Causes of Death in Down Syndrome: Accuracy of British Columbia Death Certificate Data, *Canadian Journal of Public Health* 81(6), pp. 456-461.

[27] Gardner, M.J. et al., *op. cit.*, p. 3; Charlton, J.R.H., *op. cit.*, p. 468; and Health and Welfare Canada, Vol. 2, *op. cit.*, p. 8.

[28] Kemp, I. and Boyle, P., *op. cit.*, p. 11.

[29] Walter, S.D. and Birnie, S.E., *op. cit.*, p. 17.

[30] Pickle, L.W., Mason, T.J., Howard, N., Hoover, R., and Fraumeni, J.F. 1987. *Atlas of U.S. Cancer Mortality Among Whites (1950-1980).* Washington, D.C.: DHHS Publication (NIH) 87-2900, U.S. Government Printing Office.

[31] Pickle, L.W. et al., *op. cit.*; Danderfer, R. and Foster, L., 1989, *op. cit.*; Danderfer, R. and Foster, L., 1990, *op. cit.*

[32] Band, P.R. et al., p. 6.

[33] Pickle, L.W. et al., *op. cit.*, p. 19.

[34] Gardner, M.J. et al., *op. cit.*, p. 1.

[35] *Ibid.*, p. 3.

[36] Charlton, J.R.H., *op. cit.*, p. 470.

[37] Non census year population estimates were provided by the Planning and Statistics Division of the Ministry of Finance and Corporate Relations.

[38] National Centre for Health Statistics (NCHS). 1989. *Advance Report of Final Mortality Statistics, 1987.* Washington, D.C.: DHHS Publication, Vol. 38, No. 5 (Suppl.).

[39] Pickle, L.W. et al., *op. cit.*

[40] Ghadirian, P. et al., *op. cit.*

[41] Breslow, N.E. and Day, N.E. 1987. *Statistical Methods in Cancer Research: Volume II - The Design and Analysis of Cohort Studies.* Lyon: International Agency for Research on Cancer.

[42] Band, P.R. et al., *op. cit.*

[43] Kemp, I. and Boyle, P., *op. cit.*

[44] Romeder, J.M. and McWhinnie, J.R. 1977. Potential Years of Life Lost Between Ages 1 and 70: An Indicator of Premature Mortality for Health Planning, *International Journal of Epidemiology* 6(2), pp. 146-149.

[45] Tan, J., Li, R. and Zhu, W. 1990. *Recent Development of Geographical Science in China.* Beijing: Science Press.

[46] Institute for Health Care Facilities of the Future. 1990. *A View of the Horizon: Regional Trends.* Ottawa: Canadian Hospital Association.

[47] Bliss, G., Foster, L.T. and Ho, K., *op. cit.*, p. 41.

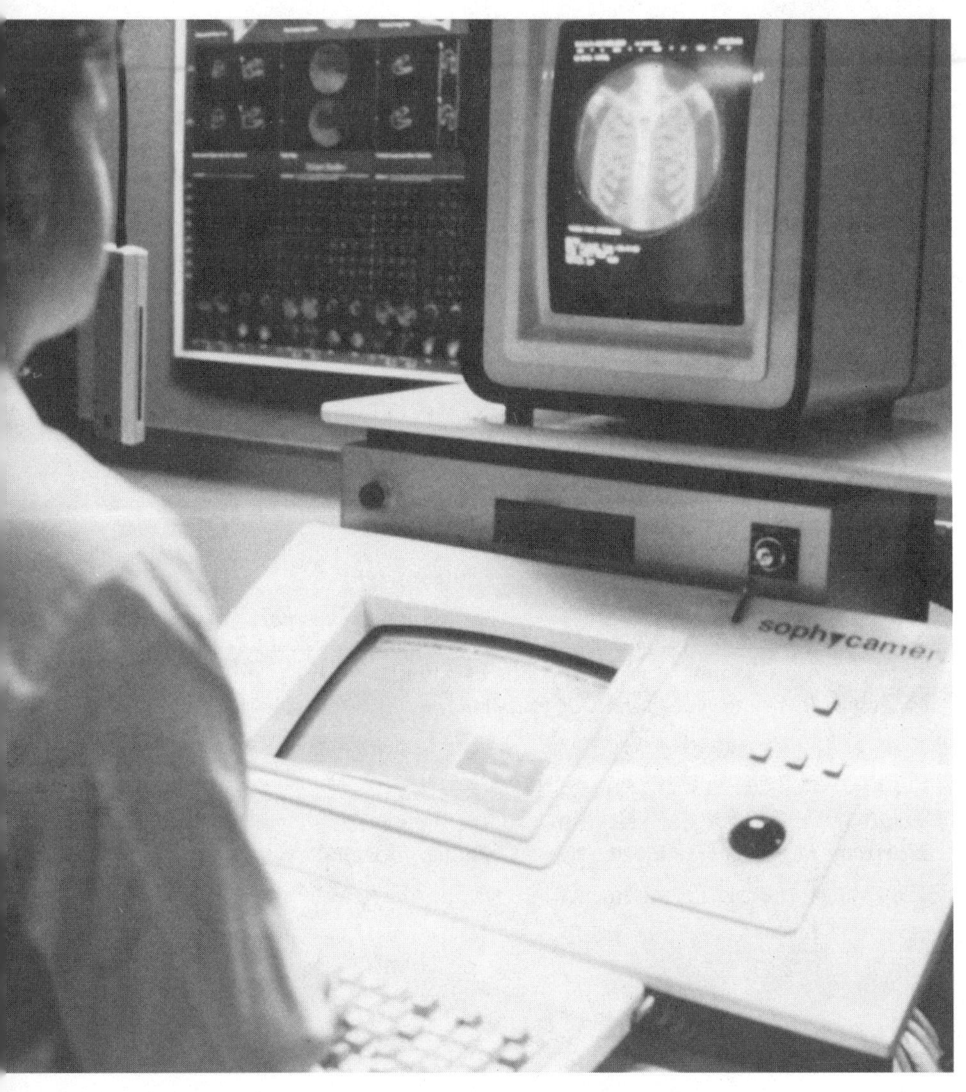

2 CANCER MAPPING IN BRITISH COLUMBIA

John J. Spinelli
Pierre R. Band
Richard P. Gallagher
William J. Threlfall
Vincent Ng

Division of Epidemiology, Biometry and Occupational Oncology
British Columbia Cancer Agency

INTRODUCTION

The study of the geographic distribution of cancer has proved useful in identifying high risk areas to be studied further using analytical epidemiological methods. The British Columbia Cancer Agency (BCCA) has developed a program to examine regional differences in cancer rates within British Columbia. This chapter will outline the geographic mapping program of the BCCA, explore some of the methodological issues and present selected results.

METHODS AND LIMITATIONS

Mortality

In a previous mortality study, data on all deaths occurring in British Columbia from 1956 to 1983 inclusive were obtained from the Division of Vital Statistics. Information on each death included the birth date, sex, place of residence and underlying cause of death. Deaths were mapped by school district; a total of 52,891 male and 40,708 female cancer deaths were included. A detailed description of the methodology can be found in the monograph, *Atlas of Cancer Mortality in British Columbia: 1956-1983.*[1]

Mortality rates have been used in preference to incidence rates for the geographic mapping of cancer for practical reasons. Death certificates are readily available and complete in most countries of the world, whereas cancer incidence statistics rely on the registration of newly diagnosed cases, which is not available in many areas and may vary in accuracy between, and even within, countries. The information on death certificates, however, presents well recognized inaccuracies and inherent limitations, specifically inaccuracies in cause of death coding and mobility after diagnosis of disease.

Cause of death coding may lack precision, and determination of the primary tumour may be difficult in patients dying of widespread metastases. Also, cancer patients may die of other causes, without the diagnosis of cancer being recorded on the death certificate.[2] This is particularly true of tumours with good prognosis, including bladder cancer and Hodgkin's disease. Moreover, studies have indicated that 15 to 20 percent of deaths certified as being due to cancer were probably attributable to other causes. The site specification was found to be particularly questionable for cancer of the stomach and pancreas.[3] In a comparison of death certificates with autopsy results, it was found that 25 percent of the cancers documented at autopsy were either not recorded on the death certificate, or ascribed to an incorrect primary site.[4] The unavailability of histologic information on death certificates makes the identification of some cancers (eg. mesotheliomas, angiosarcomas) impossible.

The decedent's residence at time of death is indicated on the death certificate, but about 10 percent of cancer patients, particularly from rural areas, will relocate between diagnosis and death to urban areas where more adequate treatment facilities may be available.[5] Thus, geographic studies of cancer mortality patterns may overestimate urban and underestimate rural cancer rates.

Incidence

For the above reasons, an examination of cancer incidence by region in British Columbia was undertaken. All cases (excluding non-melanoma skin cancer cases) ascertained by the B.C. Cancer Registry from 1983 to 1989 inclusive were studied. Since 1970, pathology reports on every newly diagnosed cancer case are submitted to the Registry from every pathology service in the province. Reciprocal agreements also exist with the three other western provinces, where B.C. residents ascertained in those provinces are reported. Ascertainment for most sites is considered to be virtually complete.

It was necessary to acquire information on address and postal code of residence at time of diagnosis for all patients registered between 1983 and 1989. For cases seen at the B.C. Cancer Agency and participants in an epidemiologic study of most males diagnosed during this time period, postal code information was available. For cases diagnosed at time of death, the school district and sometimes postal code at time of death was available from the Division of Vital Statistics. Address and postal code information was obtained from the referring physician or pathology laboratory for the remaining cases.

The use of incidence data also has some inherent limitations. Although the problem of mobility after diagnosis is eliminated by the use of incidence data, mobility prior to diagnosis is still a major concern. No information is available on the duration of residence prior to diagnosis, and in the case of retirement areas it is likely that an elderly individual would have spent the majority of his or her life in another region. Therefore, the incidence rates in those regions may not reflect exposure to risk factors in those areas, thus obscuring real geographic differences. Mapping studies also do not usually include data on lifestyle information, such as cigarette smoking, alcohol consumption or dietary information. These lifestyle factors, which account for many cancers, may influence incidence rates in certain areas.

Statistical Limitations

Two final limitations of geographic mapping studies that will be presented are statistical in nature. The first is known as the multiple comparison problem. In an analysis where many statistical tests are performed, it is likely that some significant differences, regions with an incidence rate different from the provincial rate, will be observed by chance. A significance test is made at a particular level, usually 0.05; meaning that there is a 5 percent chance that a significant difference will be observed even if there is no true difference. In a mapping project with approximately 75 regions and 40 cancer sites, 3000 significance tests are made. On average, therefore, 60 significantly elevated or lowered incidence rates will be observed simply by chance. Thus, an observed significantly elevated incidence rate cannot alone be interpreted as a true elevated rate.

The converse limitation is lack of power. Power is the probability that a significant difference will be observed given that a true difference exists. In a mapping study, the power depends on two factors: the magnitude of the true difference and the expected number of cases. The expected number of cases is related to the population and age structure of a region, and the provincial incidence rate. An elevated cancer rate may be difficult to ascertain when the cancer is rare and the population of a region is small, as is the case for some of the B.C. school districts.

Interpretation

Due to the inherent methodological limitations of geographic mapping studies, the interpretation of regional variation in cancer rates is extremely difficult. In an attempt to minimize the reporting of false positive associations, guidelines to eliminate spurious results were initiated. The criteria for further examination are as follows:

1. School districts with a significantly elevated cancer rate for both males and females, or
2. School districts with a significantly elevated cancer rate on both the mortality and incidence analyses.

METHODOLOGICAL CONSIDERATIONS

Despite the inherent limitations, the geographic mapping of cancer has been helpful in generating hypotheses for further epidemiologic research. Even in the absence of the hypothesis generating function, geographic mapping offers an attractive method to represent incidence and mortality data. Once it is decided to undertake a mapping project, several methodologic considerations need to be addressed, including: the procedure for estimating risk; the choice of geographic unit; and the choice of values to be mapped.

The desired outcome of any mapping project is to examine regional variation in disease rates. A rate is the number of cases with the disease in a given time period divided by population at risk for the disease during that time period. The following notation will be helpful to the discussion:

I = number of age groups,
n_{ij} = number of cases of disease in age group i and region j,
p_{ij} = population at risk in age group i and region j,
N_i = number of cases of disease in the entire province in age group i,
P_i = population at risk in the entire province in age group i,
$r_{ij} = n_{ij} / p_{ij}$ = the disease rate in age group i and region j.

The population at risk is calculated by summing the annual population figures, or by multiplying a mid-period population by the number of years in the time period. For example, over a 10 year period, a region with a mid-period population of 10,000 people will have a population at risk of 100,000.

The simplest comparison of regions is by use of a crude rate. The crude rate in region j is defined as follows:

$$r_j = \sum_{i=1}^{I} n_{ij} / \sum_{i=1}^{I} p_{ij}$$

Since cancer incidence and mortality rates are age dependent, the crude rate can only be used to compare two regions if the age distributions in the two regions are similar. If the two regions have different age distributions, the comparison of rates between the two regions will be confounded by age. Since the age distributions of regions vary greatly in British Columbia, the crude rate is not a useful measure. A method which is used to adjust rates to compare different regions is referred to as age-standardization or age-adjustment.

There are two methods of adjusting for age: direct and indirect. Both methods have advantages and disadvantages. A directly standardized rate is a weighted sum of stratum (usually age) specific rates, with the weights generally based on the population of one of the two regions being compared or on some "standard population". A crude rate can be thought of as a directly standardized rate with weights proportional to the strata population of that region. A directly standardized rate is referred to as an age standardized mortality (or incidence) rate (ASMR or ASIR). The ASMR for region j is defined as:

$$ASMR_j = \sum_{i=1}^{I} w_i \frac{n_{ij}}{p_{ij}}$$

where w_i is the weight for age-stratum i. The ratio of two directly standardized rates is referred to as the comparative mortality (or incidence) figure (CMF or CIF). For the mapping of cancer in B.C., each school district is compared to the entire B.C. population. The CMF for region j is thus defined as follows:

$$CMF_j = \frac{\sum_{i=1}^{I} w_i \frac{n_{ij}}{p_{ij}}}{\sum_{i=1}^{I} w_i \frac{N_i}{P_i}}$$

The method of indirect standardization is a comparison of a region with a standard population, and is referred to as the standardized mortality (or incidence) ratio (SMR or SIR). The SMR is a ratio of observed to expected deaths, defined:

$$SMR_j = \frac{\sum_{i=1}^{I} n_{ij}}{\sum_{i=1}^{I} p_{ij} \frac{N_i}{P_i}}$$

An important property of directly standardized rates is that if the rate in each stratum is greater in region A than in region B, then the directly standardized rate for region A will be larger than the directly standardized rate for region B. This property is known as consistency and will generally not be true for two indirectly standardized rates.

The major advantage of indirectly standardized rates is that the stability of the SMR is higher than that of the CMF. This means that an indirectly standardized rate is less affected by small populations in any of the strata, and thus more stable. In fact, under reasonable assumptions, the SMR is the minimum variance estimate of the common rate ratio.[6]

Both directly and indirectly standardized rates have been used in the production of cancer atlases. Due to the small populations of many regions of B.C., the SMR and SIR were the measures utilized in the B.C. mortality atlas and incidence mapping study.

The second methodologic issue to be discussed is the choice of the geographic unit. Ideally, geographic units should be chosen to reflect areas which are most homogeneous within areas and most heterogeneous between areas - for example, with respect to the environment, industry and lifestyle. Usually, however, units must be based on administrative areas which have no identifiable common characteristics. Within this constraint, the investigator can only make a choice as to the size of the unit. Smaller units are preferred on the basis of interpretability, as larger regions will generally include more heterogeneous populations. Amalgamation of areas may also obscure the identification of high risk areas by diluting the area of higher incidence with areas of no increased incidence. The use of smaller units will, however, lead to a study with lower power to detect significant differences due to smaller populations.

In the B.C. mortality atlas, school districts were chosen for the geographic unit because they were the only regions for which population figures existed over the entire time span, and whose boundaries remained relatively constant during that time period. For the incidence mapping project the postal code is available, so that any geographic unit with available population data for the time period can be examined. For the purpose of comparison with the mortality data, school districts were the first unit examined.

The choice of class intervals for display on maps is one of the most important and difficult steps in constructing maps, as there is no consensus concerning the best procedure.[7] The primary value to be indicated on a cancer map is the incidence rate or relative incidence rate of a particular region. Because the estimates of incidence rate are more precise in areas with larger populations, the display of classes should somehow reflect this. Usually, this is accomplished by indicating statistical significance as an interval discriminator. That is, classes are chosen based both on the absolute incidence rate and the statistical significance of the rate. In the B.C. mortality atlas, four intervals were used: significantly high, SMR in the highest decile but not statistically significant, significantly low and other. The same intervals are used in the cancer incidence maps, except that SIRs were used instead of SMRs.

Advanced Statistical Techniques

Some recent advances in statistical methodology have presented alternatives to the traditional estimation procedures and class intervals for display.[8] This new methodology is summarized in a recent monograph.[9] Ideally, the risk estimates to be mapped should be of approximately the same precision for each region. The empirical Bayes estimate does this by utilizing the data in all the regions.

The rationale is as follows. If there are no data in a region, the best estimate of the incidence rate in that region is the incidence rate of the entire population. As the data in the region increases, the estimate is based more on the regional estimate of incidence and less on the overall incidence rate. If we consider the empirical Bayes estimator (EB) as an extension of the standardized incidence ratio (SIR), the EB for region j is defined as follows:

$$EB_j = \frac{\sum_{i=1}^{I} n_{ij} + \alpha}{\sum_{i=1}^{I} p_{ij} \frac{N_i}{P_i} + \beta}$$

where α / β is an estimate of the SIR based on the entire population. Clearly, as the observed and expected number of cases increases, the EB is based more on the data for the region and less on the data for the entire population. The estimates of α and β, and thus the EB, will depend on the particular model chosen for the variation in SIR across regions. An extension of this idea is to use the incidence of areas near to the region of interest rather than the whole population, or to weight the influence of other regions by the distance from the region of interest. This extension can also be handled by the empirical Bayes approach.

RESULTS

Table 1,2 shows the proportion of cases for whom address information was successfully determined, by sex and status at diagnosis (alive or deceased). A total of 78,378 cases, 41,282 male and 37,096 female, were available for analysis. The all cancer incidence maps for males and females are presented in Figures 1,2 and 2,2. The maps illustrate some of

Table 1,2 Number of Cases Registered by the B.C. Cancer Registry During the Years 1983-1989, and the Number and Percent of Cases for whom School District at Diagnosis was Obtained, by Sex and Status at Diagnosis

			School District at Diagnosis Obtained (%)
Total Cases	80,122	78,378	(97.8)
Males	41,833	41,282	(98.7)
Alive	37,595	37,149	(98.8)
Diagnosis at Death	4,238	4,133	(97.5)
Females	38,289	37,096	(96.9)
Alive	34,289	33,854	(96.9)
Diagnosis at Death	3,336	3,242	(97.2)

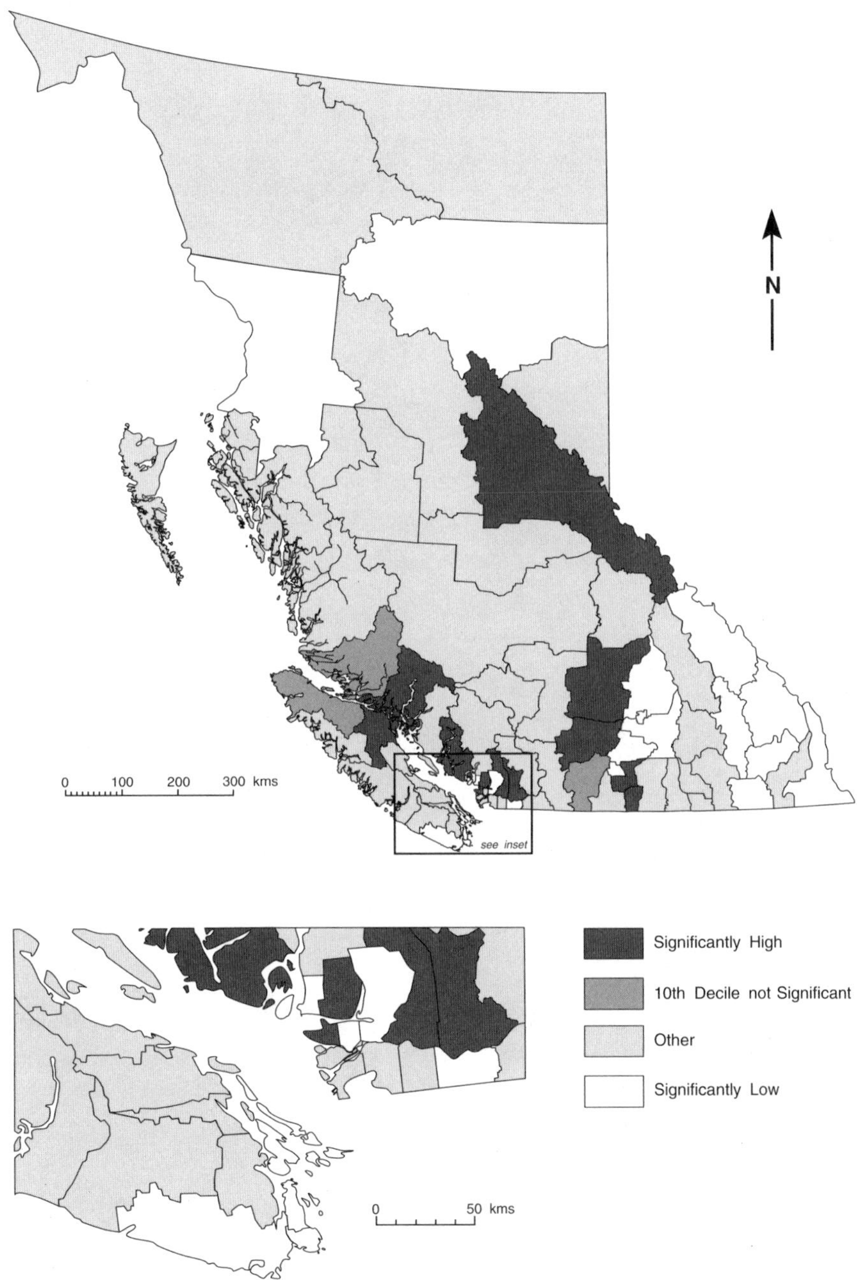

Figure 1,2 All Cancers (Male)

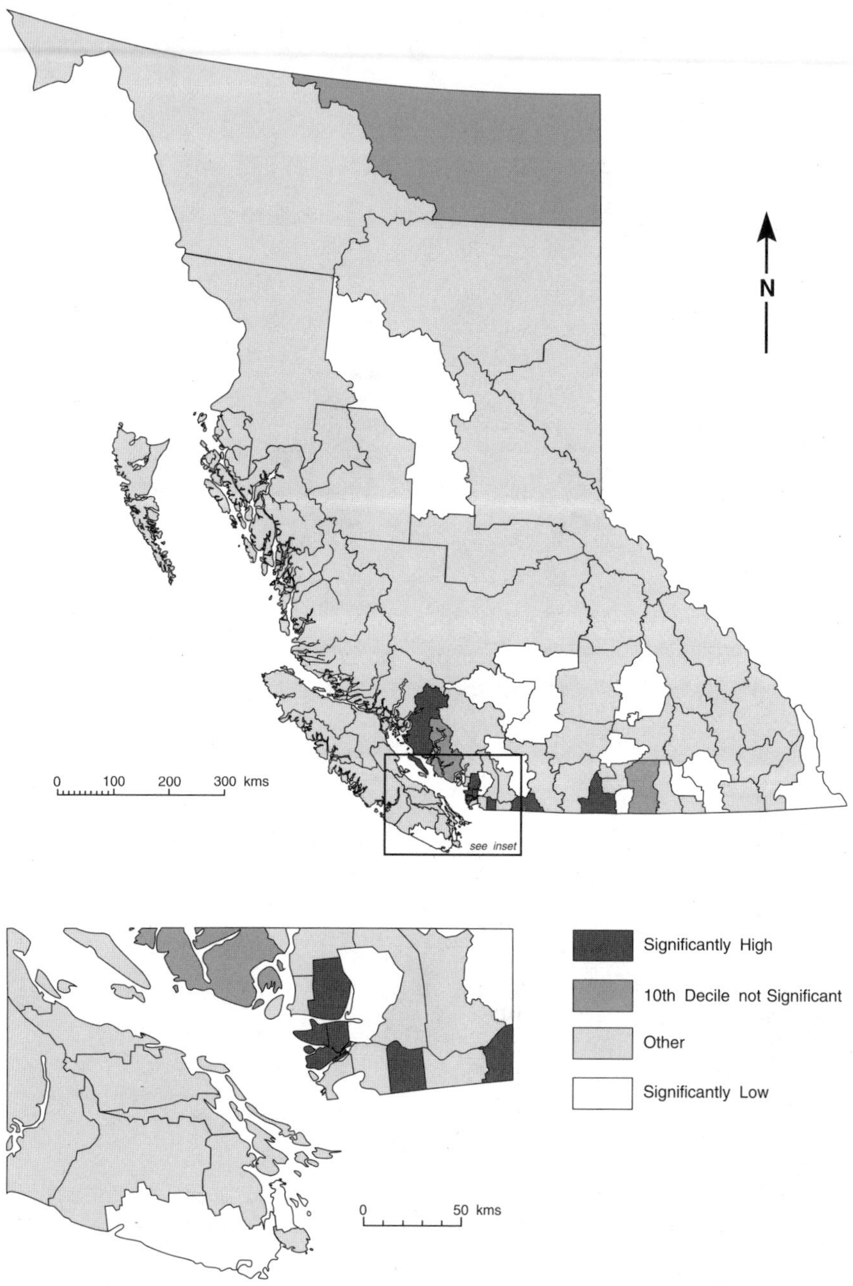

Figure 2,2 All Cancers (Female)

the difficulties of interpretation in cancer maps. The areas in the northern part of the province are visually most striking, but are also the areas with the least population. Also, there is little overlap in high risk areas between the cancer maps for males and females. Only three school districts were significantly high in all cancer incidence for both males and females; Vancouver, the largest city in British Columbia, and two school districts in the greater Vancouver area - New Westminster and North Vancouver.

Several school districts did have specific cancer sites with elevated SIRs for both males and females, and these are presented in Table 2,2. The results from the mortality mapping are also included for comparison. The Vancouver school district had elevated incidence and mortality of cancers at three sites, the mouth/pharynx, esophagus and liver. The incidence of cancers at all three of these sites has been related to alcohol consumption. Lung cancer incidence was elevated for both males and females in three school districts, Kamloops, New Westminster and Courtenay, none of which had elevated lung cancer mortality.

Table 2,2 School Districts with a Cancer Site with an Elevated Standardized Incidence Ratio (SIR) for Both Males and Females and the Corresponding Standardized Mortality Ratio (SMR)

School District/			*Males*			*Females*	
Cancer Site		*No.*	*SIR/SMR*	*Rank+*	*No.*	*SIR/SMR*	*Rank*
Kamloops (24)							
Lung	I	182	1.21**	15	98	1.31**	11
	M	250	0.97	26	68	1.08	33
Vancouver (39)							
Mouth/Pharynx	I	314	1.63**	7	139	1.28**	13
	M	389	1.58**	6	154	1.14	29
Esophagus	I	128	1.34**	23	72	1.39**	15
	M	388	1.29**	10	201	1.27**	26
Liver	I	127	1.84**	7	60	1.44**	13
	M	245	1.51**	8	105	1.04	36
New Westminster (40)							
Lung	I	166	1.28**	8	98	1.30**	12
	M	339	1.11	6	106	1.11	23

Table 2,2 continued

School District/ Cancer Site		*Males* No.	*SIR/SMR*	*Rank+*	*Females* No.	*SIR/SMR*	*Rank*
West Vancouver (45)							
Melanoma	I	31	1.53*	4	41	1.82**	4
	M	10	1.31	14	12	1.58	16
Cowichan (65)							
Pancreas	I	28	1.90**	5	20	1.63**	8
	M	52	1.14	12	31	1.10	26
Courtenay (71)							
Lung	I	119	1.22*	14	70	1.45**	5
	M	159	0.98	21	45	1.14	20
Campbell River (72)							
Stomach	I	17	1.64*	7	9	1.92*	5
	M	31	1.08	19	8	0.75	55

I = Incidence M = Mortality

* p<.05 ** p<.01

\+ Rank of the SIR/SMR for that school district relative to the SIR/SMR for the other 73 districts. A rank of 1 indicates the largest SIR/SMR.

School districts with significantly elevated SIRs for specific cancer sites on both the mortality and incidence analyses are presented in Tables 3,2 and 4,2, for males and females respectively. For males, the Vancouver school district had five sites with elevated incidence and mortality, cancers of the mouth/pharynx, esophagus, liver, larynx, and lung. Cancer of the esophagus also showed elevated incidence and mortality for females. Other notable results are an elevated incidence and mortality from lung cancer in males in the Prince George school district. For females, Prince George also showed an significantly elevated lung cancer mortality and a non-significantly elevated lung cancer incidence. The Victoria school district showed a small but significantly elevated incidence and mortality from breast cancer, the cancer site with the largest incidence for females.

Table 3,2 School Districts with a Cancer Site with Elevated Incidence and Mortality for Males

School District	*Site*	*Incidence*			*Mortality*		
		No.	*SIR*	*Rank+*	*No.*	*SMR*	*Rank*
Armstrong -Spallumcheen (21)	Myeloma	6	4.27**	3	5	2.34*	2
Kamloops (24)	Prostate	228	1.27**	13	110	1.28**	12
Vancouver (39)	Mouth/Pharynx	314	1.63**	7	389	1.58**	6
	Esophagus	128	1.34**	23	388	1.29**	10
	Liver	127	1.84**	7	245	1.51**	8
	Larynx	138	1.24*	24	197	1.51**	7
	Lung	1408	1.10**	26	4230	1.19**	2
Prince George (57)	Lung	146	1.33**	7	230	1.15*	4
Peace River South (59)	Prostate	70	1.38**	7	47	1.54**	4

* p<.05 ** p<.01

\+ Rank of the SIR/SMR for that school district relative to the SIR/SMR for the other 73 districts. A rank of 1 indicates the largest SIR/SMR.

Table 4,2 School Districts with a Cancer Site with Elevated Incidence and Mortality for Females

School District	*Site*	*Incidence*			*Mortality*		
		No.	*SIR*	*Rank+*	*No.*	*SMR*	*Rank*
Vancouver (39)	Esophagus	72	1.39**	15	201	1.27**	15
	Cervix	268	1.46**	12	459	1.14**	29
New Westminster (59)	Rectum	48	1.33*	13	61	1.45**	10
Victoria (61)	Breast	1009	1.07*	15	1022	1.13**	10

* p<.05 ** p<.01

\+ Rank of the SIR/SMR for that school district relative to the SIR/SMR for the other 73 districts. A rank of 1 indicates the largest SIR/SMR.

Table 5,2 shows the results of the empirical Bayes estimation for male lung cancer and a comparison with the classical estimation procedures. The school district with the largest standardized incidence ratio (SIR) is Fort Nelson, with an SIR of 2.40, which is statistically significant but based on only 10 cases. The empirical Bayes (EB) estimate, however, is 1.21, the seventh largest, indicating a large degree of uncertainty in the SIR. More striking are the results for Vancouver Island West, which had the largest comparative incidence figure (CIF), 3.50 based on only four cases. The CIF is the estimate most affected by small population. The SIR is much smaller, 1.16 (20th largest), reflecting the fact that it is a more stable estimate. The EB estimate is 1.02 (34th largest), indicating that very little weight can be placed on the data for the individual school district due to the small number of deaths. The school district with the second largest SIR, Merritt, has the largest EB. Vancouver, with the largest population and thus the most stable estimates, shows almost no difference among the three estimates. However, the relative rankings do change as Vancouver moves from 26th with the SIR to 18th with the EB.

Table 5,2 Comparison of the Standardized Incidence Ratio (SIR), Comparative Incidence Figure (CIF) and Empirical Bayes Estimate (EB) for Male Lung Cancer in Selected School Districts

School District	*No.*	*SIR*	*Rank+*	*Sig**	*CIF*	*Rank*	*EB*	*Rank*
Fort Nelson (81)	10	2.40	1	.004	2.64	2	1.21	7
Merritt (31)	36	1.85	2	<.001	1.90	3	1.39	1
Revelstoke (19)	26	1.70	3	.006	1.61	5	1.27	3
Campbell River (72)	78	1.42	4	.002	1.40	7	1.30	2
Kitimat (80)	20	1.42	5	.112	1.80	4	1.16	11
Queen Charlotte (50)	9	1.37	6	.340	1.54	6	1.08	21
Prince George (57)	146	1.33	7	.001	1.37	8	1.27	4
New Westminster (40)	166	1.28	8	.001	1.30	9	1.24	5
Vancouver Is. West (84)	4	1.16	20	.767	3.50	1	1.02	34
Vancouver (39)	1408	1.10	26	<.001	1.11	27	1.10	18

+ Rank of the SIR/SMR for that school district relative to the SIR/SMR for the other 73 districts. A rank of 1 indicates the largest SIR/SMR.

* Significance level

SUMMARY

Several important methodological issues and solutions have been presented with examples from the mapping program of the B.C. Cancer Agency. Clearly, great care is needed in the production and interpretation of cancer atlases. However, new developments in statistical methodology may alleviate some of the problems.

REFERENCES

[1] Band, P.R., Spinelli, J.J., Gallagher, R.P. et al. 1989. *Atlas of Cancer Mortality in British Columbia*. Statistics Canada, Hull.

[2] Heasman, M.A. and Lipworth, L. 1966. Accuracy of Certification of Causes of Death. *General Register Office Studies on Medical and Population Subjects*, No. 20. HMSO, London.

[3] Moriyama, I.M., Baun, W.S., Haenszel, W.M. et al. 1958. Inquiry into Diagnostic Evidence Supporting Medical Certifications of Death. *American Journal of Public Health* 48, pp. 1376-1387.

[4] Engel, L.W., Strauchen, J.A., Chiazze, L. et al. 1980. Accuracy of Death Certification in an Autopsied Population with Specific Attention to Malignant Neoplasms and Vascular Lesions. *American Journal of Epidemiology* 111, pp. 99-112.

[5] Fritchman, K. and Wallace, R.B. 1979. Geographic Mobility of Cancer Patients Between Incidence and Death (abstract). *American Journal of Epidemiology* 110, p. 370.

[6] Breslow, N.E. and Day, N.E. 1987. *Statistical Methods in Cancer Research: Volume II - The Design and Analysis of Cohort Studies*. Lyon: IARC.

[7] Schmid, C.F. 1983. *Statistical Graphics: Design Principles and Practices*. New York: Wiley.

[8] See for example: Tsutakawa, R.K., Shoop, G.L. and Marienfeld, C.J. 1985. Empirical Bayes Estimation of Cancer Mortality Rates. *Statistics in Medicine* 4, pp. 201-212; Clayton, D. and Kaldor, J. 1987. Empirical Bayes Estimates of Age-standardized Relative Risks for Use in Disease Mapping. *Biometrics* 43, pp. 671-681; Tsutakawa, R.K. 1988. Mixed Model for Analyzing Geographic Variability in Mortality Rates. *Journal of the American Statistical Association* 83, pp. 37-42; Manton, K.G., Woodbury, M.A.,

Stallard, E. et al. 1989. Empirical Bayes Procedures for Stabilizing Maps of U.S. Cancer Mortality Rates. *Journal of the American Statistical Association* 84, pp. 637-650; Mollie, A. and Richardson, S. 1991. Empirical Bayes Estimates of Cancer Mortality Rates Using Spatial Models. *Statistics in Medicine* 10, pp. 95-112; Marshall, R.J. 1991. Mapping Disease and Mortality Rates Using Empirical Bayes Estimators. *Applied Statistics* 40, pp. 283-294.

[9] Kaldor, J. and Clayton, D. 1989. Role of Advanced Statistical Techniques in Cancer Mapping, in Boyle, P., Muir, CS. and Grundmann, E. (eds.), *Cancer Mapping*. Berlin: Springer Verlag.

MORTALITY AND SOIL GEOCHEMISTRY: Tentative Working Hypotheses

Harold D. Foster

Department of Geography
University of Victoria

INTRODUCTION

In 1974, recognizing that societal health did not reflect merely the availability of medical care, the Canadian Department of Health and Welfare[1] introduced the "Health Field Concept". This model demonstrated that health was the function of a great diversity of factors that could be conveniently aggregated into four distinct categories, namely human biology, lifestyle, environment and health care organization. Each of these major areas of impact could be further subdivided. Environment, for example, includes physical dimensions, such as altitude and climate; social factors, like population density and industrial and agricultural activities and associated pollutants and biological attributes, exemplified by viruses, bacteria, pollen and fungal spores. This chapter focuses on one relatively neglected dimension of the "health field model", the impact on human well-being of soil quality.

The scientific method involves a series of steps that, when taken in sequence, after rigorous testing, permit the adoption of new ideas.[2] The generation of an initial hypothesis is the crucial first stage in this process, one that can often be stimulated through the use of statistical correlation. While such correlations, in themselves, are only suggestive and can never be used to prove causal relationships, they are sometimes capable of promoting rapidly new avenues of research.[3]

DATA BANKS

The author had been using such an approach to study the spatial distribution patterns of disease, both in North America and elsewhere.[4] To this end, a series of disease and environmental data banks have been established, that permit large scale correlation to be conducted. The most comprehensive of these has been developed for the United States. Ideally, to conduct such analysis correctly, it is necessary to have medical and environmental data that relate to both the same spatial units and time periods. Even in the United States, most environmental data have been collected by water and fisheries managers, geologists, geographers, foresters and agriculturalists for reasons completely unrelated to human health. As a result, no such synchronous environmental data set exists. The author searched the relevant literature, therefore, and accepted for use any data collected before, during or immediately after the period covered by the bulk of the medical statistics, that is 1950 to 1975.

The most comprehensive source of environmental data was that published by the United States Geological Survey.[5] In 1961, the Geological Survey began a soil and regolith sampling program that was designed to give estimates of the range of element abundance in surficial materials that were as unaltered as possible by human activity. These were to represent the natural geochemical environment of the United States. Samples were taken at a depth of approximately 20 centimetres below the surface, from locations about 80 kilometres apart, throughout the conterminous United States, and were then analyzed to determine their mineral content. In this way, 863 sample sites were chosen and the results of the geochemical analyses for 35 elements plotted on maps. On the published maps, the abundance of each element was represented by a symbol that indicated whether, at that site, its level was very high, high, medium, low or very low, when compared with the geometric mean for the nation as a whole. These figures were used by this author to generate percentage values for each state, for each level and element. For instance, in Florida 100 percent of the samples were found to be low in barium. In contrast, in Louisiana 25 percent were very low, 8.33 percent low, while 33.33 percent contained medium levels of barium and 33.33 percent had high barium levels. There were no very high barium concentrations recorded in this state. Similar percentage values were calculated, for this study, for all 35 elements, for all conterminous states. The elements involved were aluminium, arsenic, barium, beryllium, boron, calcium, cerium, chromium, cobalt, copper, fluorine, gallium, iron, lanthanum, lead, lithium, magnesium, manganese, mercury, molybdenum, neodymium, nickel, niobium, phosphorus, potassium, scandium, selenium, sodium, strontium, titanium, vanadium, ytterbium, yttrium, zinc and zirconium (Table 1,3).

In summary, the environmental data base included details of a wide range of air and water pollutants, climatic data; and a variety of industrial, commercial and agricultural activities. In addition, as previously described, geochemical information, that appeared to reflect, as closely as is possible, the natural chemical environment of the United States, was also utilized. In total, information on 221 geographical variables was available for use in correlation. This data bank is described in more detail in a previous publication.[6]

This chapter examines only associations between variations in health, in the United States, and measured levels of 35 soil elements. The author fully realizes that it is a gross oversimplification to correlate disease patterns with the total amount of any soil element. This is because the availability to plants and water supply of many such elements, and hence to the food chain, reflect soil pH (which influences processes like dissolution and precipitation reactions and ion exchange and complexing) and the redox potential (a measurement of a solution's oxidising or reducing capacities). Indeed, the behaviour of soil elements is not only controlled by mineral-solution equilibrium, but also reflects co-precipitation, adsorption, ion exchange and interactions with organic phases.[7] Clearly, the results that follow are tentative and designed to promote discussion and further research.

It should be noted that the state was used as the unit analysis rather than the county. There were several reasons for this, the most compelling of which was the absence of so much of the environmental data, at the county level. While detailed information on some of the environmental variables is available for certain counties, it is still impossible to conduct a holistic study, of the whole of the United States, at this scale. The present examination, therefore, is designed to generate hypotheses, that can subsequently be explored further, in more detail, anywhere the necessary environmental data are available. It is of interest, however, that many of the most heavily populated counties produce little of their own food or water and, as a consequence, are less likely to demonstrate strong correlations between illnesses and the local geochemistry. For such densely populated areas, therefore, there may be certain advantages of analysis at the state level. However, ultimately health-environment relationships should be examined at a variety of scales.

Table 1,3 Soil Elements Used in Pearson Correlations with Medical Data

Soil Element	*How Used*	*Data Source*
ALUMINIUM (Al)	Five Classes (very high, high, median, low, very low)	1
ARSENIC (As)	Five Classes	5
BARIUM (Ba)	Five Classes	1
BERYLLIUM (Be)	Three Classes (very high, median, very low)	1
BORON (B)	Five Classes	1
CALCIUM (Ca)	Five Classes	1
CERIUM (Ce)	Three Classes (very high, median, very low)	1
CHROMIUM (Cr)	Five Classes	1
COBALT (Co	Five Classes	1
COPPER (Cu)	Five Classes	1
FLUORINE (F)	Five Classes	5
GALLIUM (Ga)	Five Classes	1
IRON (Fe)	Five Classes	1
LANTHANUM (La)	Four Classes (very high, high, median, very low)	1
LEAD (Pb)	Five Classes	1
LITHIUM (Li)	Five Classes	3
MAGNESIUM (Mg)	Five Classes	1
MANGANESE (Mn)	Five Classes	1
MERCURY (Hg)	Five Classes	4
MOLYBDENUM (Mo)	Two Classes	1
NEODYMIUM (Nd)	Three Classes	1
NICKEL (Ni)	Five Classes	1
NIOBIUM (Nb)	Four Classes (very high, median, low, very low)	1
PHOSPHORUS (P)	Five Classes	1
POTASSIUM (K)	Five Classes	1
SCANDIUM (Sc)	Five Classes	1
SELENIUM (Se)	Five Classes	2
SODIUM (Na)	Five Classes	1
STRONTIUM (Sr)	Five Classes	1
TITANIUM (Ti)	Five Classes	1

Table 1,3 continued

Soil Element	*How Used*	*Data Source*
VANADIUM (V)	Five Classes	1
YTTERBIUM (Yb)	Five Classes	1
YTTRIUM (Y)	Five Classes	1
ZINC (Zn)	Five Classes	1
ZIRCONIUM (Zr)	Five Classes	1

SOURCES:

1. Shacklette, Hamilton, Boerngen and Bowles (1971)
2. Keller (1979)
3. Shacklette, Boerngen, Cahill and Rahill (19730
4. Shacklette, Boerngen and Turner (1971)
5. Shacklette, Boerngen and Keith (1974)

To date, the author has collected data on the spatial distributions of 127 diseases, or for differing time periods for the same illnesses. This information has been abstracted from a variety of published sources. Data from *Vital Statistics of the United States*, volumes of which are produced annually by the U.S. Department of Health and Human Services, National Center for Health Statistics, have been extensively used.[8] These publications, at the state scale, provide data on the United States death rate per 100,000 population for diseases of the heart, malignant neoplasms, cerebrovascular diseases, pneumonia, diabetes mellitus, cirrhosis of the liver, arteriosclerosis, suicide and early infant diseases (Table 2,3).

To test some of the hypotheses generated by analysis of these data, far more detailed information on spatial variations in mortality from neoplasms was taken from *Patterns in Cancer Mortality in the United States: 1950-1967*, authored by Burbank.[9] This volume, Monograph 33 of the United States National Cancer Institute, contains age adjusted death rates per 100,000 for 65 specific or sub-groups of cancer, and for malignant neoplasms as a whole. This information is also presented at the state scale. It was derived from death certificates for the period 1950 through 1967, and includes the cause of mortality, identified by the International Classifications of Diseases code, and the sex, race, state of residence and age at death.

METHODOLOGY

In an effort to stimulate initial hypotheses that might be valuable in the search for possible causes of disease, Pearson correlation coefficients were used to explore relationships between "natural" geochemical soil and regolith conditions and the spatial distributions of major causes of death for the year 1976. This analysis generated 1,467 Pearson correlation coefficients, only the most statistically significant (P = 0.005) of which are illustrated in Table 2,3. This process was then repeated using medical data abstracted from Burbank's[10] publication, so allowing the identification of possible links between the geochemical environment and a wide variety of specific cancers to be examined in more detail (Table 3,3).

RESULTS

As can be seen from Table 2,3, there were 15 positive, or negative, correlations at the p = 0.0001 level, between the spatial distributions of the major causes of death in 1976, in the United States, and the previously described 35 soil elements. To illustrate, the strongest negative associations for diseases of the heart were with soils that were very high in calcium (r = -0.62884), strontium (r = -0.53034) and barium (r = -0.52198), and low in arsenic (r = -0.53108). Mortality from malignant neoplasms showed the highest negative correlations with soils that were also very enriched with calcium (r = -0.56191), or had low arsenic levels (r = -0.53868). In contrast, death from cerebrovascular disease, in 1976, appeared to have occurred most often in states having very low levels of soil magnesium (r = 0.60663), strontium (r = 0.52869) and gallium (r = 0.52102), but very high levels of soil zirconium (r = 0.54548). No highly significant statistical links (p = 0.0001) were found between pneumonia, cirrhosis of the liver, arteriosclerosis and early infancy diseases and soil characteristics. However, diabetes mellitus mortality was also depressed in states where soils were very calcium enriched (r = -0.53936) and suicide had occurred more often where these were very high in strontium (r = 0.60459), sodium (r = 0.58275), potassium (r = 0.57203) and barium (r = 0.57042). In all the instances cited, p = 0.0001.

Examination of mortality data for 1977 to 1980 confirmed that many of these highly significant correlations (p = 0.0001) are repetitive. To illustrate, there is only one instance from these four additional years in which very high soil levels of calcium, strontium and barium are not significantly correlated with mortality from diseases of the heart, at the p = 0.0001 level. This occurred in 1980 with very high barium soil content (r = -0.51014, p =

0.0002). In addition, in each of these four years, death rates from diseases of the heart were also negatively correlated with low soil arsenic at the p = 0.0001 level. Similar highly significant correlations were also found to occur, each year, between very low soil magnesium and cerebrovascular disease. This was true of links between the spatial distributions of cerebrovascular mortality and soils that were very low in strontium and gallium, for all years except 1980.

It is not claimed, of course, that these strong correlations imply causal links. However, they suggest that such possibilities ought to be examined further. In an effort to take the enquiry process an step further, Pearson correlations were used to explore associations between the 66 cancers and groups of cancers for which data are available in Burbank's[11] *Patterns in Cancer Mortality in the United States: 1950-1967* and the 219 geographical variables previously described. The analysis of possible links between cancers and geochemical soil and regolith conditions alone yielded 4,620 correlations. These have already been examined in detail elsewhere.[12] For the purpose of this chapter, attention is drawn to some of the most intriguing associations, those between soil calcium, mercury and selenium levels and cancers of the digestive tract, in White males (Table 3,3). Although similar associations were noted for White females and Non-white males and females, they are not as pronounced as in White males.

The significant negative correlations between very high soil calcium and mortality from both diseases of the heart and diabetes mellitus (Table 2,3) have already been discussed. Although the selenium-mercury association noted in many cancers is not as obvious in either of these two diseases, there is some evidence of it in heart disease. To illustrate, in 1980, the correlations between high soil selenium and very high soil mercury levels and diseases of the heart were $r = -0.47121$ ($p = 0.0007$) and $r = 0.33864$ ($p = 0.0186$) respectively. In 1978, the same relationships were $r = -0.43308$ ($p = 0.0012$) and $r = 0.32997$ ($p = 0.0220$). However, correlations between mortality from diabetes mellitus and soil selenium and mercury levels, for the years 1976 to 1980, are not suggestive of any possibility of a causal relationship.

Soil elements do not occur at random but, in large measure, reflect geological and geomorphological processes. As a consequence, they often appear together in particular associations. This tendency, therefore, complicates the interpretation of correlations, frequently making collinearity a significant problem. To examine this issue further, Tables 4,3 and 5,3 present Pearson correlation coefficients for those soil elements that have similar spatial distributions to diseases of the heart and to cancer of the digestive tract, in White males, in the United States.

Table 2,3 Pearson Correlation Coefficients Illustrating Relationships Between Soil Levels of Various Elements and the Major Causes of Death in the United States in 1976 (p = 0.0050)

	Diseases of the Heart	*Malignant Neoplasms*	*Cerebro-vascular*	*Pneumonia*	*Diabetes Mellitus*	*Cirrhosis of Liver*	*Arterio-sclerosis*	*Suicide*	*Early Infant Diseases*
Arsenic	-0.53108 0.0001 (L)	-0.53868 0.0001 (L)	0.45313 0.012 (VH)						
Barium	-0.52198 0.0001 (VH)				0.51398 0.0002 (L)			-0.41714 0.0032 (L)	
								0.57042 0.0001 (VH)	
Beryllium				0.42216 0.0028 (M)					0.41028 0.0038 (VL)
Boron			0.45101 0.0031 (H)	0.41706 0.0032 (L)					0.45371 0.0012 (H)
				0.43187 0.0022 (M)					

Calcium	-0.50758 0.0002 (H)	-0.56191 0.0001 (VH)	0.47915 0.0006 (VL)	0.43944 0.0018 (M)	0.42225 0.0028 (VL)			0.44438 0.0016 (VH)
	-0.62884 0.0001 (VH)				0.41376 0.0035 (L)			0.51858 0.0002 (H)
					-0.46507 0.0009 (H)			
					-0.53936 0.0001 (VH)			
Fluorine				0.48263 0.0005 (H)	-0.44562 0.0015 (VH)		0.41891 0.0030 (H)	0.42204 0.0028 (VH)
Gallium			0.52102 0.0001 (VL)					
Iron			0.43742 0.0019 (VL)					
Lanthanum							0.42196 0.0028 (H)	
Lithium							0.41935 0.0030 (M)	

Table 2,3 continued

	Diseases of the Heart	*Malignant Neoplasms*	*Cerebro-vascular*	*Pneumonia*	*Diabetes Mellitus*	*Cirrhosis of Liver*	*Arterio-sclerosis*	*Suicide*	*Early Infant Diseases*
Magnesium			0.60663 0.0001 (VL)				0.43165 0.0022 (H)		
Mercury	-0.41484 0.0034 (VL)	-0.46425 0.0009 (L)						0.40204 0.0046 (VL)	
	-0.41787 0.0031 (L)	0.45724 0.0011 (VH)							
Nickel	-0.42704 0.0025 (L)					-0.42602 0.0025 (H)			
Niobium				0.49027 0.0004 (L)					
Phosphorus			0.41950 0.0030 (VL)						0.40742 0.0041 (VL)
									0.46489 0.0009 (L)

Potassium			0.45053 0.0013 (VL)					0.57203 0.0001 (VH)	
Selenium	-0.50167 0.0003 (H)	-0.45487 0.0012 (H)	0.42626 0.0025 (L)					0.42841 0.0024 (H)	0.39939 0.0049 (L)
	-0.42715 0.0025 (VH)							0.40330 0.0045 (VH)	
Sodium	-0.46510 0.0009 (VH)		0.40176 0.0046 (VL)	0.46438 0.0009 (M)	-0.50525 0.0002 (VL)			0.58275 0.0001 (VH)	0.44894 0.0014 (VL)
Strontium	-0.53034 0.0001 (VH)	-0.40522 0.0043 (H)	0.52869 0.0001 (VL)	0.48738 0.0004 (H)	0.46905 0.0008 (L)			0.60459 0.0001 (VH)	
					-0.51447 0.0002 (VH)				
Titanium								0.45684 0.0011 (L)	
Zirconium			0.54548 0.0001 (VH)		-0.46886 0.0008 (L)			0.40203 0.0046 (L)	

Soil Levels of Elements: VL = Very Low L = Low M = Medium H = High VH = Very High

Table 3,3 Pearson Correlation Coefficients Illustrating Relationships Between Cancer Mortality, in White U.S. Males for the Years 1950-1967, and Soil Levels of Selenium, Mercury and Calcium

Cancer type	*High soil selenium*	*Very high soil mercury*	*Very high soil calcium*
Tongue	-0.46827	0.58782	-0.51676
	0.0008	0.0001	0.0002
Floor of mouth	-0.46709	0.60977	-0.42650
	0.0008	0.0001	0.0025
Oral mesopharynx	-0.55596	0.50743	-0.50314
	0.0001	0.0002	0.0003
Pharynx	-0.36509	0.47415	-0.33892
	0.0107	0.0007	0.0185
Esophagus	-0.55505	0.59005	-0.48143
	0.0001	0.0001	0.0005
Large intestine	-0.61023	0.56623	-0.42010
	0.0001	0.0001	0.0030
Rectum	-0.55264	0.61850	-0.32785
	0.0001	0.0001	0.0229
Bladder	-0.47663	0.53598	-0.28283
	0.0006	0.0001	0.0514
All malignant neoplasms	-0.52248	0.58946	-0.46634
	0.0001	0.0001	0.0008

Tables 4,3 and 5,3 provide some evidence of collinearity. Very high levels of soil calcium, strontium and barium, for example, appear to occur together in many states. Nevertheless, this factor does not appear to be of overriding importance in this analysis. Certainly, it cannot be used as a complete explanation for the calcium-selenium-mercury relationships discussed with respect to many cancers.

Table 4,3 Pearson Correlation Coefficients Illustrating the Strength of the Associations Between those Soil Elements that are Linked Strongly Spatially, Either Positively or Negatively, with U.S. Mortality from Diseases of the Heart (1976-1980)

	Very high calcium	*Very high strontium*	*Very high barium*	*Low arsenic*
Very high calcium		0.55775 0.0001	0.56775 0.0001	0.35180 0.0142
Very high strontium	0.55775 0.0001		0.36523 0.0107	0.40943 0.0039
Very high barium	0.56775 0.0001	0.36523 0.0107		0.25426 0.0812
Low arsenic	0.35180 0.0142	0.40943 0.0039	0.25426 0.0812	

Table 5,3 Pearson Correlation Coefficients Illustrating the Strength of the Associations Between those Soil Elements that are Linked Strongly Spatially, Either Positively or Negatively, with U.S. Mortality from Cancer of the Digestive Tract (1950-1967), in White U.S. Males

	Very high calcium	*High selenium*	*Very high mercury*
Very high calcium		0.41636 0.0032	-0.31414 0.0297
High selenium	0.41636 0.0032		-0.40481 0.0043
Very high mercury	-0.31414 0.0297	-0.40481 0.0043	

CONCLUSION

Clearly, in such large scale analyses, some apparently significant correlations are likely to be generated by chance. Others may illustrate collinearity, or may not reflect a causal relationship. In addition, some of the stronger correlations, such as those between highly calcareous soils and low mortality from diseases of the heart, might have been anticipated, given the longstanding and widespread debate about this possibility.[13] However, other groups of associations were less expected, and clearly warrant further examination. These include the selenium-mercury-calcium associations with cancer of the digestive tract (Table 3,3).

The Chinese have traditionally used concretions from loess to treat cancer of the esophagus. These nodules, termed *jiang-shi*, are now being used in an experiment to modify drinking water in parts of Xingtai county, Hebei Province.[14] In 1974, for example, new wells were constructed in the five villages of Baian. These either had jiang-shi incorporated into their structure, or had numerous concretions dropped into them. In the past, this region had an extremely high esophageal cancer mortality rate. Prior to 1975, at least one death from this disease was expected every year in each of the five villages. Since this date, none has been recorded. The Chinese have been piling loess nodules, calcium enriched and containing a variety of trace elements including selenium, at the bottoms and sides of such drinking water wells, using between 0.2 and 0.4 tons per cubic metre of standing water. These concretions are replenished annually. While this experiment does not prove a causal relationship between calcium and selenium deficiency and cancer of the esophagus, it is extremely suggestive, given other very positive results recorded from China.[14]

In conclusion, therefore, it is possible that some of the relationships illustrated in Tables 2,3 and 3,3 are causal. Certainly, several require more attention. Included among these are the increase in diabetes mellitus mortality that seems to occur with declining soil calcium, and the decrease in death from diseases of the heart as soil mercury levels fall. This drop also appears associated with a decline in mortality from heart diseases as soil selenium levels rise. The possibility that strontium might play some role in protection against diseases of the heart and cerebrovascular disease is also intriguing. Clearly the hypothesis that selenium and calcium together may be protective against numerous cancers is worthy of additional attention.

REFERENCES

1 Lalonde, D. 1974. *A New Perspective on the Health of Canadians: A Working Document*. Ottawa: Canadian Department of Health and Welfare.

2 Wightman, W.P.D. 1964. *The Growth of Scientific Ideas*. New Haven: Yale University Press. See also Barnes, B. (ed.) 1972. *Sociology of Science*. Harmondsworth: Penguin.

3 Jones, K. and Moon, G. 1987. *Health, Disease and Society: An Introduction to Medical Geography*. London: Routledge and Kegan Paul.

4 Foster, H.D. 1986. *Reducing Cancer Mortality: A Geographical Perspective*. Western Geographical Series Vol. 23, Victoria: Department of Geography, University of Victoria.

5 Shacklette, H.T., Boerngen, J.G., Cahill, J.P.and Rahill, R.L. 1973. Lithium in Surficial Materials of the Conterminous United States and Partial Data on Cadmium, *Geological Survey Circular* 673, 8 pp.; and Shacklette, H.T., Boerngen, J.G. and Keith, J.R. 1974. Selenium, Fluorine and Arsenic in Surficial Materials of the Conterminous United States, *Geological Survey Circular* 692, 14 pp. See also Shacklette, H.T., Boerngen, J.G. and Turner, R.L. 1971. Mercury in the Environment in Surficial Materials of the Conterminous United States, *Geological Survey Circular* 644, 5 pp. Literature used also included Shacklette, H.T., Hamilton, J.C., Boerngen, J.G. and Bowles, J.M. 1971. Elemental Composition of Surficial Materials in the Conterminous United States, *Geological Survey Professional Paper* 574-D, 71 pp.

6 Foster, *op. cit.*

7 Bowie, S.H.U and Thornton, I (eds.) *Environmental Geochemistry and Health: Report to the Royal Society's British National Committee for Problems of the Environment*. Dordrecht: D. Reidel.

8 See, for example, U.S. Department of Health and Human Services, Public Health Service, National Center for Health Statistics Vital Statistics of the United States 1980, Volume II - Mortality. Part A. Hyattsville, Maryland, 1985.

9 Burbank, F. 1971. *Patterns in Cancer Mortality in the United States: 1950-1967*. National Cancer Institute Monograph 33, United States Department of Health, Education and Welfare.

10 *Ibid.*

11 *Ibid.*

12 Foster, *op. cit.*

13 See, for example, Jones, K. and Moon, G. *op. cit.*; and Safe Drinking Water Committee, National Research Council, Drinking Water and Health. Washington, DC: National Academy of Sciences, 1977 for conflicting views on this topic.

[14] Zhu, Cheng and An, Yonglu. 1988. The Effect of Jiang-shi (stong [sic] drugs) in Prevention and Treatment of Esophageal Cancer in the High Incident District, *Abstracts, International Symposium on Environmental Life Elements and Health*, 1-5 November, Beijing, China: Chinese Academy of Sciences, p. 225.

[15] Gao, Zhenwie et al. 1984. Observation on the Effect of Preventing of Esophageal Cancer Through Changing the Quality of Drinking Water by Jiangshi in the Western Mountain Area of Xingtai County, Hebei Province. *Proceedings of the First National Symposium on Trace Elements and Human Health.* Nanjing, People's Republic of China.

4

SELENIUM AND CANCER: Overview and Data From Alberta

J. Berkel
G. Bako

Division of Epidemiology and Preventive Oncology
Alberta Cancer Board

INTRODUCTION

Cancer is the second most frequent cause of mortality in Canada, accounting for the death of about 50,000 Canadians every year. The leading cause of death is still cardiovascular diseases (CVD). This pattern is well known and has existed for the last several decades. However, when examining the trends in mortality, in more detail, it becomes clear that the relative importance of cancer is increasing: in the 1960s the mortality rate ratio of CVD and cancer was 3.2, decreasing to 1.4 in the late 1980s. This decrease is due in part to a decline in death from CVD, but also to an increase in mortality from cancer. If these trends continue, cancer will be the leading cause of death in our society before the turn of the century. The impact of cancer on society is even more significant if one looks at the "potential years of life lost" (PYLL) caused by several diseases. Cancer is the leading cause of premature mortality, that is it is responsible for most PYLL, in persons under age 75. Figures 1,4 and 2,4 illustrate both the leading causes of death and the PYLL in Canada.[1]

From these and many other descriptive data, it is obvious that at present cancer is a major health care problem in our society and if progress is not made in reducing its impact, it will soon be a catastrophe. The search for the causes of cancer, therefore, is extremely important. In their

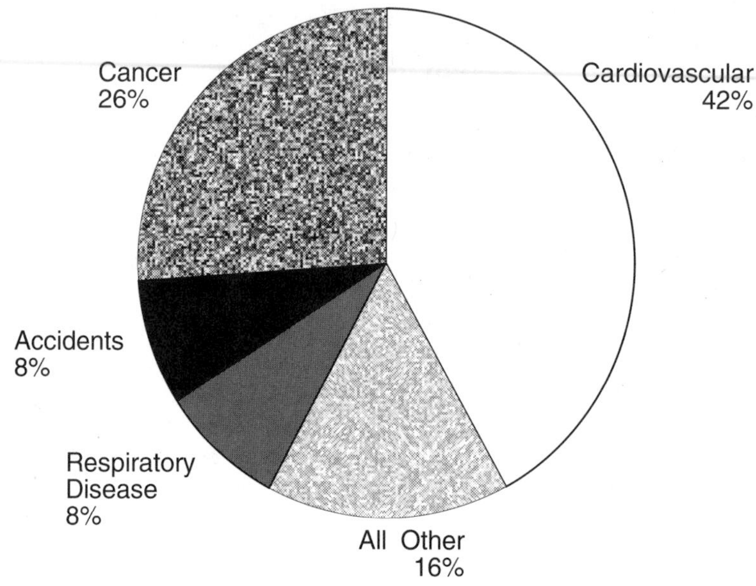

Figure 1,4 Leading Causes of Death, Canada, 1987.

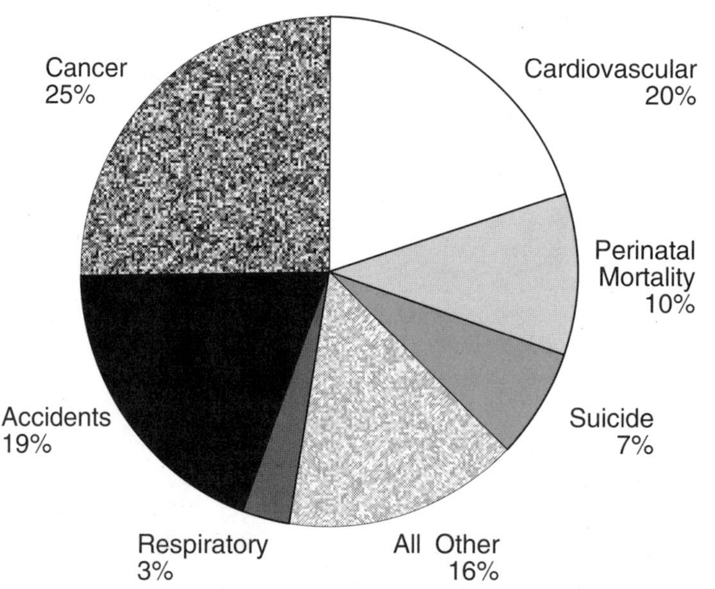

Figure 2,4 Potential Years of Life Lost, Canada, 1987.

classical monograph on the "causes of cancer" Doll and Peto calculated the attributable risk of various known risk factors for mortality of cancer.[2] Their estimates are shown in a slightly modified way in Table 1,4. This table clearly shows the involvement of both the macro- and micro-environments. The macro-environment includes those factors that are really beyond the influence of a single individual. The micro-environment, in contrast, is composed of those factors over which the individual virtually has complete control. Table 1,4 ignores hereditary, or genetically determined, cancers. In the authors' opinion, however, heredity most likely contributes relatively little to the total cancer problem.

Table 1,4 Proportion of Cancer Deaths Attributable to Various Factors (modified from Doll, Peto)[2]

MACRO-ENVIRONMENT		
Pollution	2	(0-5)
Industrial products	<1	(0-2)
Geophysical factors	3	(2-4)
Occupation	4	(2-8)
MICRO-ENVIRONMENT		
Tobacco	30	(25-40)
Alcohol	3	(2-4)
Diet	35	(10-70)
Reproductive and sexual behaviour	7	(1-13)

The distinction between macro- and micro-environment is not always complete, one of the prime examples being exposure to selenium and the risk of cancer. While selenium (Se) intake occurs mainly from food, the Se content of food-items differs considerably, reflecting the geographic area in which the food was grown. Thus Se-exposure depends not only on individual decisions about what kind of food and how much should be eaten, but also on environmental circumstances such as the Se-content of the soil, over which the individual has little influence. The remainder of this chapter is concerned with a discussion about the relationship between this micronutrient and cancer incidence.

SELENIUM

Selenium (Se) is a component of the enzyme glutathione peroxidase which protects cell membranes and tissues against damage caused by oxygen-radicals, generated as by-products of the lipid peroxidation process. Several reviews of the physiological role and the biochemical functions of Se have recently been published, from which it is clear that Se is an essential constituent of the glutathione peroxidase enzyme.[3,4,5] Se, therefore, is an essential micronutrient, a deficiency of which is known to be associated with at least two disease-syndromes in humans, those being Keshan disease, a cardiomyopathy, and Kashin-Beck syndrome, an osteoarthropathy.

Selenium appears to be ubiquitous, but rock and soil concentrations vary geographically. This results in areas with either very low, or very high natural levels of Se in the environment. Depending on the natural availability of selenium in the environment, the range of Se concentrations in different foods also varies greatly. Food is the main source of Se for humans, the intake from drinking water being much less, with a concentration of only a few µg/liter of water being typical. Because of such geochemical differences in Se availability, estimates of adult human exposure from diet range from 11 to 5000 µg/day, in different parts of the world. Table 2,4 shows estimated dietary intake of selenium in various countries.

Table 2,4 Estimated Dietary Intake (µg/day) of Selenium in Various Countries

New Zealand[6]	28-32
Finland[7]	50-60
Netherlands[8]	50
United Kingdom[9]	60
Japan[10]	80-90
Canada[11]	90-224
U.S.A.[12]	132-216
Venezuela[13]	326
China[14]	11-5000

Metabolism

Absorption of soluble Se compounds through the gastro-intestinal tract in animals seems to be highly efficient. In various species absorption rates of 92 percent, 91 percent and 81 percent have been described for selenite, selenomethionine and selenocystine respectively.[15,16] Most of the absorption occurs in the duodenum, with less absorption occurring in the jejunum and ileum. The absorption of Se from natural foods has been studied to a much lesser extent, but seems less efficient, and quite a variation of bio-availability of Se from different food sources has been described.[17]

Studies conducted with ^{75}Se-selenomethionine in humans showed that Se from selenomethionine is better (±90 percent) absorbed than the Se originating from selenite-selenium (±60 percent). After adjustment for the endogenous fecal excretion of Se, the average total true absorption of naturally occurring Se in women, freely consuming "normal diets", was 79 percent.

The absorbed Se is rapidly distributed among various tissues. In humans, the highest levels of Se are found in descending order in the following organs: kidney, liver, spleen, pancreas, testes, heart, intestine, lung and brain.[18,19] A close correlation exists between blood-selenium levels and the amount of Se in the diet. This correlation has also been described for blood-Se levels and Se-content of the soil. Superimposed on these geographical variations in blood-Se level are the effects of general nutritional status and/or the effects of certain diseases. In cancer patients, for example, a lower serum-Se concentration has been described repeatedly.[20,21]

The excretion of Se occurs mainly with the urine, and urinary Se-levels have long been used to monitor occupational Se-exposure. The importance of the kidneys in the homeostasis of Se was further demonstrated by the results of studies in persons with a low Se-status. These people were found to have a low renal-plasma clearance of Se.

Selenomethionine seems to be retained in the body to a greater extent than Se in the form of selenite or selenate, with only ±25 percent of the absorbed dosage of Selenomethionine being excreted in urine and faeces, compared to ±60 percent and ±90 percent of the dosage of selenite and selenate respectively.

Physiologic Effects

The biological functions of selenium relate for the most part to its role as an essential constituent of the enzyme glutathion-peroxidase (GSHpx). This enzyme functions in the cell as part of a defense system against oxygen-induced damage. Free radicals originate from various (metabolic) sources, but they all result in lipid peroxidation. Uncontrolled lipid peroxidation damages the structural integrity of the cell membranes and thus can be very detrimental to the cell. Se-GSHpx serves as an anti-oxidant defense system by reducing the amount of lipid hydroperoxides. Secondly, Se-GSHpx exerts influence on the prostaglandin metabolism by protecting the enzyme prostaglandin dehydrogenase, a key enzyme in the degradation pathway of prostaglandins.[22]

As well as lowered Se-GSHpx levels, Se-deficiency results in changes in liver enzymes and substrates involved in the metabolism and detoxification of drugs and other foreign compounds through a variety of pathways. Finally Se has been shown to have an effect on both B-cell (humoral) and T-cell (cellular) dependent immune functions as well as on the immune functions of phagocytic cells, potentially resulting in a lower resistance to infectious diseases in Se-deficient persons.

SELENIUM AND CANCER

Animal-experimental Studies

Selenium-containing compounds were suggested for chemotherapy of cancer as early as 1915, long before the essential role of Se in normal metabolism was recognized. The first study that demonstrated that a high level of dietary Se prevented the development of colon tumors, in rats, was published in 1949.[23] Although early animal experimental studies were not able to resolve the question whether Se was in fact carcinogenic or anticarcinogenic, studies in the late 1960s and early 1970s showed, in an unequivocal way, that there was no evidence of carcinogenicity. In the last two decades a great number of animal experimental studies have been conducted, the results of which have been discussed in several excellent recent reviews.[4,5] Of the 56 studies that have evaluated the effects of pharmacological, that is elevated levels of Se on experimental carcinogenesis in more than two dozen animal models, 49 (88 percent) have found that high level Se treatment reduced the development of tumors by at least 15 percent. In 66 percent of these "positive" studies, a very substantial (>50 percent) decrease in tumorigenesis was observed. Only seven studies did not find any effect of Se treatment on tumor outcome, and two of these studies used doses of Se that are probably within the nutritional range. One other observation derived from these studies is of importance: reduction in tumor outcome was accomplished at levels of Se intake that are not toxic. If the results of all the animal experimental studies are combined, a dose-response relationship between dietary Se intake and reduction in tumor incidence becomes apparent, as is shown in Table 3,4.

Table 3,4 Dose-response Relation Between Dietary Se Intake and Tumor Incidence (combined data from animal experimental studies)

Se-intake	*Reduction in Tumor Load (%)*	*Number of Studies*
<1 ppm	19	8
1-2 ppm	42	12
2-4 ppm	45	18
4-8 ppm	41	24
>8 ppm	58	8

Relatively few studies (n=24) have evaluated the effect of nutritionally deficient Se levels on carcinogenesis in experimental models. Unfortunately many of these studies did not also measure specific indicators of Se status (e.g. how deficient were the animals?), and the statistical methods were not always appropriate (e.g. how valid are the results?). Four animal tumor model studies showed that high levels of Se might enhance tumorigenesis. All these studies, however, showed a differential effect on various organs; for example, in one study small intestinal tumors seemed to increase in frequency, while colorectal tumors showed a decreased incidence. Overall conclusions that can be drawn from animal experimental studies are:

1. Se treatment decreased tumor incidence (in a dose response fashion) in animal experiments; and

2. Although the available information is suggestive to this effect, it is not yet possible to definitively say that Se-deficiency exacerbates carcinogenesis.

This impressive body of data from experimental studies in animals necessitates an evaluation of the role of Se as a potential preventive factor in human cancer etiology.

Ecological Studies

In 1969 Shamberger first described the inverse relationship between Se levels in crops and cancer mortality rates in various regions in the USA.[24] Subsequent papers from an associated group of researchers expanded on this concept and confirmed earlier results. Essentially the same results were obtained by Schrauzer, who found that in 27 developed countries, the overall mortality rate from cancer was inversely associated with the estimates of per capita intake of Se.[20] These investigators also reported that age-adjusted mortality rates in the United States for breast, colon, rectum and lung cancer were negatively associated with whole blood Se concentration. An ecological association between blood Se level and cancer mortality was also reported by Yu, in a study of 24 regions of China.[21] An inverse relation between geographic distribution of liver cancer incidence and blood Se level and/or Se content of local foods was also described.

In general, correlational studies have found a high mortality from cancers of the tongue, oesophagus, stomach, colon, rectum, liver, pancreas, larynx, lung, kidneys, bladder and for malignant lymphomas to be associated with low selenium areas.

The above mentioned correlational (ecological) studies can be criticized because of the use of sometimes less appropriate methodology. For example, in some studies mortality rates were used that were not age-adjusted, some locations were excluded from the analysis and estimates of national per capita Se intake from "typical food contents" were used, rather than the result of actual analyses of the food composition, in the various geographical areas. However, although the individual evidence derived from ecological studies might not be very strong, all correlational studies *consistently* suggest an inverse relation between Se intake and cancer mortality. Moreover this "indication" was reinforced by the results of a well-designed correlational study done by Clark, who used the forage crop Se-levels per county (thereby minimizing the chance of misclassification of per capita Se intake) and excluded counties with more than 50,000 residents (reducing confounding effect of urbanization). Adjustments were made for urbanization, ethnicity, presence of major industrial activities and annual sunlight exposure. The Se intake was compared to U.S. age-adjusted cancer mortality data for 1950-1969, which were also adjusted for regional differences. A state-of-the-art analysis was performed using least-squares regression weighted in inverse proportion to the variance of the site- and sex-specific cancer mortality rate. The results of this study show a significantly lower mortality rate in intermediate- and high-

Se consuming counties, compared to low Se areas. This was the case, for males and females, for all cancers combined as well as for cancers of the lung, colon, rectum, bladder, oesophagus and pancreas separately. In females, an inverse association also existed for mortality from malignant tumors from the breast, ovary and cervix. Contrasting to the "protective" effect of Se intake for these sites, a positive association was found between forage crop Se level and mortality from tumors from liver and stomach, as well as from M. Hodgkin and the leukaemias.

In conclusion, although it is not within the power of this type of study to prove causality of associations, *all* the ecological studies to date indicate that the hypotheses that a low Se status is indeed a risk factor for cancer in humans is plausible. To test this hypotheses more sophisticated epidemiological study designs are needed. The results of these studies will be described below.

Case-Control Studies

In most case-control studies addressing a "diet-cancer" question, the intake of the micronutrients of interest is estimated by obtaining a dietary history. Methods of obtaining a dietary history include diet records, or food frequency questionnaires. Unfortunately this approach is not reliable for estimating Se intake, since the Se content of the same foods can vary considerably depending on the Se levels in the soil, in which they were grown. Consequently nutrient composition tables upon which evaluation of the intake of nutrients depends are not useful in studies regarding Se intake. As a proxy measure for Se intake, Se levels in blood, serum, hair or toenails need therefore to be used as the exposure variable.

Another concern related to this is the fact that alterations in blood Se level, observed in cases compared to controls, might be the (metabolic) consequence rather than the cause of the cancer. Se is known to be concentrated in high levels in certain tumors and a large tumor mass could, therefore, have a substantial impact on blood levels. In fact, several studies have shown that cancer patients with low serum Se levels had more recurrences, wider dissemination of the disease, more multiple primaries and a shorter survival.

The strength of any evidence obtained from "cross-sectional" case-control studies needs therefore to be weighted carefully. However, of 21 studies of this type 13 (62 percent) show that Se status in cancer patients was lower compared to healthy controls. In six studies (28 percent) the differences found were not statistically significant, although Se levels in these cases was virtually always lower than the control group. It has to be noted that a wide variety of tumor sites has been studied, but that the studies that specifically looked also at GI tract cancer all found a decreased Se status in patients, as opposed to controls.

Although, as mentioned before, the strength of the evidence of "cross-sectional case-control studies" is not sufficient to prove causality, one can summarize the results of these studies by saying that they support and are consistent with the hypothesis that low Se status increases cancer risk in humans.

More convincing evidence comes from "prospective" case-control studies in which the design most often used is a so called "nested case-control" approach. In a cohort of two healthy persons, apparently free of the disease of interest, parameters are measured that might influence the disease outcome. The cohort is followed up prospectively and in time some members of the cohort may develop the disease under study. These cases are then compared with healthy members of the cohort with regard to initial exposure levels. The results of 16 of these prospective case control studies are summarized in Table 4,4.

Table 4,4 Results of 16 Prospective Case-control Studies on the Relation Between Se Status and Cancer Risk

Country	*Cancer Site - Result*	*Comments*
USA	All sites - RR=2.0 (low vs high) GI tract - signifantly lower serum Se	Hypertension detection and follow-up study. Total of 111 cases and 210 controls, GI tract cancer n=13 only. Outcome: incidence.
Finland	All sites - RR=3.1 (low vs high) GI tract - } signif. lower serum Se Lung - } signif. lower serum Se Haematol. - } signif. lower serum Se	North Karelia Project. 128 cases/128 controls. GI tract n=21 only. Outcome: incidence.
Finland	All sites - RR=5.8 (low vs high) GI tract - not significantly lower serum Se Lung - significantly lower serum Se	Eastern Finland Heart Survey. GI tract cancer n=18 Outcome: mortality.
USA	All sites - *no* significant difference	Evans County Study. 130 cases/130 controls.
USA	Colon - RR=1.4 (low vs high; not signif.)	72 cases/143 controls.
USA	Lung - no difference	99 cases/196 controls.

Table 4,4 (continued)

Country	*Cancer Site - Result*	*Comments*
Netherlands	All sites - RR=2.7	Zoetermeer-EPOZ cohort. 69 cases/164 controls. Significant in males only. Outcome: mortality.
Netherlands	Breast - RR=1.1 (not significant)	DOM (breast screening cohort)-study: premenopausal women. Outcome: incidence.
Netherlands	Lung - no difference	Zutphen study: 63 cases.
Hawaii	Colon - RR=1.8 Bladder - RR=3.1 Lung - RR=1.1 (not significant) Rectum - RR=1.6 Stomach - RR=0.9	Males of Japanese ancestry. Outcome: incidence.
Sweden	All sites - RR=3.8 (low vs high)	Outcome: mortality.
Finland	All sites - RR=1.34 (not significant)	109 cases.
USA	All sites - no difference GI tract - no difference Breast - RR=3.4 (not significant)	156 cases/287 controls. Outcome: incidence.
USA	Breast - no difference	393 cases/393 controls. Outcome: incidence.
USA	Skin - RR=0.8 (high vs low) non-melanoma	177 cases. Outcome: incidence.
Finland	GI tract - Upper GI tract: males RR=3.3 females RR=2.4 (not significant) Colorectal: males RR=1.7 (not significant)	Upper GI tract n=86. Colorectal n=57. Outcome: incidence.

RR = Relative Risk

None of the studies found a negative effect of high Se level on cancer (neither on incidence nor on mortality). In seven studies (47 percent) a significantly increased risk for cancer was found associated with a low Se status. The other studies, however, did not show a significant association between Se and cancer risk. Several comments can be made to explain this seeming lack of consistency. First of all, many of the studies suffer from the "small-number-problem". The total number of cases (cancers of all sites) exceeded 200 in only one study. When this number is broken down for specific sites, very small groups remain (usually <50). Secondly, some studies used incidence as outcome measure, while others used mortality. It is not necessary that when Se is related to cancer incidence it is also related to cancer mortality. Thirdly, some studies adjusted for confounding or effect modifying factors in the analysis, while others did not. Finally, Se is not necessarily related to *all* cancer sites. Evaluation of the seven studies that specifically looked at gastrointestinal (GI) tract cancers shows that only one did not reveal any difference in Se level in these patients. The other six all showed a decreased Se status in GI tract cancer cases, reaching statistical significance in three studies. Therefore, it is justified to conclude that most well designed prospective studies to date confirm an association between low Se status and increased GI tract cancer risk. That not all investigations have demonstrated this relationship might indicate that the role of a low Se status is that of an effect modifier, rather than an effector *per se* in colon carcinogenesis.

MECHANISM OF ANTICARCINOGENIC ACTION OF SE

Possible pathophysiologic mechanisms explaining the apparent protective effect of Se on (colon-) carcinogenesis are:

1. Se status influences the metabolism of (chemical) carcinogens.
2. Se protects against oxidative damage through its role as an essential component of the enzyme glutathione peroxidase (SeGSHpx).
3. Se-metabolites are toxic to proliferating cells.
4. Se inhibits cell proliferation associated with decreased mitochondrial oxidative metabolism, increased levels of cAMP and reduced synthesis of RNA, DNA and protein.
5. Se affects the immune response.

At present it is not completely clear which (combination of?) the above mentioned mechanisms relates to the antitumorigenic effect that has been shown to exist. In itself all the mechanisms have been experimentally shown to be plausible. What is clear, however, is that high levels of Se can impair metabolism in a fundamental way, having its effects on proliferating cells.

ALBERTA - DATA

The authors used data from the population-based Alberta Cancer Registry to evaluate the possible correlation between environmental distribution of Se and various cancer sites. From the cancer registry, information about the place-of-residence-at-diagnosis was retrieved. All patients diagnosed between 1976-1985 were included. Age-standardized incidence rates for each census division (CD) were calculated using the place-of-residence-at-diagnosis as the determinant of the CD. The 1981 census data were used to establish the reference population. The various tumors were combined into 12 major groups.

Information about the selenium distribution in Alberta was provided by Alberta Agriculture, which had collected 7,539 samples of animal feed, greenfeed, grains, hay, silage and straw in the 15 Census Divisions of Alberta, during 1971-1986. Feed samples are used as proxi for soil samples for their greater reliability in selenium estimation. The selenium levels by census division in Alberta are shown in Figure 3,4. Not included are samples of undetermined census divisions, indian reserves, provincial parks and so-called "urban lands", or city properties. Generally, Se levels in Alberta were found to increase from west to east. Variation in Se was quite large, with the lowest Se level occurring in the mountain areas and the highest being in the south-eastern part of the province. The authors then estimated the correlation-coefficient between environmental Se-level and census-specific incidence rates for all cancers combined, as well as for the various site groups. The correlation-coefficients for Se-levels and incidence of cancer (-groups) in the 15-CDs are shown in Table 5,4.

This analysis only revealed one statistically significant correlation ($p < 0.05$): the incidence of malignant melanoma of the skin in males was negatively/inversely related to the Se level in the different CDs (Figure 4,4). However, a non-significant correlation in the opposite direction was present in females with the same tumor. As has been argued above, data from the literature suggest that Se deficiency is related to cancers specifically of the digestive tract, mainly the large intestine. In the overall analysis, digestive tract cancers did not show a significant correlation with Se level.

Table 5,4 Correlation Coefficients of Se-level and Cancer Incidence in Alberta (by sex)

Tumor Group	*Sex*	*Correlation Coefficients*	*p-value*
Head and Neck	M	0.1659	0.555
	F	-0.0036	0.990
Digestive Tract	M	0.3537	0.196
	F	-0.0956	0.735
Lung	M	-0.3720	0.172
	F	0.1564	0.578
Bone and Soft Tissue	M	0.1265	0.653
	F	-0.2416	0.386
Melanoma	M	-0.5150	0.049
	F	0.3974	0.142
Skin	M	0.2008	0.473
	F	0.3496	0.202
Genital Organs	M	-0.1768	0.528
	F	-0.1946	0.487
Bladder and Kidney	M	0.3090	0.262
	F	-0.4606	0.084
CNS and Eye	M	-0.0512	0.856
	F	-0.2778	0.316
Thyroid and Endocrine	M	0.2161	0.439
	F	0.0195	0.945
Lymphoma and Leukemia	M	0.4621	0.083
	F	-0.1902	0.497
Other	M	-0.0601	0.831
	F	-0.5062	0.054

When analyzing only the data for colon cancer (males and females combined) a correlation coefficient $r = 0.31$ was found, which was not statistically significant.

The apparent limitation of this analysis is the geographical unit. The authors used census divisions for this purpose. The Province of Alberta is

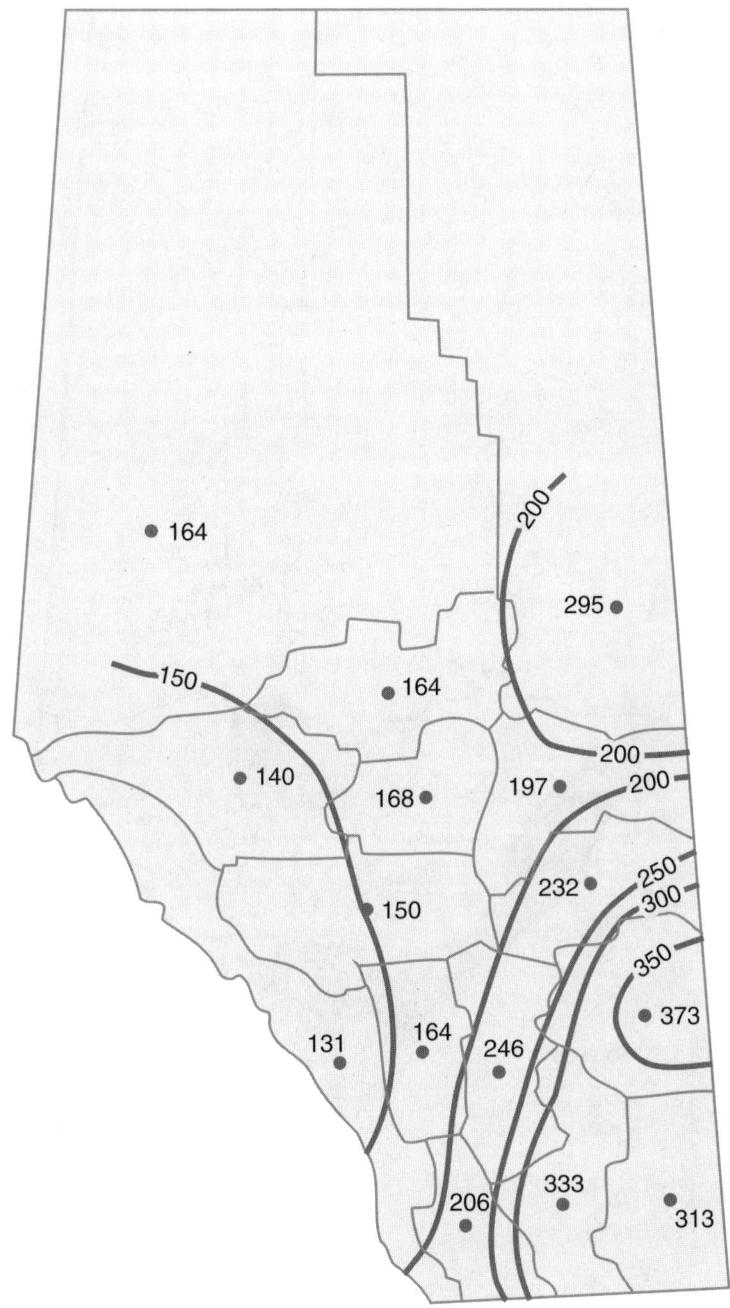

Figure 3,4 Average Selenium (PPB) Distribution in Alberta 1971-1989

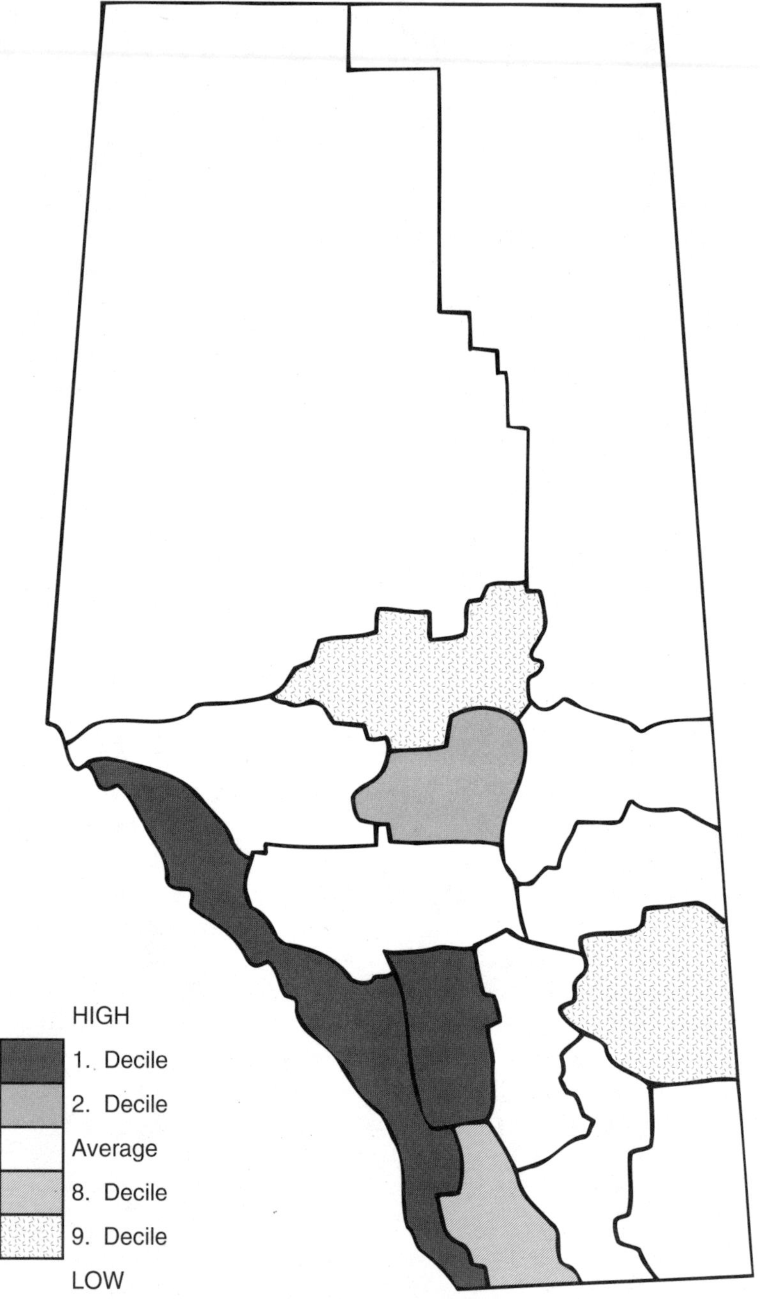

Figure 4,4 Age-Adjusted Incidence Rates/10^5 Population for Melanoma - Males 1971-1985 (N=523)

not very heavily populated and the population density varies widely across the province. This restricts the use of small geographical units because of the "small-number-problem" that would arise. The use of larger geographical areas, however, could easily mask local variation in environmental Se levels. This dilemma is a useful example of the advantages and limitations of descriptive geographical analyses.

REFERENCES

[1] Health Reports, Vol. I, No. 1, 1989. Published by Canadian Centre for Health Information (Statistics Canada).

[2] Doll, R. and Peto, R. 1981. *The Causes of Cancer.* Oxford University Press.

[3] *Ibid.*

[4] Combs, G.F. Jr. and Combs, S.B. 1986. *The Role of Selenium in Nutrition.* Academic Press Inc.

[5] *Selenium.* W.H.O. Environmental Health Criteria #58 (1987).

[6] Thomson, C.D. and Robinson, M.F. 1980. Selenium in Human Health and Disease with Emphasis on Those Aspects Peculiar to New Zealand. *American Journal of Clinical Nutrition* 303(33).

[7] Varo, P. and Koivistoinen, P. 1981. Annual Variations in the Average Selenium Intake in Finland: Cereal Products and Milk as Sources of Selenium in 1979/80. *International Journal of Vitamin and Nutrition Research* 51(62).

[8] Ockhuizen, T.L., Cardinaals, H., Bos, P. et al. 1985. Seleen u de preventie van kanker. *Voeding* 46(416).

[9] Thorn, J., Robertson, J., Buss, D.H. et al. 1978. Trace Elements. Selenium in British Food. *British Journal of Nutrition* 39(391).

[10] Sakurai, H. and Tsuchiya, K. 1975. A Tentative Recommendation for Maximum Daily Intake of Selenium. *Environmental Physiology and Biochemistry* 5(107).

[11] Thompson, J.N., Erdody, P. and Smith, D.C. 1975. Selenium Content of Food Consumed by Canadians. *Journal of Nutrition* 112(274).

[12] U.S. Food and Drug Administration, Bureau of Foods Compliance Program. 1975. *Evaluation Report: Total Diet. Studies Fiscal Year 1974.* Washington, D.C.: U.S. Government Printing Office.

[13] Mondragen, M.C. and Jaffe, W.A. 1976. Consumo de selenio en la cividad de caracas en comparacion von el de otras cividades del mundo. *Arch. Latin American Nutrition* 26(31).

[14] Yang, G. 1986. Research on Se-related Problems in Human Health in China. *Proceedings, Third International Symposium on Selenium in Biology and Medicine*. Westport, Conn.

[15] Thomson, C.D. and Stewart, R.D.H. 1973. Metabolic Studies of (^{75}Se) Selenomethionine and (^{75}Se) Selenite in the Rat. *British Journal of Nutrition* 30(139).

[16] Thomson, C.D., Robinson, B.A., Stewart, R.D.H. et al. 1975. Metabolic Studies of (^{75}Se) Selenocystine and (^{75}Se) Selenomethionine in the Rat. *British Journal of Nutrition* 43(501).

[17] Combs, *op. cit.*, Chapter 4.

[18] Schroeder, H.A., Frost, D.V. and Balassa, J.J. 1979. *Essential Trace Metals in Men: Selenium.* Journal of Chronic Disease 23(227).

[19] Diskin, C.J., Tomasso, C.L., Alper, J.C. et al. 1979. Long Term Selenium Exposure. *Arch. Int. Medicine* 139(824).

[20] Schrauzer, G.N., White, D.A. and Schneider, C.J. 1977. Cancer Mortality Correlation Studies III. Statistical Association with Dietary Selenium Intake. Biology and Inorganic Chemistry 7(23).

[21] Yu, S.Y., Chu, Y.J., Gong, X.L. et al. 1985. Regional Variation of Cancer Mortality and its Relation to Selenium Levels in China. *Biol. Trace Elem. Res.* 7(21).

[22] North, L.N., Mathias, M.M. and Schatte, C.L. 1984. Effect of Dietary Vitamin E or Selenium on Prostaglandin Dehydrogenase in Hyperoxic Rat Lung. *Aviat. Space Environm. Med.* (July).

[23] Clayton, C.C. and Baumann, C.A. 1949. Diet and Azo Dye Tumors: Effects of Diet During a Period When the Drug is Not Fed. *Cancer Research* 9(575).

[24] Shamberger, R.J. and Frost, D.V. 1969. Possible Protective Effect of Selenium Against Human Cancer. *Canadian Medical Association Journal* 104(82).

[25] Clark, L.C. 1985. The Epidemiology of Selenium and Cancer. *Feder. Proc.* 44(2584).

WATER HARDNESS AND THE DIGESTIVE CANCERS

Ian H. Norie

Department of Geography, University of Victoria

CANCER AND OUR PHYSICAL ENVIRONMENT

Introduction

In the Developed World, cancer is the second leading cause of death after diseases of the heart.[1] In its many forms, it is thought to be responsible for an estimated 4.3 million global deaths annually, with 5.9 million additional cases being diagnosed each year.[2] However, the spatial distribution of cancer shows major international variations. To illustrate, although 75 percent of the world's population live in the Developing World, it accounts for only 2.3 million annual cancer deaths. In contrast, in the Developed World, with only some 25 percent of the global population, the mortality rate is roughly triple, with 2.0 million cancer deaths occurring annually.[3] It is clear, however, that cancer incidence does not merely rise as a simple consequence of industrialization. As can be seen in Figure 1,5, the world's highest age adjusted incidence rate for esophageal cancer, in males, occurs in India, while the highest incidence rates for cancer of the esophagus and liver, in females, are found in China.

Evidence is accumulating to support the view that variations in cancer incidence are often related to characteristics of the physical environment, namely its climate, geology, soils and water supply. There is a considerable

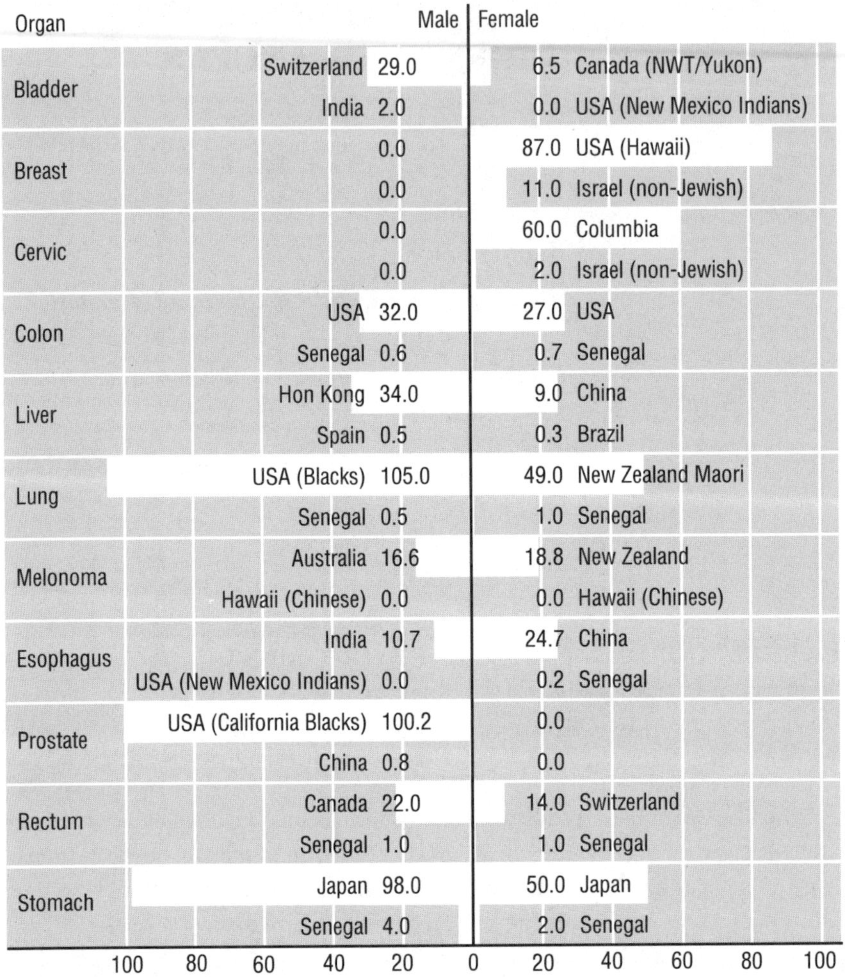

Organ	Male		Female	
Bladder	Switzerland	29.0	6.5	Canada (NWT/Yukon)
	India	2.0	0.0	USA (New Mexico Indians)
Breast		0.0	87.0	USA (Hawaii)
		0.0	11.0	Israel (non-Jewish)
Cervic		0.0	60.0	Columbia
		0.0	2.0	Israel (non-Jewish)
Colon	USA	32.0	27.0	USA
	Senegal	0.6	0.7	Senegal
Liver	Hon Kong	34.0	9.0	China
	Spain	0.5	0.3	Brazil
Lung	USA (Blacks)	105.0	49.0	New Zealand Maori
	Senegal	0.5	1.0	Senegal
Melonoma	Australia	16.6	18.8	New Zealand
	Hawaii (Chinese)	0.0	0.0	Hawaii (Chinese)
Esophagus	India	10.7	24.7	China
	USA (New Mexico Indians)	0.0	0.2	Senegal
Prostate	USA (California Blacks)	100.2	0.0	
	China	0.8	0.0	
Rectum	Canada	22.0	14.0	Switzerland
	Senegal	1.0	1.0	Senegal
Stomach	Japan	98.0	50.0	Japan
	Senegal	4.0	2.0	Senegal

100 80 60 40 20 0 20 40 60 80 100

Figure 1,5 Highest and Lowest Global Age-Adjusted Incidence Rates, per 100,000 Population, for Specific Cancer Sites, 1984 (after Howe, 1986)

literature that suggests that there may be associations between variations in drinking water quality and spatial differences in mortality from cancers.[4-7] This chapter sets out to provide an international overview of the evidence both for, and against, such potential relationships, especially with respect to digestive cancers. Emphasis is then placed on such relationships in Canada, and especially in British Columbia.

Searching for Causal Variables

It is extremely important to identify why such spatial variations occur in the incidence of cancer, at the international and national levels. Interestingly enough, epidemiological evidence in the United States does not tend to support the widely held belief that cancer is primarily due to industrial pollution. To illustrate, in 1981, an extremely comprehensive overview of the epidemiology of cancer, in the United States, was undertaken by Doll and Peto.[8] They claim that the epidemiological evidence suggests that, with the exception of tobacco-related lung cancer, the causes of cancer are generally traditional, not modern. They concluded that, since there is no rapid increase in the probability of an individual of a given age developing most specific types of cancer, current United States mortality is far more likely to be due to long established aspects of American lifestyle and environment, than it is to the modern environment.

Drinking Water and Soils

Since 1968, there has been much interest in studying geographical patterns of site-specific cancer incidence and how these might be explained by the regional variability in the levels of natural bulk and trace elements in soils and drinking water supplies. Several researchers have suggested, for example, that the mineral content of either, or both, drinking water and soils may influence the spatial distribution of cancer. To illustrate, as early as 1952, Legon[9] argued that stomach cancer rates were highest among those inhabitants of North Wales and Cheshire, who lived in areas of highly organic soils. Since that period, the general focus of interest has narrowed to that of studying very specific environmental conditions as they relate to present cancer incidence or mortality. For example, in 1960 Allen-Price[10] provided evidence to support Tromp's[11] earlier view that there was a link between water quality and cancer incidence. In a study of stomach cancer incidence in West Devon, Allen-Price found this disease to be most common among those drinking soft, highly mineralized ground water, derived from Devonian bedrock. This relationship is illustrated in Figure 2,5, which shows both the geology of West Devon, England, and high and low stomach cancer incidence zones. As can be seen, the highest rates of stomach cancer appear to be associated with drinking water derived from highly mineralized Devonian aquifers. In contrast, the lowest incidence rates for this cancer seem to be associated with drinking water derived from Carboniferous sediments. The poorly mineralized granites of Devon also appear to be associated with low stomach cancer incidence.

GASTRIC CANCER

Standardized Mortality Ratios

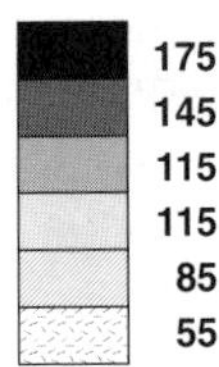

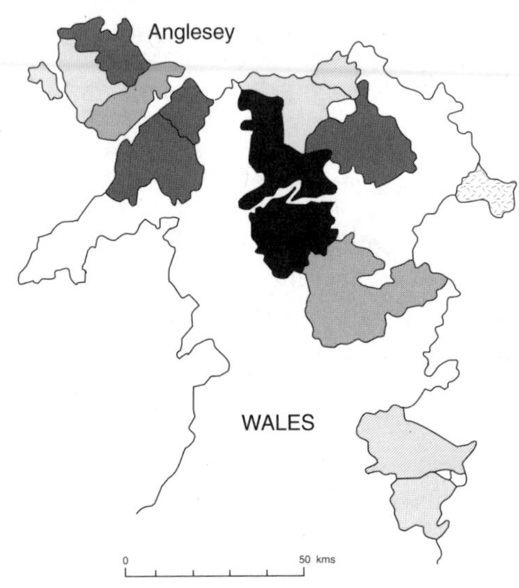

IGNITION LOSS OF CULTIVATED SOIL

(% dry weight)

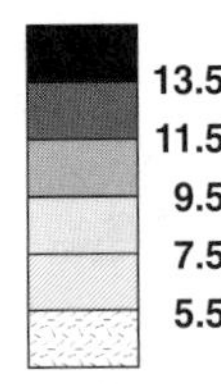

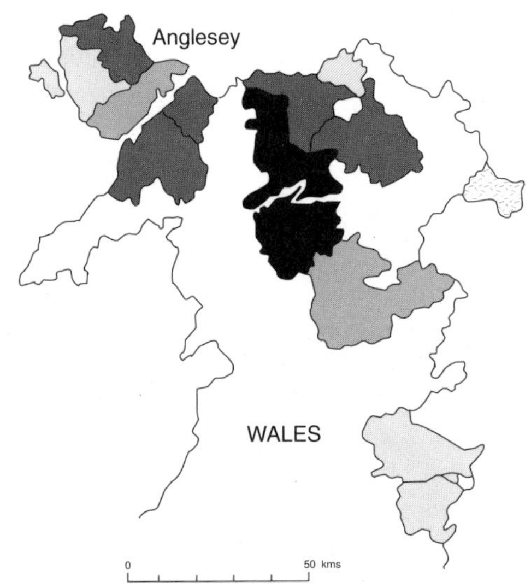

Figure 2,5 Comparison of Cancer Mortality Ratios and Ignition Loss of Cultivated Soils in North Wales (after Legon, 1952)

Haenszel et al.,[12] in a study of stomach cancer in Japan, found that consumers of leafy vegetables had lower rates of mortality than those eating western diets. He postulated that the high levels of calcium in leafy vegetables might have had a beneficial effect in reducing the incidence of disease. Reicher[13] argued that, in the United States, cancer appeared to be much more common in those communities using poor quality acidic drinking water. Similar conclusions were reached in the same year by Kuzma and his colleagues,[14] who studied the relationship between cancer incidence and drinking water quality in Ohio. These researchers concluded that cancer rates were highest among those consuming acidic surface water. Similarly Tuthill and his colleagues[15] argued that in Massachusetts there were important associations between cancer incidence and the pH, alkalinity, chlorine and nitrate content of drinking water. Further support for the view that spatial variations in the incidence of cancer reflected geographical variables came from Armstrong,[16] who studied disease patterns in Southeast Asia. He concluded that both water quality and diet were of major significance in understanding spatial variations in the incidence of cancer.

In 1980, Kendrick,[17] published a major review of the relationships between socio-economic variables and cancer incidence in New Zealand. His correlations indicated that environmental and socio-economic factors were strongly involved in the etiology of the disease. Kendrick argued that, not only did water quality parameters give more meaningful correlations, but that water provided a useful means of measuring the impact of local geochemical variations. This was because, unlike foods which may have been derived from many locations, water is generally taken from the immediate vicinity and utilized by the total population. Interestingly, unlike the previous research workers quoted, Kendrick found no relationship between the hardness of water and colonic cancer incidence in New Zealand. In contrast, Zeilhuis and Haring,[18] in their study of water hardness and mortality in the Netherlands, found strong negative correlations between water hardness and both cancer and heart disease.

Later, Thouez, Beauchamp and Simard[19] attempted to identify links between cancer and the physio-chemical quality of drinking water in Quebec. A major objective of this study was to examine possible relationships between drinking water conditions and the incidence of site-specific cancers. Statistics presented by these authors indicated that the cancer risk in Canadian males was greatest in Quebec. The results of their study indicated a negative association between water hardness and the incidence of stomach cancer, and other organs of the digestive system. Thouez and his colleagues pointed out that these results confirmed, in part, the results and theories of Crawford et al.[20] and Stocks.[21]

In 1986, Foster[22] undertook a major statistical study of cancer mortality in the United States. His results suggested that the distribution of deaths from cancers of the digestive system, the respiratory system and the reproduction system are strongly influenced by soil bulk and trace element excesses, or deficiencies. Foster[23] also examined these relationships, in a less rigorous manner, in a variety of other countries, demonstrating that such links appear to hold true at the international scale.

WATER QUALITY AND DIGESTIVE CANCER IN CANADA

It is clear from the preceding discussion of the literature that there is growing evidence to suggest that, in both the Developed and Developing Worlds, soil and water quality may play a key role in determining variations in cancer incidence. For example, several research workers, including Crawford et al.,[24] Kendrick,[25] Zeilhuis and Haring,[26] and Foster,[27] have suggested negative links between water hardness and the incidence of cancers of the digestive tract. It was hypothesized, therefore, that Canadian and British Columbian mortalities from digestive cancers might also be related to the elemental quality of drinking water. This chapter explores possible links between Canada's and British Columbia's mortality, from cancers of the digestive tract and certain aspects of drinking water quality, the research plan being shown as Figure 3,5.

Data Sources

Two sets of data were used in this analyses. The first, undertaken at the national scale, used published mortality and water quality data, at the census division level. The second British Columbia data set consisted of mortality data, provided by the British Columbia Cancer Control Agency, and comprehensive water quality data collected and assembled by the author, at the school district level.

Canadian Study by census division

To conduct such an analysis effectively, it is necessary to have environmental and medical data that relate to both the same spatial units and time periods. Variations in the Canadian death rate from cancers of the digestive tract, that is for the tongue, mouth and pharynx; stomach, large intestine and rectum are illustrated in the *Mortality Atlas of Canada, Vol-*

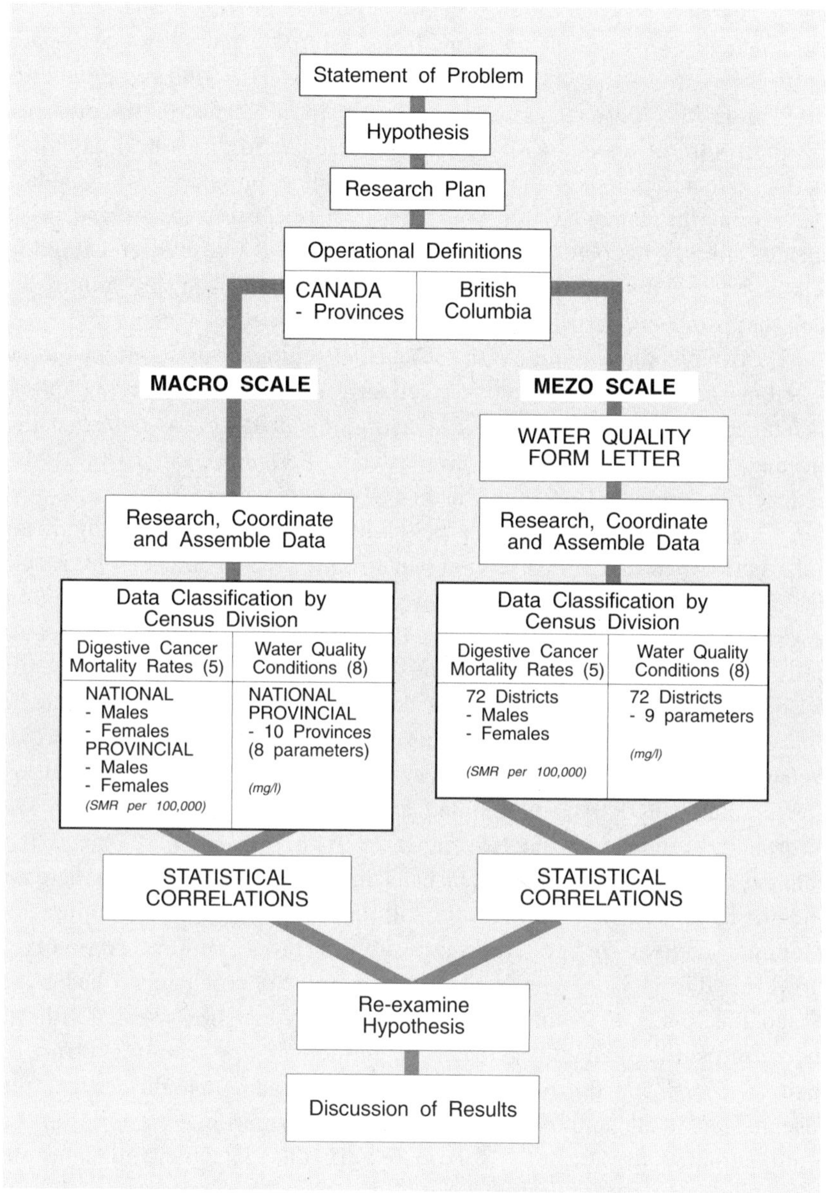

Figure 3,5 Research Plan for Assembly of Data and Testing the Hypothesis

ume 1: Cancer, produced jointly by Health and Welfare Canada and Statistics Canada. This volume was published in 1980 and shows the spatial distribution of cancer mortality, by census division, during the period 1966 to 1976. Maps of 16 cancers, or groups of cancers, are provided, together with an appendix of age standardized mortality rates, calculated by the direct method, using the 1971 Canadian population as standard. Data from this appendix, for both sexes, were abstracted for use in the current study. It consisted of mortality rates for five cancer categories related to the digestive tract; namely those of the tongue, mouth and pharynx; stomach, large intestine, rectum and colorectal cancer.

To explore the possibility that Canadian regional variations in mortality from cancers of the digestive tract were due, in part, to water quality differences, it was necessary to use hydrological data, also collected during the period 1966 to 1976, in any analysis. Fortunately, in 1970, a group of research workers from the University of Ottawa and Health and Welfare Canada began to explore the possibility of the existence of significant links between the chemical content of drinking water and cardiovascular health. To achieve this objective they undertook a nation-wide sampling of drinking water. This was collected from 526 communities with populations of at least 1,000, between June 1970 and December 1972, generally during the winter of 1971. Of these locations, 152 were then resampled at random to gauge seasonal fluctuations. In addition, all settlements with a water hardness of over 100 parts per million and cities with populations greater than 100,000 were sampled again in 1972 in greater detail. All information collected from November 1970 to December 1972 was then collated and published. This data bank included details of water hardness and the total calcium, magnesium, lithium, copper, zinc, chromium and cadmium content of the drinking water of the communities involved. However, during laboratory analysis, Neri and his colleagues[28] had established the levels of some elements, especially magnesium and lithium, more frequently than others. The magnitude of the sampling effort, its national scope and the time period involved, meant that this survey represented an ideal source of water quality information. Permission was kindly granted by Dr. L.C. Neri to rework this data bank to determine whether there were any significant links between these measured water quality parameters and cancer mortality rates. Since the location of each water supply had been identified by latitude and longitude, it was possible to establish from which census division it had been taken. The age standardized mortality rates for various cancers of the digestive tract were also known for each of these census divisions. As a result, Pearsons correlation could be used to establish the strength of the associations between those two groups of variables.

British Columbia by School District

The major objective of this study, however, was to test the hypothesis for conditions in British Columbia. To do this, it was necessary for the author to assemble a comprehensive set of medical and water quality data for the province. Mortality rates for site-specific digestive cancers were obtained from the Statistics Division of the British Columbia Cancer Control Agency, Vancouver. Data on all cancer-related deaths, in British Columbia for the period 1956 to 1978, at the school district level, were provided. Standardized mortality rates (SMR) per 100,000 population, by age and sex were provided for 75 school districts. These data were originally collected from death certificates by the Division of Vital Statistics, Ministry of Health, Province of British Columbia. Certificates included the deceased's age, sex, place of residence and principal cause of death. This information was subsequently coded and classified according to the International Classification of Diseases.

The school districts for which mortality rates are available are shown in Figure 4,5. The B.C. Cancer Control Agency used school districts rather than health districts, since school district boundaries have remained virtually unchanged for some 30 years. Health district boundaries, in contrast, have altered regularly as new population centres have developed. It should be noted, however, that although mortality rates for 75 school districts were provided by the B.C. Cancer Control Agency, only 72 were used for this study. School districts 81 (Fort Nelson), 87 (Stikine) and 92 (Nisga) were omitted from the analysis because these districts are very sparsely populated and, as a result, medical data are misleading. The Cancer Control Agency's data were then sorted by site-specific designation and classified by school district. This provided the basis from which the distribution of standardized digestive cancer mortality rates for the period 1956-1978 were derived.

To test the hypothesis, at a provincial scale, it was also necessary to have a very comprehensive water quality data set for British Columbia. Although governed under the Ministries of Environment and Health, most of the water district distribution systems within the province are administered by Regional District authority. The operation and quality control of the water distribution systems are undertaken by the various cities, towns or villages within each district. For this reason, no comprehensive set of water quality data for the province is available in published form.

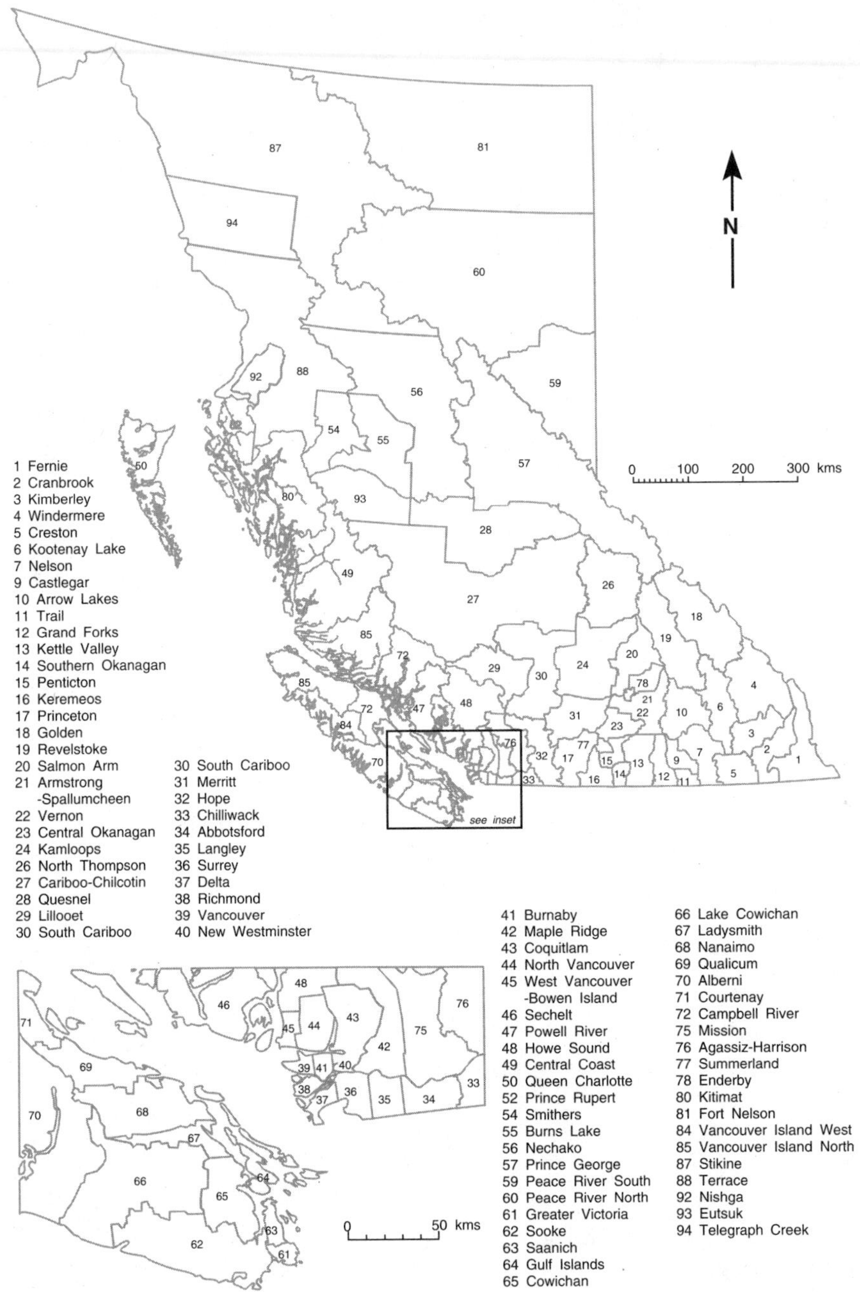

Figure 4,5 School District Boundaries in British Columbia

In order to develop one, it was necessary for the author to contact all administrative bodies responsible for water distribution and quality control in the province by form letter. These included cities, towns and villages, company towns, private subdivisions, irrigation and water improvement districts, as well as the Federal Department of Indian Affairs.

Extensive water quality information was received in response to these form letters, although it was not always complete. Fortunately, information was generally available on those parameters that were of particular importance to this study. These included the source of the water, its pH, hardness, and calcium, copper, iron, fluoride, magnesium, nitrogen, sodium and zinc levels. It should be noted that, in all cases, analysis was conducted on primary water sources, before any treatment had taken place.

A very select and efficient system of data management is required to handle such an enormous amount of information. Since medical data were available at the school district level, it was decided that the study would include the identification of all data under a field system by school district, as shown in Figure 4,5. Data collected also included the school district name, its water source, latitude, longitude, and the 10 water quality parameters utilized for this study.

Coding sheets where then developed to assist in the organization of these data and to provide an efficient way of loading it into the computer system by field. In most cases, respondents to the form letter asking for water quality information sent several year's records, including the most recent. In an effort to keep data from this study synchronized, only information collected during the period 1975 to 1978 was coded. In a few instances, however, the only records available were more recent, and where relevant, this updated information was used.

One difficultly involved in this process was determining the location of the distribution system within school district boundaries. A second problem encountered in this process involved establishing representative water quality levels when two or more population centres, with differing water sources, were situated within a single school district boundary. The author aggregated, or weighted, the elemental listings for different water sources, as a percent of the total population within a school district. As a result, aggregate listings were developed for each district that included pH, hardness, calcium, copper, iron, fluoride, magnesium, nitrogen, sodium, and zinc. Once medical and water quality data for British Columbia had been assembled in appropriate order by school district, Pearson correlation was used to establish the strength of association between the two variables.

STATISTICAL CORRELATIONS

As has been described, Canadian water quality information by provincial census divisions and British Columbia water quality by school district boundaries, were assembled into two separate data sets. Age-standardized digestive cancer mortality rates, at both the national and provincial levels, along with that of British Columbia by school district, were also incorporated into the same data system. As a result, it was then possible, using Pearson correlation coefficients, to test the hypothesis that there may be an association between cancers of the mouth, esophagus, stomach, small intestine, colon and rectum and elevated or depressed levels of water hardness, and pH, calcium, cadmium, chromium, copper, fluoride, iron, lithium, magnesium, nitrates, sodium and zinc.

To achieve this objective, Neri's Canadian water quality data were first correlated with Canadian cancer mortality data, derived from the *Mortality Atlas of Canada, Volume 1: Cancer.* Secondly, mortality data, provided by the British Columbia Cancer Control Agency, were correlated with the detailed water quality information from British Columbia.

Canada

Two methods of analysis were used in the national Canadian study. These procedures were chosen because of the nature of water sampling procedure and differences in the size of census divisions, which resulted in variations in the availability of water quality data. In the first set of analyses, all water data, in the form provided by Neri[29] and his associates, were utilized in Pearson correlations. In the second set, an arithmetic mean was calculated for each water quality parameter, for each census division for which such data were available. This mean was then used to represent this parameter when calculating Pearson correlation coefficients. While these two approaches resulted in minor variations in the strength and statistical significance of resulting correlations, both methods identified the same major trends. For this reason only the results obtained using the first method of analysis will be presented in this chapter. As there were fairly large samples, a significance level of $p = <0.0001$ was used.

The results gave little evidence of statistically significant ($p = <0.0001$) correlations between mortality from cancer of the digestive system and levels of copper, chromium, cadmium and zinc in Canadian drinking water. In contrast, as shown in Tables 1,5 and 2,5, the analysis established repeated significant negative correlations between the hardness and lithium, calcium and magnesium content of drinking water and cancers of the tongue, mouth and pharynx; stomach, large intestine, rectum and colorectal cancer. Although some exceptions to this generalization occurred, a preponderance of significant negative associations were established at both national and provincial scales. To illustrate, in males, there were repetitive negative correlations between drinking water magnesium levels and cancer of the tongue, mouth and pharynx. This can be seen for Canada as a whole (r = -0.30440, p = 0.0001), and also in British Columbia (r = -0.52745, p = 0.0001); Alberta (r = -0.25901, p = 0.0001); Manitoba (r = -0.44400, p = 0.0001); Ontario (r = -0.22669, p = 0.0001); Quebec (r = -0.15014, p = 0004) and New Brunswick (r = -0.32619, p = 0.0008). No analysis could be undertaken for Prince Edward Island because of the very limited water quality data available from that province.

As can be seen in Tables 1,5 and 2,5, the strongest negative correlations, at the national level, were between colorectal cancer and magnesium in both males and females. In addition, with one exception, all correlations between digestive cancer mortality and water hardness, lithium, calcium and magnesium content were negative, for both males and females, at the national scale. However, it should be noted that the strength of the correlations appear generally higher for males than for females. Similarly, the strongest negative correlations tend to be with magnesium, followed in declining order by those with hardness, calcium and lithium. To illustrate, in cancer of the large intestine in Canadian males, the correlations, all significant at the 0.0001 level, were as follows: magnesium (r = -0.34180), hardness (r = -0.29368), calcium (r = -0.26297) and lithium (r = -0.21742).

While such statistical correlations, in and by themselves, can never prove causal relationships, these results are suggestive. However, interpretation is complicated by the fact that hardness, calcium, magnesium and lithium in drinking water are themselves strongly positively correlated. In Canadian water supplies, for example, lithium content correlated at the 0.0001 level with magnesium (r = 0.58691), calcium (r = 0.74801), and hardness (r = 0.70111). Similarly there are marked positive correlations, at the same level of significance, between magnesium and both calcium (r = 0.68231) and water hardness (r = 0.83200). There is also a very strong positive link between calcium and water hardness (r = 0.93448). These correlations reflect geological realities. Water flowing over, or through, sedimentary rocks, as in the Canadian Rockies, tends to be hard and relatively enriched in these three elements. In contrast, soft water, from igneous or metamorphic aquifers or drainage basins, for example from the Canadian Shield, is relatively deficient in these three substances.

Table 1,5 Pearson Correlation Coeffecients and Level of Statistical Significance for Cancers of the Digestive Tract in Canadian MALES and the Hardness and Calcium, Magnesium and Lithium Content of Drinking Water

	Tongue Mouth Pharynx	Stomach	Large Intestine excluding Rectum	Rectum	Large Intestine and Rectum	Tongue Mouth Pharynx	Stomach	Large Intestine excluding Rectum	Rectum	Large Intestine and Rectum	Tongue Mouth Pharynx	Stomach	Large Intestine excluding Rectum	Rectum	Large Intestine and Rectum
	CANADA					**BRITISH COLUMBIA**					**ALBERTA**				
Calcium	-0.16421	-0.26537	-0.26297	-0.07704	-0.24542	-0.44264	-0.19848	-0.24726	-0.02749	-0.33866	-0.13589	-0.31028	-0.09306	0.12977	0.01416
	0.0001	0.0001	0.0001	0.0374	0.0001	0.0042	0.2195	0.1240	0.8663	0.0326	0.2264	0.0048	0.4086	0.2482	0.9002
	730	730	730	730	730	40	40	40	40	40	81	81	81	81	81
Hardness	-0.24759	-0.30928	-0.29368	-0.10594	-0.28308	-0.58692	-0.23520	-0.23674	-0.03263	-0.34324	-0.23632	-0.25377	-0.01451	0.21884	0.11703
	0.0001	0.0001	0.0001	0.0031	0.0001	0.0001	0.1001	0.0979	0.8220	0.0147	0.0266	0.0170	0.8933	0.0405	0.2775
	780	780	780	780	780	50	50	50	50	50	88	88	88	88	88
Lithium	-0.09417	-0.21464	-0.21742	-0.01758	-0.18500	0.44270	-0.14673	-0.22171	-0.07044	-0.27485	-0.15301	-0.35232	-0.08252	0.16915	0.04489
	0.0001	0.0001	0.0001	0.4023	0.0001	0.0001	0.0971	0.0609	0.4276	0.0016	0.0151	0.0001	0.1916	0.0071	0.4781
	2272	2272	2272	2272	2272	129	129	129	129	129	252	252	252	252	252
Magnesium	-0.30440	-0.29556	-0.34180	-0.16284	-0.34462	-0.52745	-0.11370	-0.22171	0.04857	-0.27924	-0.25901	-0.05972	-0.13512	0.22846	0.22294
	0.0001	0.0001	0.0001	0.0001	0.0001	0.0001	0.1995	0.0116	0.5847	0.0014	0.0001	0.3780	0.0453	0.0006	0.0009
	2130	2130	2130	2130	2130	129	129	129	129	129	220	220	220	220	220
	SASKATCHEWAN					**MANITOBA**					**ONTARIO**				
Calicum	-0.20569	0.12993	-0.08077	0.05038	-0.02633	-0.28712	0.26821	-0.35203	-0.18337	-0.35134	-0.02576	-0.17393	-0.14904	0.10542	-0.07783
	0.1214	0.3310	0.5467	0.7072	0.8445	0.0805	0.1035	0.0302	0.2705	0.0305	0.7085	0.0110	0.0297	0.1251	0.2581
	58	58	58	58	58	38	38	38	38	38	213	213	213	213	213
Hardness	-0.26747	-0.05411	0.02449	0.11322	0.07397	-0.40655	0.21213	-0.43678	-0.20310	-0.43622	-0.09566	-0.19565	-0.10737	0.09407	-0.04335
	0.0326	0.6711	0.8477	0.3730	0.5613	0.0102	0.1948	0.0054	0.1588	0.0055	0.1555	0.0034	0.1106	0.1625	0.5205
	64	64	64	64	64	39	39	39	39	39	222	222	222	222	222
Lithium	-0.17041	0.18312	-0.05700	-0.13817	-0.10604	-0.12713	0.26282	-0.10658	-0.16192	-0.00785	-0.22669	-0.19436	-0.04485	-0.14197	-0.13637
	0.0233	0.0147	0.4511	0.0667	0.1601	0.1836	0.0053	0.2655	0.0895	0.9349	0.0259	0.0184	0.0001	0.0002	0.0003
	177	177	177	177	177	111	111	111	111	111	697	697	697	697	697
Magnesium	-0.17338	0.10422	-0.07428	0.13851	0.02567	-0.44400	0.13490	-0.39838	-0.15735	-0.37449	-0.22669	-0.19436	-0.04508	-0.04485	-0.00586
	0.0283	0.1897	0.3505	0.0807	0.7473	0.0001	0.1659	0.0001	0.1055	0.0001	0.0001	0.0001	0.2628	0.2652	0.8843
	160	160	160	160	160	107	107	107	107	107	619	619	619	619	619

	Tongue Mouth Pharynx	Stomach	Large Intestine excluding Rectum	Rectum	Large Intestine and Rectum
QUEBEC					
Calcium	-0.16983	-0.03697	-0.10437	-0.11197	-0.13352
	0.0215	0.6193	0.1597	0.1313	0.0716
	183	183	183	183	183
Hardness	-0.20380	-0.07950	-0.10465	-0.17176	-0.16015
	0.0047	0.2743	0.1497	0.0175	0.0269
	191	191	191	191	191
Lithium	-0.09081	-0.01059	-0.14853	-0.14853	-1.19540
	0.0324	0.8033	0.0001	0.0004	0.0001
	555	555	555	555	555
Magnesium	-0.15014	-0.07204	-0.11677	-0.18843	-0.17627
	0.0004	0.0932	0.0064	0.0001	0.0001
	544	544	544	544	544
NOVA SCOTIA					
Calcium	'0.30217	-0.06313	-0.25750	-0.27318	-0.37514
	0.0223	0.6408	0.0531	0.0398	0.0040
	57	57	57	57	57
Hardness	0.32304	-0.06944	-0.23173	-0.28231	-0.36473
	0.0142	0.6078	0.0828	0.0334	0.0053
	57	57	57	57	57
Lithium	0.31009	0.00776	-0.12659	-0.24243	-0.24152
	0.0001	0.9232	0.1141	0.0022	0.0023
	157	157	157	157	157
Magnesium	0.17906	0.01456	-0.26084	-0.24472	-0.36612
	0.0248	0.8564	0.0010	0.0020	0.0001
	157	157	157	157	157

Tongue Mouth Pharynx	Stomach	Large Intestine excluding Rectum	Rectum	Large Intestine and Rectum
NEW BRUNSWICK				
-0.52356	-0.20409	-0.23563	-0.54950	-0.12771
0.0018	0.2546	0.1868	0.0009	0.4788
33	33	33	33	33
-0.48303	-0.15414	-0.29142	-0.40306	-0.01803
0.0018	0.3488	0.0001	0.0719	0.9133
39	39	39	39	39
-0.31017	-0.06934	-0.29142	-0.25835	-0.17967
0.0014	0.4864	0.0001	0.0084	0.0694
103	103	103	103	103
-0.32619	-0.10198	-0.25429	-0.21173	-0.09715
0.0008	0.3053	0.0095	0.0318	0.3289
103	103	103	103	103
NEWFOUNDLAND				
-0.01231	-0.23406	-0.46487	-0.40341	-0.57518
0.9534	0.2601	0.0192	0.0455	0.0026
25	25	25	25	25
0.08186	-0.20009	-0.18726	-0.16278	-0.24269
0.6786	0.3073	0.3400	0.4079	0.2134
28	28	28	28	28
-0.00490	-0.14630	-0.51130	-0.19115	-0.53424
0.0683	0.1953	0.0001	0.0894	0.0001
80	80	80	80	80
-0.14339	-0.39386	-0.68162	-0.21958	-0.69628
0.2045	0.0003	0.0001	0.0503	0.0001
80	80	80	80	80

British Columbia

As the preceding national study by census division has shown (Tables 2,5 and 3,5) there was little evidence in British Columbia, of statistically significant ($p = 0.0001$) correlations between mortality from male and female digestive cancers and the levels of cadmium, chromium, copper and zinc in drinking water. In contrast, however, there were several significant ($p = 0.0001$) negative correlations for British Columbia males, between levels of hardness, magnesium and lithium and cancers of the tongue, mouth and pharynx ($r = -0.58692$, $r = -0.52745$ and $r = -0.44270$ respectively) suggesting that these elements may provide a protective

Table 2,5 Pearson Correlation Coeffecients and Level of Statistical Significance for Cancers of the Digestive Tract in Canadian FEMALES and the Hardness and Calcium, Magnesium and Lithium Content of Drinking Water

	Tongue Mouth Pharynx	Stomach	Large Intestine excluding Rectum	Rectum	Large Intestine and Rectum	Tongue Mouth Pharynx	Stomach	Large Intestine excluding Rectum	Rectum	Large Intestine and Rectum	Tongue Mouth Pharynx	Stomach	Large Intestine excluding Rectum	Rectum	Large Intestine and Rectum
	CANADA					**BRITISH COLUMBIA**					**ALBERTA**				
Calcium	-0.00360	-0.22765	-0.15345	-0.07567	-0.15712	0.04824	0.17296	-0.43184	-0.05503	-0.47890	-0.06713	-0.15195	-0.26480	-0.26243	-0.30269
	0.9227	0.0001	0.0001	0.0410	0.0001	0.7675	0.2858	0.0054	0.7359	0.0018	0.5515	0.1757	0.0169	0.0179	0.0060
	730	730	730	730	730	40	40	40	40	40	81	81	81	81	81
Hardness	-0.03413	-0.23991	-0.19684	-0.12956	-0.21354	0.12566	0.10759	-0.30242	-0.06026	-0.35493	0.15266	-0.15830	-0.34067	-0.22986	-0.34733
	0.3412	0.0001	0.0001	0.0003	0.0001	0.3846	0.4571	0.0328	0.6776	0.0114	0.1556	0.1407	0.0012	0.0312	0.0009
	780	780	780	780	780	50	50	50	50	50	88	88	88	88	88
Lithium	0.03759	-0.17607	-0.13830	-0.04802	-0.13507	0.14917	0.22140	-0.31866	0.07718	-0.30397	-0.00220	-0.25726	-0.31387	-0.32882	-0.34661
	0.0733	0.0001	0.0001	0.0221	0.0001	0.0916	0.0117	0.0002	0.3846	0.0005	0.9723	0.0001	0.0001	0.0001	0.0001
	2272	2272	2272	2272	2272	129	129	129	129	129	252	252	252	252	252
Magnesium	-0.02377	-0.23586	-0.27701	-0.15216	-0.29142	0.21316	-0.05047	-0.30807	0.03822	-0.35792	0.40291	-0.14978	-0.29262	-0.08340	-0.25381
	0.2728	0.0001	0.0001	0.0001	0.0001	0.0153	0.5700	0.0004	0.6672	0.0001	0.0001	0.0263	0.0001	0.2179	0.0001
	2130	2130	2130	2130	2130	129	129	129	129	129	220	220	220	220	220
	SASKATCHEWAN					**MANITOBA**					**ONTARIO**				
Calcium	-0.14466	-0.20425	0.03306	0.18945	0.12012	-0.32616	0.04131	-0.01249	0.12169	0.04066	-0.10649	0.01143	0.13664	0.12092	0.16271
	0.2786	0.1241	0.8054	0.1544	0.3691	0.0457	0.8055	0.9406	0.4667	0.8085	0.1213	0.8683	0.0464	0.0783	0.0175
	58	58	58	58	58	38	38	38	38	38	213	213	213	213	213
Hardness	-0.23538	0.11483	0.10143	0.10057	0.14080	-0.31593	0.05116	-0.01778	0.18116	0.06062	-0.10703	0.04983	0.15915	0.08034	0.16951
	0.6212	0.3662	0.4252	0.4291	0.2671	0.0501	0.7571	0.9145	0.2697	0.7139	0.0112	0.4601	0.0176	0.2332	0.0114
	64	64	64	64	64	39	39	39	39	39	222	222	222	222	222
Lithium	0.01124	0.31111	-0.11744	0.26416	0.01822	-0.22714	0.06426	0.06802	0.18766	0.13214	-0.00392	0.11982	0.06209	0.01585	0.06225
	0.8820	0.0001	0.1195	0.0004	0.8098	0.0165	0.5028	0.4781	0.0486	0.1668	0.9177	0.0015	0.1015	0.6762	0.1006
	177	177	177	177	177	111	111	111	111	111	697	697	697	697	697
Magnesium	-0.14866	0.17072	0.03116	0.23832	0.13908	-0.22907	0.06066	0.01087	0.28803	0.13199	-0.29224	0.08164	0.13232	0.02706	0.13037
	0.0606	0.0309	0.6957	0.0024	0.0794	0.0176	0.5348	0.9115	0.0026	0.1754	0.0001	0.0423	0.0008	0.5016	0.0012
	160	160	160	160	160	107	107	107	107	107	619	619	619	619	619

	QUEBEC Tongue Mouth Pharynx	QUEBEC Stomach	QUEBEC Large Intestine excluding Rectum	QUEBEC Rectum	QUEBEC Large Intestine and Rectum	NEW BRUNSWICK Tongue Mouth Pharynx	NEW BRUNSWICK Stomach	NEW BRUNSWICK Large Intestine excluding Rectum	NEW BRUNSWICK Rectum	NEW BRUNSWICK Large Intestine and Rectum
Calcium	0.00583	-0.12114	-0.00206	-0.23729	-0.09027	-0.39099	-0.10974	-0.42815	-0.30409	-0.54703
	0.9376	0.1023	0.9779	0.0012	0.2243	0.0245	0.2877	0.0129	0.0853	0.0010
	183	183	183	183	183	33	33	33	33	33
Hardness	0.05073	-0.13790	-0.02059	-0.19949	-0.09324	-0.18746	0.07361	-0.19191	-0.30375	-0.30986
	0.4859	0.0571	0.7774	0.0057	0.1995	0.2531	0.6561	0.2419	0.0601	0.0549
	191	191	191	191	191	39	39	39	39	39
Lithium	-0.09627	-0.07431	-0.06679	-0.11602	-0.10408	0.02071	0.07135	-0.16723	-0.19503	-0.23374
	0.0233	0.0803	0.1160	0.0062	0.0142	0.8355	0.4739	0.0913	0.0484	0.0175
	555	555	555	555	555	103	103	103	103	103
Magnesium	0.00825	-0.12785	-0.02952	-0.03696	-0.03979	0.00266	0.10202	-0.04576	-0.30713	-0.15942
	0.8478	0.0028	0.4920	0.3896	0.3543	0.9787	0.3051	0.6463	0.0016	0.1077
	544	544	544	544	544	103	103	103	103	103

	NOVA SCOTIA Tongue Mouth Pharynx	NOVA SCOTIA Stomach	NOVA SCOTIA Large Intestine excluding Rectum	NOVA SCOTIA Rectum	NOVA SCOTIA Large Intestine and Rectum	NEWFOUNDLAND Tongue Mouth Pharynx	NEWFOUNDLAND Stomach	NEWFOUNDLAND Large Intestine excluding Rectum	NEWFOUNDLAND Rectum	NEWFOUNDLAND Large Intestine and Rectum
Calcium	0.30232	-0.19439	0.17965	-0.35181	0.05478	0.03011	-0.14980	0.04122	0.16500	0.06980
	0.0223	0.1474	0.1811	0.0073	0.6857	0.8864	0.4748	0.8449	0.4306	0.7402
	57	57	57	57	57	25	25	25	25	25
Hardness	0.30877	-0.19561	0.22075	-0.34829	0.09539	-0.02984	-0.15137	0.12553	0.22561	0.15848
	0.0194	0.1448	0.0989	0.0079	0.4803	0.0802	0.4420	0.5245	0.2484	0.4205
	57	57	57	57	57	28	28	28	28	28
Lithium	0.34173	-0.09476	0.17885	-0.30555	0.07300	-0.08854	-0.08787	-0.05104	0.18659	-0.00629
	0.0001	0.2378	0.0250	0.0001	0.3636	0.4348	0.4383	0.6530	0.0975	0.9559
	157	157	157	157	157	80	80	80	80	80
Magnesium	0.21967	-0.09081	0.04122	-0.15294	-0.01042	-0.07176	-0.35279	-0.26634	0.05761	-0.22571
	0.0057	0.2580	0.6083	0.0558	0.8970	0.5270	0.0013	0.0169	0.6117	0.0441
	157	157	157	157	157	80	80	80	80	80

effect against cancer. This was also evident for British Columbia females where a significant correlation was found between magnesium and colorectal cancer (r = -0.35792, p = 0.0001) and at a less significant level (r<0.0005) between female colonic cancer and lithium (r = -0.31866, p = 0.0002); and magnesium (r = -0.30807, p = 0.0004) and between female colorectal cancer and lithium (r = -0.30397, p = 0.0005). Again the negative correlations suggest the protective effect of these elements.

In contrast, as can be seen from Tables 3,5 and 4,5, analysis at the *provincial school district level* produced far fewer significant correlations (p<0.01) than the national study. These weak associations did, however, indicate similar relationships to those found in the Canadian study, and

Table 3,5 Pearson Correlation Coeffecients and Level of Statistical Significance for Cancers of the Digestive Tract in British Columbia MALES and Nine Water Quality Parameters

	Esophagus Cancer	Stomach Cancer	Small Intestinal Cancer	Colon Cancer	Rectal Cancer
pH	0.03182	-0.00582	0.03863	-0.21201	-0.31828
	0.7907	0.9613	0.7473	0.0738	0.0064
	72	72	72	72	72
Calcium	-0.04618	-0.04404	-0.01831	-0.14690	-0.31251
	0.7001	0.7133	0.8787	0.2182	0.0075
	72	72	72	72	72
Copper	-0.04170	0.29371	-0.06499	0.14030	0.05952
	0.7280	0.0123	0.5876	0.2398	0.6195
	72	72	72	72	72
Fluoride	0.04534	0.00947	0.28515	-0.23000	-0.11781
	0.7053	0.9371	0.0152	0.0519	0.3243
	72	72	72	72	72
Iron	0.22156	-0.03406	0.33585	0.15091	-0.07313
	0.0614	0.7764	0.0039	0.2057	0.5416
	72	72	72	72	72
Hardness	-0.07747	-0.07138	-0.00708	-0.14097	-0.29605
	0.5178	0.5513	0.9529	0.2376	0.0116
	72	72	72	72	72
Magnesium	-0.13284	-0.11872	-0.02009	-0.11409	-0.31554
	0.2660	0.3206	0.8670	0.3399	0.0069
	72	72	72	72	72
Nitrates	-0.08785	0.01353	0.00003	-0.33873	-0.06959
	0.4631	0.9102	0.9998	0.0036	0.5614
	72	72	72	72	72
Sodium	0.36508	0.01610	0.00870	-0.09657	-0.18340
	0.0016	0.8932	0.9422	0.4197	0.1231
	72	72	72	72	72
Zinc	-0.03534	0.02860	-0.06331	-0.05887	-0.18340
	0.7682	0.8115	0.5973	0.6233	0.0538
	72	72	72	72	72

Table 4,5 Pearson Correlation Coeffecients and Level of Statistical Significance for Cancers of the Digestive Tract in British Columbia FEMALES and Nine Water Quality Parameters

	Esophagus Cancer	Stomach Cancer	Small Intestinal Cancer	Colon Cancer	Rectal Cancer
pH	0.20685	0.15854	0.16072	-0.02336	-0.02505
	0.0813	0.1835	0.1774	0.8456	0.8345
	72	72	72	72	72
Calcium	0.09317	-0.06063	0.05594	-0.06718	-0.09508
	0.4363	0.6129	0.6407	0.5750	0.4269
	72	72	72	72	72
Copper	-0.10883	0.25539	0.01029	0.27902	0.07810
	0.3628	0.0304	0.9316	0.0176	0.5144
	72	72	72	72	72
Fluoride	0.04049	-0.13663	0.03327	-0.21335	-0.12439
	0.7359	0.2525	0.7815	0.0720	0.2979
	72	72	72	72	72
Iron	0.14364	0.05215	-0.06524	0.07239	-0.16027
	0.2287	0.6635	0.5861	0.5456	0.1787
	72	72	72	72	72
Hardness	0.09036	-0.07375	0.02522	-0.05475	-0.09526
	0.4503	0.5381	0.8335	0.6478	0.4261
	72	72	72	72	72
Magnesium	0.13875	-0.13220	-0.06717	-0.06165	-0.16019
	0.2451	0.2683	0.5751	0.6069	0.1789
	72	72	72	72	72
Nitrates	-0.03433	0.27614	0.42513	-0.03892	0.04051
	0.7745	0.0189	0.0002	0.7455	0.7354
	72	72	72	72	72
Sodium	0.11625	0.02185	0.14221	0.14300	-0.15910
	0.3308	0.8554	0.2334	0.2308	0.1819
	72	72	72	72	72
Zinc	0.11472	0.12019	-0.07853	0.03695	0.12374
	0.3373	0.3146	0.5120	0.7579	0.2663
	72	72	72	72	72

cited elsewhere in the literature. For example, in males, positive associations were found between male esophageal cancer and sodium (r = 0.36508, p = 0.0016), and between small intestinal cancer and iron (r = 0.33585, p = 0.0039), suggesting an antagonistic association. In contrast, a negative association was found between male colonic cancer and nitrates (r = -0.33873, p = 0.0036). The strongest and most significant associations, in males, were the negative correlations between cancer of the rectum and pH (r = -0.31828, p = 0.0064); calcium (r = -0.31251, p = 0.0075) and magnesium (r = -0.31554, p = 0.0069).

As can be seen from Table 5,5, the analysis indicated that, for British Columbia females, meaningful associations between site-specific digestive cancers and water quality were much less frequent and, for the most part, not strongly statistically significant. The repeated negative associations between calcium and magnesium and site-specific digestive cancers, in males, was, however, notable.

Table 5,5 Summary of Significant Correlations Between Site-specific Digestive Cancers, Water Hardness and Associated Water Quality Parameters in Canada and British Columbia

	Calcium		Hardness		Lithium		Magnesium	
Tongue	Can/M	-0.16*	Can/M	-0.25*	Can/M	-0.09*	Can/M	-0.30*
Mouth							Can/F/AB	-0.40*
Pharynx							Can/F/ON	-0.29*
Stomach	Can/M	-0.26*	Can/M	-0.31*	Can/M	-0.21*	Can/M	-0.29*
	Can/F	-0.23*	Can/F	-0.23*	Can/F	-0.17*	Can/F	-0.23*
					Can/F/AB	-0.25*		
					Can/F/SK	+0.31*		
Colon	Can/M	-0.26*	Can/M	-0.29*	Can/M	-0.22*	Can/M	-0.34*
	Can/F	-0.15*			Can/F	-0.14*	Can/F	-0.28*
	Can/F/BC	-0.43#			Can/F/BC	-0.31+	Can/F/BC	-0.30+
					Can/F/AB	-0.31*	Can/F/AB	-0.29*
Rectum	BC/M	-0.31*	Can/M	-0.11*	Can/F/AB	-0.32*	Can/M	-0.16*
			Can/F	-0.13+	Can/F/NS	-0.33*	Can/F	-0.15*
							BC/M	-0.31*

Legend:
Can = Canadian Study by Province
BC = British Columbia Study
M = Male
F = Female

Significance Levels:
* = 0.0001
+ = 0.001
# = 0.01

CONCLUSIONS

While the results of these two analyses, at the national and provincial levels, do not prove causal links between specific digestive cancers and water quality, they are suggestive. The evidence presented in summary Table 6,5 tends to imply that, at the national level at least, water hardness and its major constituents, calcium and magnesium, along with lithium may play a role in reducing mortality from cancers of the tongue, mouth, stomach, colon and rectum. This association has been found elsewhere, in literature reviewed earlier in this study.[30-33] However, it has been suggested that other soil and water parameters, such as the concentrations of barium, mercury, phosphorus and selenium, may also play causal roles.[34] If this is true, then it may help to explain why some of the correlations produced by the current study are not stronger, since data were unavailable for these elements.

The hypothesis that digestive cancer mortality may be linked to elevated or depressed levels of bulk or trace elements in drinking water is suggested by the literature, which also indicates that water hardness and its major constituents, calcium and magnesium, appear to provide a protective effect from digestive cancers. Some of these negative relationships with water hardness appear to have been confirmed, in part, by the results of this study, especially with regard to cancers of the lower digestive tract, such as that of the colon and rectum.

Calcium

Given the apparent negative correlations between cancers of the digestive tract and water hardness, how likely is it that calcium and/or magnesium in drinking water can influence carcinogenesis? Calcium is the medium for all cell communication, carrying vital messages between cells. It is also known that in low serum calcium environments, cell division is stimulated. This leads to hyperplasia, which appears to be an initial precancerous stage. Certainly, an individual living in a high calcium environment, such as the typical hardwater areas of British Columbia, is less likely to be calcium deficient. They are, therefore, probably less prone to develop hyperplasia.[35] Under these circumstances, it seems logical that many cancers, especially those of the digestive tract, will be less common in such hard water areas.

There is also growing evidence to suggest a link between high fat diets and the incidence of colorectal cancer. It is thought that an elevated fat intake may increase the secretion of bile acids, which are needed to digest fat, altering bacterial populations in the large bowel, and thereby increasing the levels of secondary acids. These secondary acids, in turn, are thought to promote lesions in the bowel, so encouraging carcinogeneses. Medical research has suggested that the level of calcium in the lower digestive tract may be a key factor in the etiology of this disease. It is thought that calcium may convert fatty acids and free bile, in the colon, to insoluble soaps, so reducing this postulated carcinogenic process. Such negative links between calcium and colorectal cancer have been noted by Garland.[36]

These theories have recently been reconfirmed by Stanfield,[37] an Australian medical researcher, who found that all excitable membranes have voltage-dependant calcium channels that provide a link between membrane changes and cellular responses, being controlled by cytosolic Ca^{2+}. The channel openings are activated by modulated changes in Ca^{2+}, signalled by hormones and neutral transmitters. Calcium channels appear to select Ca^{2+} over monovalent cations such as sodium and potassium by momentarily binding Ca^{2+} as it permeates. Significantly, it was also found that divalent cations like Cd^{2+} block channels, as do other organic calcium antagonists.

While the potential role of calcium in reducing carcinogenesis appears relatively easy to justify from the medical literature, it is more difficult to account for the relatively strong negative correlations between digestive cancer mortality and the magnesium content of drinking water. However, such protective relationships seem apparent from the proceeding geographical analyses.

Selenium

There is also considerable epidemiological evidence to suggest that cancer mortality appears to be depressed in regions where levels of environmental selenium are high. This apparently protective effect of selenium has been well documented by animal experiments. Carcinogenesis is most likely to occur in animals that are selenium deficient. Protection appears to increase with selenium dose until toxic levels are reached.[38,39] In the People's Republic of China, researchers found that the mean selenium content of the hair of control patients was far higher than that of 54 patients suffering from cancer of the digestive tract.[40] It is also thought that elevated levels of serum selenium discourage the accumulation, in the body, of harmful elements such as arsenic and mercury.

It is possible, therefore, that the negative correlations between hard water and digestive cancer mortality, described in this chapter, occur simply because high calcium and magnesium environments elevate selenium absorption by crops and increase its presence in local food and water supplies.

FURTHER RESEARCH

The role of selenium in drinking water should not be overlooked in future research, since this element may be a key catalyst in many biochemical processes and deficiencies may ultimately lead to poor health. It is further evident from the medical literature that there is a need to clearly assess the metabolic role of dietary selenium and its apparently protective effects in digestive cancers.[41-47] Perhaps investigations should also be undertaken to determine whether digestive cancer mortality can be reduced by treating soft drinking water with calcium and magnesium. This process might be required in very soft water areas, such as Victoria and Vancouver metropolitan areas.

The addition of bulk or trace elements to drinking water would at first glance appear to be a rather drastic move to prevent the occurrence of disease. In the People's Republic of China, however, it has been found that the incidence of digestive cancer can be reduced by such supplementation to natural drinking water. For example, high incidence rates of esophageal cancer in the Xingtai, Hebei Province, were dramatically reduced by the addition of jiang-shi to the well water of several villages. Jiang-shi are natural loess concretions that contain high levels of calcium, selenium and a variety of other trace elements and have been used as a treatment for esophageal cancer for centuries. During 1974, new wells were constructed in the five villages of Baian. Jiang-shi were either incorporated into their structures, or added to the ground water. Since that date no new cases of esophageal cancer have been reported in an area that had formerly experienced one of the world's highest incidence rates of this disease. This Chinese evidence obviously strongly supports the hypothesis that water hardness, and perhaps selenium, have a protective effect against certain digestive cancers.[31]

It should be recognized, however, that the intake of minerals from drinking water can have subtle, unpredictable outcomes. For this reason, alteration or enrichment of raw water with calcium, magnesium or selenium, should be carried out over a lengthy time period, with great caution, so as to be able to clearly identify total health implications.

REFERENCES

[1] Hayakawa, N. and Kurihara, M. 1981. International Comparison of Trends in Cancer Mortality for Selected Sites. *Social Science and Medicine* 15D, pp. 245-249.

[2] Howe, G.M. (ed.) 1986. *Global Geocancerology: A World Geography of Human Cancers*. Edinburgh: Churchill Livingstone.

[3] *Ibid.*

[4] Reicher, N.A. 1977. *An Epidemiologic Investigation of the Relationship Between Chemical Contaminants in Drinking Water and Cancer Mortality.* Ph.D. Thesis. Department of Preventative Medicine, The Ohio State University.

[5] Zielhuis, R.L. and Haring, B.J.A. 1981. Water Hardness and Mortality in the Netherlands. *The Science of the Total Environment* 18, pp. 35-45.

[6] Durlach, J. and Bara, M. 1985. Magnesium Levels in Drinking Water and Cardiovascular Risk Factor: A Hypotheses. *Magnesium* 4, pp. 5-15.

[7] Foster, H.D. 1986. *Reducing Cancer Mortality. A Geographic Perspective.* Western Geographic Series, Vol. 23. Victoria: Department of Geography, University of Victoria.

[8] Doll, R. and Peto, R. 1981. The Causes of Cancer: Quantitative Estimate of Avoidable Risk of Cancer in the United States Today. *Journal of National Cancer Institute* 66B, pp. 1193-1308.

[9] Legon, C.D. 1952. The Etiological Significance of Geographical Variations in Cancer Mortality. *British Medical Journal*, pp. 700-702.

[10] Allen-Price, E.D.1960. Uneven Distribution of Cancer in West Devon. *Lancet* 1, pp. 1235-38.

[11] Tromp, S.W. 1955. Possible Effects of Geophysical and Geochemical Factors on Development and Geographic Distribution of Cancer. *Schweiz Z. Path. Bakt.* 81, p. 929.

[12] Haenszel, W., Kurihara, M., Locake, F.B., Shimuzu, K. and Sego, M. 1976. Stomach Cancer in Japan. *Journal of the National Cancer Institute* 56, pp. 265-274.

[13] Reicher, *op. cit.*

[14] Kuzma, R.J., Kuzma, C.M. and Buncher, C.R. 1977. Ohio Drinking Water Source and Cancer Rates. *American Journal of Public Health* 67, p. 725.

[15] Tuthill, R.W. and Moore, G. 1978. *Chlorination of Public Drinking Water Supplies and Subsequent Cancer Mortality on Ecological Time-lag Study.* Office of Research and Development, U.S. Environmental Protection Agency, Cincinnati, OH.

[16] Armstrong, R.W. 1980. Geographical Aspects of Cancer Incidence in Southeast Asia. *Social Science and Medicine* 14D, pp. 299-306.

[17] Kendrick, B.L. 1980. A Spatial Environmental and Socioeconomic Appraisal of Cancer in New Zealand. *Social Science and Medicine* 14D, pp. 205-214.

[18] Zielhuis and Haring, *op. cit.*, pp. 35-45.

[19] Thouez, J.P., Beauchamp, Y. and Simard, A. 1981. Cancer and the Physiochemical Quality of Drinking Water in Quebec. *Social Science and Medicine*.15D, pp. 213-223.

[20] Crawford, M.D., Gardner, M.J. and Morris, J.N. 1968. Mortality and Hardness of Local Water Supplies. *Lancet*. 547, pp. 827-831.

[21] Stocks, P. 1973. Mortality from Cancer and Cardiovascular Diseases in the County Boroughs of England and Wales, Classified According to the Sources and Hardness of their Water Supplies, 1958-1967. *Journal of Hygiene*. 71, pp. 237-252.

[22] Foster, *op. cit.*

[23] *Ibid.*

[24] Crawford et al., *op. cit.*

[25] Kendrick, *op. cit.*

[26] Zielhuis and Haring, *op. cit.*

[27] Foster, *op. cit.*

[28] Neri, L.C. et al. 1977. *Chemical Content of Canadian Drinking Water Related to Cardiovascular Health*. Department of Energy, Mines and Resources, Reproduction Services, University of Ottawa.

[29] *Ibid.*

[30] Reicher, *op. cit.*

[31] Armstrong, *op. cit.*

[32] Thouezet al., *op. cit.*

[33] Foster, *op. cit.*

[34] *Ibid.*

[35] Garland, C. and Garland, F. 1989. *The Calcium Connection*. New York: Fireside Books.

[36] *Ibid.*

[37] Stanfield, P.R. 1986. Voltage Dependant Calcium Channels of Excitable Membranes. *British Medical Bulletin* 42-44, pp. 359-367.

[38] Shamberger, R.J. 1970. Relation of Selenium to Cancer. Inhibitory Effect of Selenium on Carcinogenesis. *Journal of the National Cancer Institute*. 44, pp. 931-936.

[39] Shamberger, R. and Willis, C. 1980. CRC Critical Reviews in Clinical Laboratory Sciences, 211-221, 1971, cited in Passwater, R.A. *Selenium as Food and Medicine*. New Canaan: Keats.

[40] Foster, H.D. 1989. Selenium and Health: Insights from the People's Republic of China. *The Journal of Orthomolecular Medicine*. 4(3), pp. 123-135.

[41] Warren, H.V. 1965. Medical Geology and Geology, *Science*. 148, pp. 534-539.

[42] Shamberger, *op. cit.*

[43] Shamberger et al., *op. cit.*

[44] Passwater, R.A. 1980. *Selenium as Food and Medicine*. New Canaan, Conn.: Keats.

[45] Schrauzer, G.N., White, D.A. and Schneider, C.J. 1977. Cancer Mortality Correlations Studies IV. Associations with Dietary Intakes and Blood Levels of Certain Trace Elements, Notably Se-antagonists. *Bioinorganic Chemistry* 7, pp. 35-36.

[46] Peng, A. and Langqiu, X. 1987. The Effects of Humic Acid on the Chemical and Biological Properties of Selenium in the Environment. *The Science of the Total Environment*. 64, pp. 89-98.

[47] Foster, *op. cit.*

[48] Zhu, C. and An, Y. 1988. The Effect of Jiang-shi in Preventation and Treatment of Esophageal Cancer in the High Incidence District, *Abstracts, International Symposium on Environmental Life Elements and Health*, November 1-5, 1988, Beijing, People's Republic of China, p. 225.

SCHIZOPHRENIA AND THE ENVIRONMENT

Donald I. Templer

California School of Professional Psychology - Fresno

INTRODUCTION

Schizophrenia is the most common of the serious mental disorders. Probably about one percent of the general population, in North America and Europe, develop schizophrenia. Perhaps the most cardinal of the symptoms of schizophrenia is what is called a thinking disorder or looseness of associations. The schizophrenic's thinking is disorganized and, at times, it can be difficult to follow the patient's discourse. Auditory hallucinations, usually the hearing of voices, are typically found in schizophrenics, and they commonly have delusions, especially of a persecutory nature. Schizophrenia usually begins in adolescence, or early adult life, and is ordinarily regarded as a chronic disorder. Social and occupational incapacitation or impaired functioning is the rule.

Whether or not the hallucinations are primarily auditory is one of the more important factors in the differentiation between schizophrenia and other psychoses. A predominance of visual, or tactile, hallucinations suggest some sort of other state, such as a toxic psychosis, a drug induced psychosis, or delirium tremens. Another critical element in such differential diagnosis is the clearness of the sensorium and adequate orientation. The schizophrenic is usually oriented with respect to time, person and place.

In regard to etiology, there is a diversity of established facts, a diversity of speculation, and a diversity of uncertainties. This chapter postulates that known and possible causes can be divided into those inherent in the person and those contained in a broadly conceptualized environment.

In the 1940s and 1950s and even in the 1960s, North American clinical psychologists viewed the etiology of schizophrenia in terms of psychodynamic and psychosocial forces. The forces within the person included such constructs as unconscious wishes, sexual urges, intrapsychic conflict, and regression to earlier psychosexual stages of development. The hypothesized forces outside of the person were said to be related to significant others, in particular, parents, especially the mother. Much was written about the seemingly disturbed and wicked "schizophrenogenic mother". There was also considerable literature about the "double bind hypothesis"[1] which postulates a duplicity of messages from the mother, on an overt level demonstrating positive regard toward the child, but on a more subtle level somehow communicating hostility and rejection. There was less said about the father, who was usually regarded as passive and ineffectual, but less malignant than the mother.

Although all of this large volume of literature on the psychological etiology of schizophrenia provides very interesting reading, the supportive research evidence has been rather meager. On the other hand, the research evidence for brain abnormality in schizophrenia has been impressive and growing in the last two decades. One should not become excessively enthused regarding these findings, however, because there is no biological anomaly that all schizophrenics have, and because the overlap in comparison to non-schizophrenics is considerable. Nevertheless, the Templer and Cappelletty[2] primary-secondary conceptualization does attempt to integrate these brain abnormalities in a meaningful fashion. This theoretical and empirically based formulation can be viewed as a more contemporary and more justifiable conceptualization of etiology inherent in the person as opposed to an etiology that is environmental. In contrast to the old viewpoint of psychological and psychosocial forces constituting the intrinsic and extrinsic etiological elements, the etiological ingredients are of a biological sort.

The primary schizophrenic is likely to have more genetic predisposition, a more insidious onset at a younger age, lower intellectual and psychosocial functioning, and a less favorable prognosis. The secondary schizophrenic is conceptualized as tending to have a lesser genetic component, later age of onset, better intellectual and psychosocial functioning, a less unfavorable prognosis, and some sort of environmental assault to the brain.

This environmental assault to the brain forms the bulk of this chapter and is divided into eight domains - industrialization, intrauterine and perinatal environment, head injury, nutritional and soil constituent environment, infection, alcohol, drugs, and the socio-economic environment.

THE GEOGRAPHY OF SCHIZOPHRENIA

Statements about the geography of schizophrenia should be made with caution, especially on an international level. This is due to national differences in criteria and diagnostic practice and because of possible variations in recording. Nevertheless, although one may question the schizophrenia rates in any given country and between pairs of countries, the big picture seems to indicate that rates of schizophrenia are not the same throughout the world (Table 1,6).

Templer, Hintze, Trent and Trent[4] found significant positive correlations between these schizophrenia rates[5] and both latitude and per capita income. Inverse correlations were found with January mean low temperature, July mean high temperature and latitude. The latitude and July temperature correlations with schizophrenia were still significant after controlling for per capita income.

Torrey's[6] pictorial representation is consistent with the information provided by Sheper-Hughes.[7] Torrey used four different types of dots to represent four categories of schizophrenia rates. Templer, Cappelletty and Kauffman[8] averaged the dots for each of the six continents and found relative rates of 2.69 for Europe, 2.00 for North America and Australia, 1.40 for Asia, 1.29 for Africa and 1.00 for South America. Torrey provided the most detailed information for Europe. He maintained that the rates are very high for Ireland, especially in the west part of the country. He reported high rates in the Scandinavian countries and especially Northern Sweden, and in Croatia, Poland, and the Soviet Union; and he maintained the rates are comparatively low in the Mediterranean countries.

There is an interesting, but not clearly understood, relationship between the epidemiology of schizophrenia and the epidemiology of multiple sclerosis. Templer, Regier and Corgiat[9] investigated this relationship because of the common properties of the two disorders. Each is a chronic disorder that begins in early adult life and runs an irregular course. Both are familial disorders and appear to be more common in the colder parts of the world. The possibility of slow virus etiology has been suggested for both. Although Templer et al. believed at the time they were the first to suggest the common properties of the two disorders, Torrey[10] stated:

Table 1,6 International Schizophrenia Prevalence

Country	Prevalence
Republic of Ireland	7.37
Sweden	6.97
Austria	5.29
New Zealand	4.97
Israel	4.78
Scotland	4.74
Northern Ireland	4.20
England and Wales	3.63
United States	2.93
Canada	2.39
Poland	2.38
Italy	2.06
Chile	1.62
Ceylon (Sri Lanka)	1.42
Spain	1.33
Japan	1.31
Cyprus	1.09
Greece	1.06
Portugal	1.01
Brazil	0.73
Ghana	0.52
Mexico	0.34
Kenya	0.29
Senegal	0.26
Nigeria	0.02

Source: After Sheper-Hughes.[3]

> Both are much more common in northern than southern Europe, in northern parts of the United States than in the South, in urban areas, and in technologically developed nations; and both afflictions were first described in the early years of the nineteenth century.

Templer et al. found that the states with the 10 highest schizophrenia rates in the USA also had significantly higher multiple sclerosis rates than the 10 lowest schizophrenia states. In a second study, a surprisingly high correlation of 0.81 ($p<.001$) between the rates of the two disorders was found in the districts of Italy.[11]

Templer, Cappelletty and Kauffman[12] reported correlations of 0.99 and 0.94 between the schizophrenia index based on Torrey's pictorial presentation for the six continents, and two different rates of multiple sclerosis. However, the correlations between multiple sclerosis rates and schizophrenia rates in the counties of Ireland were not significant. The association between schizophrenia and multiple sclerosis rates for the United States was also positive, but less strong than in the Templer, Regier and Corgiat[13] study that used data consisting of only the very highest and lowest schizophrenia rate states.

A study by Templer et al.[14] further extended the previous research that had found geographic similarity of the distributions of schizophrenia and multiple sclerosis. The independent variables located for both Italy and the United States were January low temperature, July high temperature, precipitation, elevation, latitude, influenza rate, sunlight, and esophageal cancer. Elevation was used because multiple sclerosis and goiter are more common in areas where soils are low in iodine.[15,16,17,18] Precipitation was chosen as an exploratory variable both because infectious disorders are more common in seasons and locations with more precipitation and because Foster[19] reported a negative correlation between sunlight and multiple sclerosis.

Esophageal cancer rate was also used as an exploratory variable because of the suggestion of Foster[20] that both esophageal cancer and schizophrenia seem to be related to selenium and calcium deficiencies, while esophageal cancer and multiple sclerosis seem to be related to selenium deficiency.[21,22]

The variables that most strongly predicted multiple sclerosis rates were greater latitude, lower January temperatures, lower July temperatures and less sunlight. The variables that most strongly predicted schizophrenic rates were elevated esophageal cancer, less sunlight, greater latitude, lower elevation, and more precipitation. Templer et al. stated:

> Perhaps the most omnibus generalization permitted by the present findings is that schizophrenia and multiple sclerosis both seem to be associated with geographical variables that at least folklore has traditionally regarded as unhealthy. The 'unhealthy' variables associated with high multiple sclerosis rates and/or schizophrenia rates include colder temperature, closer to the hemispheric poles, more precipitation, low elevation, and less sunlight. Infection and nutrition are two of the possible etiological variables that could account for this pattern.

It is well established that schizophrenics tend to be born in the colder months of the year.[23,24] There are negative correlations between temperature of month and number of schizophrenics born.[25,26] The most viable explanation for this association seems to be the "harmful influence" hypothesis of McNeil, Raff and Cromwell[27] which contends that there are harmful effects surrounding birth, or during gestation, such as infection or nutritional variables. Consistent with the harmful influence hypothesis is the finding of greater seasonality over the 20th century in Europe than the United States, a phenomenon attributed by Templer[28] to the greater prosperity and protection from the elements in the U.S. On the basis of this technology-based explanation, Templer and Austin[29] predicted and found a decrease in the seasonality of schizophrenic births from 1900 to 1960 in Missouri. To further test the harmful effects hypothesis, Templer and Veleber[30] determined that schizophrenics with presumably a greater genetic predisposition, namely hebephrenics and catatonics, displayed less seasonality than schizophrenics with lesser genetic predisposition, namely paranoid schizophrenics. Furthermore, Corgiat, Regier and Templer[31] found that schizophrenics with an older age of onset tended to have greater seasonality of birth. Apparently, with a greater genetic predisposition, less of a harmful influence is required for the development of schizophrenia.

CIVILIZATION AND INDUSTRIALIZATION

Torrey[32] has contended that schizophrenia is a product of "civilization". He presented evidence that schizophrenia did not appear to any notable extent until around 1800 when it became much more prevalent and spread globally. Torrey contended that schizophrenia is still rare in remote parts of the world. He stated that schizophrenia is more common in areas with greater population density. Research upholds his contention. Templer and Veleber[33] reported a correlation of 0.39 between schizophrenia prevalence and population density for the 50 U.S. states. Foster[34] found schizophrenia prevalence correlations of 0.54 with proportion of population employed in manufacturing, 0.54 with population density, 0.48 with population of state, and 0.55 for industrial water withdrawals. Foster stated:

> Clearly schizophrenia is particularly common in heavy manufacturing states. This tends to confirm Torrey's[35] hypothesis that the disease is essentially one of civilization. Why high schizophrenia prevalence may be related

> to manufacturing is unclear, but several hypotheses seem obvious. Firstly, some pollutant or group of pollutants may be involved. Secondly, industrialization generally results in less exposure to sunlight, as a consequence of more indoor work and shade from tall buildings. Thirdly, it is associated with major changes in diet and the greater consumption of refined, processed and canned foods, that is with alterations in the intake of various minerals and vitamins. Such a decline in the consumption of fresh foods may have various mental health related implications.[36] Of course, industrialization is also associated with enormous political, social and economic changes which may influence the prevalence of the disease.

There are a number of neurotoxins, perhaps especially metals, that industrial workers are exposed to that can cause psychosis having at least some resemblance to schizophrenia. The reader should neither attach too little nor too much importance to this resemblance. Psychosis is usually less apparent than the neurological and neuropsychological effects. And often the psychosis resembles other toxic psychoses more than schizophrenia.

Persons who made hats in England in the 19th century displayed erratic behavior which we now know was due to mercury toxicity. Thus the expression of "mad as a hatter" was obtained. Huggins[37] and Hoffer[38] suggested that "silver" dental fillings (actually 50 percent mercury) can cause psychiatric disturbances and neurological problems. Workers more vulnerable to mercury toxicity include dental office employees, photographers, photoengravers, feltmakers, tanners, embalmers, persons who make electrical switches and batteries, and those who work with disinfectants, fungicides, pesticides, wood preservatives, paper and pulp, and cosmetics. There is a considerable amount of mercury in waste water discharges.

Lead exposure can produce hallucinations.[39] Lead smelters, painters, battery workers, automobile repair workers and ceramic glazers are among the persons exposed to lead. Workers involved in ship breaking, demolition of metal structures such as bridges, and lead scrap smelting are also at a higher risk.

Manganese poisoning also can result in hallucinations and in incoherent talk and aggressive behavior.[40,41] Manganese toxicity can occur in manganese miners, welders, dry cell battery workers, steel alloy workers, and those persons who work with ceramics, fertilizer, glass, animal food additives, antiseptics, dyes, germicides, matches, oxidizing solutions, and welding rods.

Arsenic intoxication can result in a psychosis resembling paranoid schizophrenia.[42] Agricultural, pharmaceutical, and iron and steel workers are vulnerable to arsenic exposure.

Thallium is a rare metallic element that is soft and malleable. It resembles aluminum in its chemical properties and lead in its physical properties. It is used in the production of optical glass, in photoelectric cells, in making green light signals, and in rodenticides. It produces psychosis and neurological abnormalities.[43]

INTRAUTERINE AND PERINATAL ENVIRONMENT

There is an impressive array of literature to convincingly document that schizophrenics have a disproportionate number of intrauterine and perinatal complications and irregularities.

Schizophrenics tend to have "minor physical anomalies". Minor physical anomalies are minor physical defects that develop in the first trimester and include abnormal-size head, low-seated ears, adherent ear lobes, malformed asymmetrical ears, high steeped mouth, furrowed feet, webbed fifth finger, single transverse palmar crease, gap between the first and second toes and an unusually long third toe.[44] They are common in Down's Syndrome and in other genetic anomalies. Factors believed to be implicated in the development of these physical abnormalities are infection, anoxia, bleeding, fetal distress, dietary deficiency, toxemia and rubella.

Lohr and Bracha[45] suggested that some cases of schizophrenia could be related to prenatal exposure to alcohol. They pointed out that some of the physical anomalies found in schizophrenia are also part of the "fetal alcohol syndrome" observed in infants and children born to women abusing alcohol during pregnancy. Furthermore, a number of the behavioral manifestations of the fetal alcoholism syndrome, such as attention deficit, are observed in schizophrenics and in children who become schizophrenics. Lohr and Bracha also suggested that heavy beer drinking, in the hot summer months, could cause second trimester abnormality associated with the seasonality of schizophrenic birth pattern in northern Europe, Canada and the United States.

There is a good deal of animal research that could be viewed as consistent with the possibility of schizophrenia being caused by prenatal alcohol exposure. Lohr and Bracha[46] pointed out that hippocampal pathology has been reported a number of times with schizophrenia, and includes reduction of volume, neuronal loss, neuronal disarray and thinning of the parahippocampal cortex. Hippocampal abnormalities have been the most commonly reported changes in rats exposed to alcohol.

Jacobsen and Kinney[47] reported that 63 Danish schizophrenics had significantly increased perinatal complications and significantly increased severe complications than 63 Danish control persons. Mura, Mednick and Schulsinger[48] found that schizophrenics had a history of significantly lower birth weight than controls. Lane and Albee[49] found that schizophrenics had a significantly lower birth weight, and in fact had twice the rate of prematurity than their siblings. Woerner, Pollack and Klein[50] found that schizophrenics had a significantly greater number of prenatal and perinatal birth complications than their siblings. Mednick[51] reported that 39 percent of schizophrenic patients, but only 13 percent of controls, were born after prolonged labor. Jacobsen and Kinney[52] reported a prolonged labor birth for 40 percent of schizophrenics and 13 percent of normal controls. Parnas, Schulsinger, Teasdale, Schulsinger, Feldman and Mednick[53] reported prolonged labor for 33 percent of schizophrenics and 19 percent of controls.

McNeil[54] reviewed the studies with childhood psychotics and found that the most common complications were of the sort to imply anoxia, specifically toxemia, bleeding during pregnancy, threatened spontaneous abortion and asphyxia.

Reveley, Gurling and Murray[55] found that schizophrenia was more likely to be exhibited in twins when one twin died in the perinatal period, compared to when both twins survived at least until adolescence. It would appear that the same unfavorable perinatal conditions that caused the death of one twin contributed to the later development of schizophrenia in the other.

Monozygotic twins constitute fertile ground for attempting to disentangle the effects of genetics and harmful influence. This is both because monozygotic twins have identical genes and because twins are more vulnerable to perinatal brain injury than singletons. McNeil and Kaij[56] found that in 72 percent of monozygotic twins discordant for schizophrenia, the schizophrenic twin manifested more birth complications. Reveley, Reveley, Clifford and Murray[57] found that the difference in ventricular size in monozygotic normal twins was less than that in monozygotic twins who were discordant for schizophrenia. The schizophrenic twins had larger ventricle size than those of the co-twins and than those of the normal pairs of twins. Reveley et al. appropriately inferred some sort of environmental process in the etiology of the schizophrenic twins. Pollin and Staberau[58] found that, in monozygotic twins discordant for schizophrenia, the schizophrenic twin had a significantly greater proportion of asphyxia and low birth weight.

In a very interesting article, Reveley and Reveley,[59] in a research project involving CT scans with adult schizophrenics, happened to notice evidence of three of their patients having stenosis of the aqueduct and resulting hydrocephalus. Examination of previous medical records of all three revealed birth complications with hydrocephalus.

HEAD INJURIES

It has long been recognized that a disproportionate number of head injured persons develop schizophrenia, or at least a schizophrenic like condition, some time after trauma.[60] Hillbon[61] reported 1.2 percent of 1,821 head injured persons developed schizophrenia within 10 years of injury; Meinertz[62] reported 1.5 percent of 1,110 patients within 15 years of injury; Feuchtwanger and Mayer-Gross[63] 1.7 percent of 1,564 patients within 15 years; Liberman[64] 0.7 percent of 4,807 patients within 15 years; Lobava[65] 9.8 percent of 1,186 patients within 15 years; and Hillbom[66] 2.6 percent of 415 patients within 20 years. Davison and Bagley[67] stated "With an expectation of developing schizophrenia in the general population of 0.8 percent over a 25 year risk period (age 15 to 40 years) the observed incidence over 10 to 20 year periods is 2 to 3 times the expected incidence."

Research subsequent to the Davison and Bagley[68] review supports their inferences. Davison[69] found that of 291 persons who had been unconscious for at least a week and followed up from 10 to 24 years later, schizophreniform psychosis developed in 2.4 percent of the cases.

Wilcox and Nasrallah[70] found that 22 (11 percent) of 200 schizophrenic patients, 6 (5 percent) of 122 manic-depressives, 3 (1 percent) of 203 depressives, and 1 (1 percent) of 134 control surgical patients had head injuries before the age of 10. It does not appear that the bulk of these persons were already schizophrenic, or pre-schizophrenic, and the head injury was coincidental or precipitated a psychosis in persons so predisposed. One basis for arguing against this possibility is that such patients do not have a disproportionate number of schizophrenic relatives. Schultz[71] reported that in 240 cases of schizophrenic manifestations, after head injury, family history was no greater than expected in the general population. Both Hillbom[72] and Davison and Bagley[73] reported family history of schizophrenia to be "rare" in head injured patients with schizophrenic manifestations.

NUTRITIONAL AND SOIL CONSTITUENT CORRELATES

The majority of studies in the area of nutrition have been correlational. Correlation does not establish causation. Nevertheless, the composite perspective, provided by both the correlational and non-correlational studies, suggests that nutrition does play a role.

Hoffer[74] stated "If all the Vitamin B3 were removed from our food, everyone would become psychotic within a year. This pandemic psychosis would resemble pellagra and it would resemble schizophrenia". Hoffer then went on to describe the similarity of the mental disturbances of the two disorders with clinical material, most of this written in the early part of the 20th century, or before. Pellagra can be effectively treated with large quantities of vitamin B3 resulting in removal of, or at least lessening of, the physical and mental symptoms. Hoffer and other eminent psychiatrists have administered vitamin B3 to thousands of schizophrenics since the early 1950s. Their rationale, besides the analogy with pellagra, includes the assumption that schizophrenics have a greater need for vitamins than the average person. This assumption is based in part upon research indicating that schizophrenics excrete less niacinamide, pyridoxine and ascorbic acid after loading with these vitamins than normal subjects. In addition to vitamin B3, megavitamin therapists typically employ other B vitamins and ascorbic acid (vitamin C). They recommend abstinence of caffeine, avoidance of excessive sleep, and a hypoglycemic diet.

Foster[75] reported that, in the United States, schizophrenia correlated 0.58 with selenium deficient fodder crops, -0.46 with level of soil calcium, and -0.43 with soil strontium. He cautiously related his findings to the suggestions of previous authors regarding the schizophrenic's excessive dopamine activity, excess amounts of a normal opiod, hypersensitivity to wheat proteins, prostaglandin deficiency, allergic phenomenon, an inability to metabolize zinc effectively and pineal deficiency. Foster pointed out that the negative correlation of schizophrenia prevalence and soil mercury could be viewed as consistent with the fact that mercury is antagonistic with selenium and tends to reduce its biological availability.[76]

It has long been recognized that schizophrenia is a familial disorder, and evidence for genetic predisposition is now overwhelming. Foster[77] suggested that "both diseases may be in part genetic and in part geochemical". He specifically suggested that what might be inherited in both schizophrenia and multiple sclerosis is the inability to effectively absorb certain trace and bulk elements through the digestive tract, or to efficiently utilize them.

Foster[78] reported a correlation of -0.52 between schizophrenia prevalence and amount of sunlight. He said that this could account for the higher prevalence of schizophrenia in northern than in southern Europe or in many of the developing countries of the world that are close to the equator. Foster said that it is possible that reduced sunshine might bring about a vitamin D deficiency, either prior to birth or later in life.

Templer and Veleber[79] reported a marginally significant correlation (r=0.38, p=.06) between schizophrenia and wheat consumption for the 18 countries for which this information could be located. However, wheat gluten in particular could be more important than wheat. The rationale for suspecting a relationship between schizophrenia and wheat gluten is based on several different discoveries and lines of investigation.[80] These include the facts that schizophrenia is very uncommon where wheat gluten is rare;[81,82] that gluten in the diet has toxic and behavioral effects in some schizophrenics;[83-87] that the relatives of celiacs have a high prevalence of schizophrenia;[88,89] that both child and adult celiac patients have a high rate of psychiatric disorders;[90,91] that immunological tests for celiac disease are often positive in schizophrenics;[92] and that the intracerebral injection of wheat gluten has produced "catalepsy" and "chewing in air" in rats.[93]

Celiac disease is an uncommon polygenetic disorder in which predominantly gastrointestinal, but also psychotic disturbances are evoked by wheat, rye and barley glutens. Diarrhea is a very prominent symptom. Malabsorption results in poor growth in children with the disorder. Dohan[94] has contended that the schizophrenia-like psychosis is produced by neuroactive peptides, in persons with genetically defective gut and other barriers. A gluten free diet brings about improvement.

Dohan[95] reported that in 239 celiac patients nine were schizophrenic and that this is 8 to 16 times the number of persons treated for schizophrenia in any given year. Graff and Hanford[96] reported celiac disease in 4 of 37 schizophrenics, about 200 times the expected number in any given year. Dohan[97] stated that Lauretta Bender reported that at Bellevue Hospital some 30 years earlier 4 of 100 schizophrenic children, 80 times the expected number, had celiac disease.

Dohan[98] found an astonishingly high correlation of 0.98 between percent change in per capita wheat and rye consumption, after World War II, and change in schizophrenic first admissions in four countries: Canada, the United States, Norway and Sweden. Dohan pointed out that in countries that produce little cereal grain schizophrenia is rare. He went on to say that the rates of schizophrenia are highest in societies that consume mostly wheat, rye or barley; somewhat less in those eating large amounts of rice and smaller amounts of wheat and barley; and still lower in societies consuming primarily corn and millet. Wheat, rye and barley contain the glutens that are most toxic to celiac patients.

Dohan, Harper, Clark, Rodrigue and Zigas[99] reported only two overtly insane schizophrenic persons in 65,000 adults observed in remote regions of Papua New Guinea, Malaita, Soloman Islands and Yap, Micronesia. These people ate little or no grain. However, in proximate regions that became "westernized" with a consumption of wheat, barley, rice and beer, the prevalence was vastly greater, in fact, about the same as the prevalence in Europe.

Dohan and Grasberger[100] reported that schizophrenics who were placed on a wheat free and milk free diet were discharged about twice as rapidly as control patients. When wheat gluten was secretly added to the experimental patient's diet, they became worse.

Singh and Kay[101] placed schizophrenics on a cereal free and milk free diet. When a wheat gluten challenge was initiated, there was an exacerbation of the schizophrenic symptoms. Upon termination of that stage of the research improvement occurred.

However, not all research has reported confirmation of the negative effects of wheat gluten. One double-blind study found no effects of a diet free of gluten, cereal grains and milk on young chronic schizophrenics. However, the authors did not rule out the possibility that some schizophrenics might benefit from such a dietary intervention.[102] Rice, Ham and Gore[103] found in an uncontrolled trial that 13 of their 16 schizophrenics were apparently not affected, although there were 3 patients who possibly demonstrated some change as a function of diet. Rice et al. viewed their findings as basically negative, but acknowledged the possibility that some subgroups of schizophrenics could respond to gluten challenge or gluten free diets.

Vlissides, Venulet and Jenner[104] performed double-blind research with schizophrenics who received either a gluten-free, or gluten-containing diet for 14 weeks. Although there were significant reductions on five scales of the Psychotic In-Patient Profile, there were no significant differences between the experimental and control groups. Vlissides et al. attributed the improvement to the increased attention given to the patients. However, the authors did note that two patients improved on the diet, but relapsed when the gluten was reintroduced. And, Vlissides et al. did point out, like the other authors who reported negative findings, the possibility that a minority of schizophrenics could possibly profit from a gluten free diet. Nevertheless, there are other authors whose findings are negative or for the most part negative.[105,106]

As discussed by Foster,[107] there are several reasons to suspect a positive association between milk consumption and schizophrenia. Foster noted that Swank and Pullen[108] pointed out a distinction between the "beer-butter" cultures of northern Europe and the "wine-oil" cultures of southern Europe. The prevalence of both schizophrenia and multiple sclerosis is lower in the wine-oil cultures. Where soils are deficient in iodine, the milk of cows is vitamin A deficient. Selenium deficiency can also bring about vitamin A deficiency in cow's milk. Foster[109] suggested that the negative association with amount of sunlight of both schizophrenia and multiple sclerosis may be a function of less iodine in the soils further from the equator. Templer and Veleber[110] found a correlation of 0.53 between milk consumption and schizophrenia prevalence for the 18 countries for which that information could be located. Dohan[111] introduced a non-milk diet along with a grain free diet to schizophrenics based on the rationale that milk proteins seem to enhance the effect of wheat gluten in some celiac patients.

The research of Christensen and Christensen[112] in an eight nation study found that favorable prognosis in schizophrenics is negatively correlated with total fat consumption and with saturated fat consumption. The correlation with unsaturated fat was in the opposite direction but not statistically significant. Nevertheless, the authors maintained that it is consistent with an admittedly uncontrolled study that seemed to indicate that linseed oil may have a beneficial effect in schizophrenia. They recommended controlled clinical trials of low fat diets with a sufficient amount of essential fatty acids.

INFECTION

The earlier stated generalization of Templer, Hughey, Chalgujian, Lavoie, Trent, Sahwell and Spencer[113] could be viewed as consistent with some sort of infectious etiology. They inferred that the most global generalization about the geographical similarities of schizophrenia and multiple sclerosis is one of a profile of variables commonly thought of as "unhealthy": colder temperature, closer to the hemispheric poles, more precipitation, lower elevation and less sunlight. Since the Templer et al. article, Templer, Hintze, Trent and Trent[114] (in press), using a sample of 25 countries, reported schizophrenia rate to be correlated 0.69 with latitude, -0.52 with January temperature, and -0.64 with July temperature.

There is now evidence that a "slow virus" can remain dormant for years, or even decades. Torrey and Peterson[115] suggested such an etiology for schizophrenia. Both anatomical[116] and immunological[117] changes, suggestive of chronic viral infection, have been found in the central nervous system of some schizophrenics. A virus like agent has been found in the cerebrospinal fluid in one third of schizophrenics.[118,119] A slow virus possibility certainly meshes with the geographical and temporal distribution of schizophrenia elucidated by Torrey[120] and others. Torrey pointed out that the non-existence of schizophrenia before 1800 and its rapid development after 1800 are consistent with the fact that viral diseases do not exist without a critical number of people to spread them. Furthermore, infectious disease flourishes more in densely populated areas and among people in low socio-economic classes. Infectious diseases also have seasonality and viruses can alter the function of nerve cells without altering their histological structure.

King et al.[121] reported notable trends for decreased cerebrospinal fluid viral antibody tissues in six of seven common viruses in schizophrenics, when compared to orthopedic patient controls. The differences were statistically significant for mumps. They found that cerebrospinal to serum ratios were reduced in the schizophrenics, with the differences being significant for mumps, measles and rubella. King et al. contended that such findings suggest reduced immune response to common viruses, in the central nervous system of schizophrenics.

ALCOHOL

Bleuler,[122] the psychiatrist who gave us the word schizophrenia, maintained that concerning alcoholic hallucinosis he could state "with certainty or greater probability that besides the alcoholism, a long standing schizophrenia was present", and that "alcoholic hallucinosis could therefore be a mere syndrome of schizophrenia induced by alcohol". Victor and Hope[123] reported on 76 cases of alcoholic admissions with auditory hallucination but with a clear sensorium including satisfactory orientation for time, person and place. They stated "our observations indicate that a syndrome may emerge in relationship to excessive drinking which is clinically indistinguishable from schizophrenia". However, they maintained that, in 68 of the 76 cases, the hallucinations were transient and lasted no more than a few days. In eight cases the hallucinations lasted for at least a few weeks, and in four of these cases the authors stated that an apparently chronic schizophrenia picture became evident. It should be pointed out here that in the field of alcoholism there has long been clinical lore that this disorder is more common in colder climates; and that research by Templer, Hintze,

Trent and Trent,[124] using data from 25 different countries, found substantial negative correlations with both winter and summer temperature. Furthermore, Templer and Veleber[125] established a significant positive correlation between per capita alcohol consumption and schizophrenic admissions for the United States.

DRUGS

Davison[126] reviewed the large and broad array of drugs that produce psychosis that has at least some resemblance to schizophrenia. Among the categories of these drugs are the hallucinogenic drugs: LSD, mescaline (from cactus plants), and cannabis (marijuana and hashish). Of these, LSD seems most capable of producing schizophreniform psychosis. Other drugs and categories of drugs that produce a schizophreniform psychosis include anticonvulsants, disulfiram (antabuse), L-dopa, drugs for cardiac conditions, antibiotics, anesthetics, barbiturates and other sedatives, and cocaine.

SOCIO-ECONOMIC ENVIRONMENT

In a classical sociological study, Farris and Dunham[127] found that the incidence of schizophrenia was inversely related to socio-economic status. They reported that the areas in the more outward concentric circles, around the center of Chicago, experience a lesser incidence of schizophrenia than do lower socio-economic neighborhoods. Hollingshead and Redlich[128] also found that the lowest social class in New Haven, Connecticut had eight times the schizophrenia prevalence rate of that of the highest class. Such results have been confirmed by other studies. The inverse correlation between social class and schizophrenia rates seems to be greater in the larger cities.[129] At the time of the Hollingshead and Redlich and the Farris and Dunham studies, schizophrenia was widely viewed as having psychogenic origins. Thus the most common inferences, at that time, were that the unfortunate social and psychological situation of the poor caused them to regress or withdraw from the real world. An alternative explanation put forward by Goldberg and Morrison[130] was the "drift hypothesis", that sees socially and economically incapacitated schizophrenics falling in socioeconomic standing. However, the present author suggests another perspective, related to the Templer and Cappelletty[131] primary versus secondary conceptualization, in which primary schizophrenia has more of a genetic, or some sort of more endogenous etiology, and secondary schizophrenia results from damage, or some sort of assault to the brain. It is possible that in the primary form of the disorder, schizophrenics have schizophrenic children because of their genes; and that both the parents and their children remain in lower socioeconomic status because of their schizophrenia. It is also possible that persons born into poverty have secondary schizophrenia, because of inferior prenatal and perinatal care, drug and/or alcohol abuse of their mothers while pregnant, poor nutrition, and head injuries from child abuse, fighting, being assaulted, or from motor vehicle and industrial accidents. They also do not have enough money to avoid going to war and suffering from associated traumas.

The psychiatric admissions rate for schizophrenics varies as a function of the economic cycle. Brenner[132] found an impressive positive relationship between schizophrenic admissions and an index of unemployment. The present author doubts that economic hard times could cause someone to become schizophrenic who had not been already predisposed. It appears more likely that the economic stress precipitates hospitalization, in persons who were already schizophrenic or preschizophrenic. Furthermore, Brenner found that an inverse relationship between psychiatric admissions and health of the economy also exists for other disorders, including those generally recognized as being primarily biological in origin, such as general paresis (syphilic affliction of the brain), epilepsy, alcoholic psychosis and psychosis with cerebral arteriosclerosis.

CULTURAL ENVIRONMENT VARIATIONS

The specific content of schizophrenic delusions, hallucinations and behaviors differ among cultures, and in different subgroups within a culture. Torrey[133] provided a useful review of these variations. Educated schizophrenics in Brazil, for example, are more likely to have delusions pertaining to electricity, while schizophrenics from the unprivileged segments of Brazil society are more apt to have delusion focused on deities and possession by spirits.[134] Lamba[135] reported that delusions of grandeur are rare in illerate Nigerians, who are fearful of dire consequences that could result from challenging the power of the Gods. Before World War II, the delusions of Japanese schizophrenics tended to center on the Emperor, but after the war were more apt to center on the Communist Party and the United States.[136] In South Africa, Bantu schizophrenic patients report hearing the voices of their deceased relatives, but White South Africans do not.[137] Schizophrenics often exhibit behavior that is considered outrageous in their culture. In Japan, where family is regarded as very important, some schizophrenics assault their relatives.[138] In Nigeria, the schizophrenic often violates accepted religious sanctions.[139]

THE PREVENTION OF SCHIZOPHRENIA

Is schizophrenia preventable? One might think that such a question is premature, insofar as we do not have a clear understanding of its causes. We do know that a greater probability of becoming schizophrenic is associated with various factors, such as head injury and obstetrical complications. However, there is no known pre-schizophrenia risk factor present in all schizophrenics. Even though, in comparison to normal persons, schizophrenics definitely exhibit an excess of brain structural and physiological abnormalities, there is no known abnormality in the brains of all schizophrenics.

On the other hand, a number of risk factors in the environment can be reduced, possibly resulting in fewer cases of schizophrenia. Even if there were no reduction in the incidence of schizophrenia, the preventative measures would cause little harm. In fact, they would probably be good for one's general health. What is good for the health of the body is probably good for the health of the brain.

Templer, Hartlage and Cannon[140] coined three closely related terms: "brain vulnerability", "brain health" and "preventable brain damage". Their book focuses upon a variety of ways in which the brain can be injured, and is divided into two sections: impact damage and chemical damage. The impact damage includes motor vehicle and occupational and recreational accidents, assault, fighting, spousal abuse, child abuse, contact sports, non-contact sports, psychosurgery and electroconvulsive treatment. Chemical damage results from agricultural neurotoxins, industrial neurotoxins, alcohol, drugs, intrauterine damage and malnutrition. The brain is much more vulnerable to damage than most laypersons realize and even more vulnerable than most health professionals are aware.

It is now apparent that minor blows to the head often do considerable damage. Traditionally, a concussion has been defined as a brain injury that causes unconsciousness, but no permanent effects. The "post-concussion syndrome" consists of memory problems, concentration difficulty, headache, irritability, anxiety, dizziness, weakness and depression, that can occur months, or even years, after the injury. It is now known that concussions can often produce considerable and permanent impediments with respect to educational and vocational performance. Many minor blows, such as are observed in boxing, can have cumulative effects. Brain injury often makes one more vulnerable to catastrophic effects produced by a subsequent blow to the head. There is often a development of progressive atrophy following a brain injury. Both contact and non-contact sports have long been valued in building character and promoting health of the body and of the mind. Unfortunately, brain damage often results. It now appears that brain damage is the rule rather than the exception in professional boxers, even those that are young and have not lost many fights.[141] Soccer (what the British call football) is the most played international sport in the world. There is evidence that "heading" the soccer ball can cause brain damage, but probably without the major neuropsychological impairments found in boxers.[142]

Recommendations for preventing impact damage contained in the Templer, Hartlage and Cannon[143] book include the wearing of adequate headgear for transportation, work, sports and other recreation, initiating more protective rules and regulations for sports, the manufacturing of safer motor vehicles, and vigilance in the prevention of family violence.

In the Central Valley of California, where the present author lives, migrant farm workers from Mexico are especially at risk from chemical damage. The symptoms of neurotoxicity include problems with memory for recent events, concentration difficulty, mental slowness, fatigue, headache, sexual dysfunction, numbness of the hands or feet, motor incoordination and sensory disturbances.[144] Very relevant to the present focus is the fact that psychosis sometimes occurs. The chief neurotoxic culprits of industry are metals and organic solvents. Both of these categories can produce a wide array of neurological and neuropsychological deficits. Legislation and its enforcement and the implementation of safety measures in the work environment are needed to prevent neurotoxicity.

CONCLUSION

Schizophrenia appears to be a cluster of closely related brain disorders with a variety of brain injurious causes. To assert that such measures as cleaning up the environment and protecting workers would prevent schizophrenia, may or may not be too great an inductive leap. Nevertheless, even if such measures should fail to prevent one case of schizophrenia, they would protect human health in other ways, and improve the quality of life.

REFERENCES

[1] Bateson, G., Jackson, D.D., Haley, J. et al. 1956. Toward a Theory of Schizophrenia, *Behavioral Science*, vol. 1, pp. 251-264.

[2] Templer, D.I. and Cappelletty, G.G. 1986. Primary Versus Secondary Schizophrenia: A Theoretical Review, *Journal of Orthomolecular Medicine*, vol. 1, pp. 255-260.

[3] Sheper-Hughes, N. 1979. *Saints, Scholars, and Schizophrenics: Mental Illness in Rural Ireland.* Berkeley: University of California Press.

[4] Templer, D.I., Hintze, J., Trent, N. and Trent, A. 1991. Schizophrenia, Latitude and Temperature, *Journal of Orthomolecular Medicine*, vol. 6, issue 1, pp. 5-7.

[5] Sheper-Hughes, *op. cit.*

[6] Torrey, E.F. 1979. *Schizophrenia and Civilization.* New York: Jason Aronson, Inc., pp. 5-187.

[7] Sheper-Hughes, *op. cit.*

[8] Templer, D.I., Cappelletty, G.G. and Kauffman, I. 1989. Schizophrenia and Multiple Sclerosis Distribution: A State to Continent Perspective, *Journal of Orthomolecular Medicine*, vol. 4, pp. 8-10.

[9] Templer, D.I., Regier, M.W. and Corgiat, M.D. 1985. Similar Distribution of Schizophrenia and Multiple Sclerosis, *Journal of Clinical Psychiatry*, vol. 46(2), p. 78.

[10] Torrey, *op. cit.*, p. 181.

[11] Templer, D.I., Cappelletty, G.G. and Kauffman, I. 1988. Schizophrenia and Multiple Sclerosis Distribution in Italy, *British Journal of Psychiatry*, vol. 153, pp. 389-390.

[12] Templer, Cappelletty and Kauffman, *op.cit.*, pp. 8-10.

[13] Templer, Regier and Corgiat, *op.cit.*, p. 78.

[14] Templer, D.I., Hughey, B., Chalgujian, H., Lavoie, M., Trent, N., Sahwell, P. and Spencer, D.A. 1990. Multiple Sclerosis, Schizophrenia, Temperature and Latitude, *Journal of Orthomolecular Medicine*, vol. 5(3), pp. 125-127.

[15] Campbell, A.M.G., Crow, R.S. and Lang, D.W. 1960. Goitre and Disseminated Sclerosis, *British Medical Journal*, vol. 1, pp. 200-201.

[16] Foster, H.D. 1987. Disease Family Trees: The Possible Roles of Iodine in Goitre, Cretinism, Multiple Sclerosis, Amyotrophic Lateral Sclerosis, Alzheimer's and Parkinson's Diseases and Cancers of the Thyroid, Nervous System and Skin, *Medical Hypotheses*, vol. 24, pp. 249-263.

[17] Foster, H.D. 1988a. Reducing Mortality From Multiple Sclerosis: A Geographical Perspective, *Environments* 19(3), pp. 14-34.

[18] Foster, H.D. 1989. Multiple Sclerosis and Schizophrenia: Some Comments on Similarities in Their Spatial Distributions, *Journal of Orthomolecular Medicine*, vol. 4(1), pp. 11-12.

[19] Foster, H.D. 1988b. The Geography of Schizophrenia: Possible Links with Selenium and Calcium Deficiencies, Inadequate Exposure to Sunlight and Industrialization, *Journal of Orthomolecular Medicine*, vol. 3, pp. 135-140.

[20] Foster, H.D., personal communication, 1988.

[21] Foster, *op.cit.*, pp. 14-34.

[22] Foster, H.D., *op. cit.*, pp. 135-140.

[23] Torrey, *op. cit.*

[24] Templer, D. and Veleber, M.A. 1981. Schizophrenia Prevalence and Demographic Variables in the United States, *Journal of Orthomolecular Psychiatry*, vol. 10(2), pp. 74-76.

[25] Templer, D.I., Halcomb, P.H., Barthlow, V.A. et al. 1978. Month of Conception and Birth of Schizophrenics as Related to the Temperature, *Journal of Orthomolecular Psychiatry*, vol. 7, pp. 231-235.

[26] Templer, D.I. and Austin, R.K. 1980a. Confirmation of Relationship Between Temperature and the Conception of Schizophrenics, *Journal of Orthomolecular Psychiatry*, vol. 9(4), pp. 220-222.

[27] McNeil, T.F., Raff, C.S. and Cromwell, R.L. 1971. Technique for Comparing Relative Importance of Season of Conception and Season of Birth, *British Journal of Psychiatry*, vol. 118, pp. 328-335.

[28] Templer, D.I. 1978. Month of Conception and Birth of Schizophrenics as Related to the Temperature, *Journal of Orthomolecular Psychiatry*, vol. 7, pp. 231-236.

[29] Templer, D.I. and Austin, R.K. 1980b. Decreasing Seasonality of Schizophrenic Births, *Archives of General Psychiatry*, vol. 37, pp. 959-960.

[30] Templer and Veleber, *op. cit.*, pp. 74-76.

[31] Corgiat, M.D., Regier, M.W. and Templer, D.I. 1983. Seasonality of Schizophrenic Births and Age of Onset, *Journal of Orthomolecular Psychiatry*, vol. 12(4), pp. 268-269.

[32] Torrey, *op. cit.*

[33] Templer and Veleber, *op. cit.*, pp. 74-76.

[34] Foster, *op. cit.*, pp. 135-140.

[35] Torrey, *op. cit.*

[36] Hoffer, personal communication, 1991.

[37] Huggins, H.A. 1982. Mercury: A Factor in Mental Disease?, *Orthomolecular Psychiatry*, vol. 11(1), pp. 4-16.

[38] Hoffer, A. 1982. Mercury-Silver Amalgams, Editorial, *Orthomolecular Psychiatry*, vol. 11(1), pp. 2-3.

[39] Grandjean, P. 1983. Behavioral Toxicity of Heavy Metals, in Zoiden, P., Cuomo, V., Racagni, G. and Weiss, B. (eds.), *Application of Behavioral Pharmacology in Toxicology*. New York: Ranen Press, pp. 331-340.

[40] Rosenstock, H.A., Simons, D.G. and Meyer, J.S. 1971. Chronic Manganism. Neurologic and Laboratory Studies During Treatment with Levodopa, *Journal of the American Medical Association*, vol. 217, pp. 1354-1358.

[41] Politis, M.J., Schaumberg, H.H. and Spencer, P.S. 1980. Neurotoxicity of Selected Chemicals", in Spencer, P.S. and Schaumberg, H.H. (eds.), *Experimental and Clinical Neurotoxicology*. Baltimore: Williams and Wilkins, 1980, pp. 613-630.

[42] Windebank, A.J., McCall, J.T. and Dyck, P.J. 1984. Mental Neuropathy, in Dyck, P.J., Thomas, P.K., Lambert, E.H. and Bunge, R. (eds.), *Peripheral Neuropathy*, vol. 2. Philadelphia: W.B. Saunders, pp. 2133-2162.

[43] Bank, W.J. 1980. Thallium, in Spencer, P.S. and Schaumberg, H.H. (eds.), *Experimental and Clinical Neurotoxicology*. London: Williams and Wilkins, 1980, pp. 570-577.

[44] Guy, J.D., Majoriski, L.V., Wallace, C.J. and Guy, M.P. 1983. The Incidence of Minor Physical Anomalies in Adult Male Schizophrenics, *Schizophrenic Bulletin*, vol. 9, pp. 571-582.

[45] Lohr, J.B. and Bracha, H.S. 1989. Can Schizophrenia Be Related to Prenatal Exposure to Alcohol? Some Speculations", *Schizophrenia Bulletin*, vol. 15(4), pp. 595-603.

[46] *Ibid.*

[47] Jacobsen, B. and Kinney, D.K. 1980. Perinatal Complications in Adopted and Nonadopted Schizophrenics and Their Controls: Preliminary Results, *Acta Psych. Scand.*, vol. 62 (suppl. 285), 337-346.

[48] Mura, E., Mednick, S.A., Schulsinger, F. et al. 1973. Erratum and Further Analysis. Perinatal Conditions and Infant Development in Children with Schizophrenic Parents, *Social Biology*, vol. 20, p. 111.

[49] Lane, E.A. and Albee, G.W. 1966. Comparative Birth Weights of Schizophrenics and Their Siblings, *Journal of Psychology*, vol. 64, pp. 227-231.

[50] Woerner, M.G., Pollack, M. and Klein, D.F. 1973. Pregnancy and Birth Complications in Psychiatric Patients: A Comparison of Schizophrenic and Personality Disorder Patients with their Siblings, *Acta Psychiatr. Scand.*, vol. 49, pp. 712-721.

[51] Mednick, S.A. 1970. Breakdown in Individuals High at Risk for Schizophrenia: Possible Predispositional Perinatal Factors, *Mental Hygiene*, vol. 54, pp. 50-52.

[52] Jacobsen and Kinney, *op. cit.*, pp. 337-346.

[53] Parnas, J., Schulsinger, F., Teasdale, T.W. et al. 1982. Perinatal Complications and Clinical Outcome Within the Schizophrenia Spectrum, *British Journal of Psychiatry*, vol. 140, pp. 416-420.

[54] McNeil, T.F. 1987. Perinatal Influences in the Development of Schizophrenia, in Helmchen, H. and Henn, F.A. (eds.), *Biological Perspectives of Schizophrenia.* New York: John Wiley & Sons, Inc., pp. 125-138.

[55] Reveley, A.M., Reveley, M.A. and Murray, R.M. 1984. Cerebral Ventricular Enlargement in Nongenetic Schizophrenia: A Controlled Twin Study, *British Journal of Psychiatry*, vol. 144, pp. 89-93.

[56] McNeil, T.F. and Kaij, L. 1978. Obstetric Factors in the Development of Schizophrenia: Complications in the Birth of Preschizophrenics and in Reproduction by Schizophrenic Parents, in Cromwell, R.L., Wynne, L.C., Matthysse, S. (eds.), *The Nature of Schizophrenia.* New York: John Wiley, pp. 401-429.

[57] Reveley, A.M., Reveley, M.A., Clifford, C.A. and Murray, R.M. 1982. Cerebral Ventricular Size in Twins Discordant for Schizophrenia, *Lancet*, vol. 1, pp. 540-541.

[58] Pollin, W. and Stabeneau, J.R. 1968. Biological, Psychological and Historical Differences in a Series of Monozygotic Twins Discordant for Schizophrenia, in Rosenthal, D. and Kety, S.S. (eds.), *The Transmission of Schizophrenia.* London: Pergamon Press, 1968.

[59] Reveley, A.M. and Reveley, M.A. 1983. Aqueduct Stenosis and Schizophrenia, *Journal of Neurology, Neurosurgery, and Psychiatry*, vol. 46, pp. 18-22.

[60] Davison, K. and Bagley, C.R. 1969. Schizophrenia-like Psychoses Associated with Organic Disorders of the Central Nervous System: A Review of the Literature, in Herrington, R.N. (ed.), *Current Problems in Neuropsychiatry, Special Publication No. 4.* Ashford, Kent: Headley, pp. 113-184.

[61] Hillbom, E. 1951. Schizophrenia-like Psychoses After Brain Trauma, *Acta. Psychiat. Neurol. Scand.*, vol. 60, pp. 36-47.

[62] Meinertz, F. 1957. Schizophreniealiche Psychosen und Grenzzustande bei Hirnverletzen, *Report of Second International Congress of Psychiatry, Zurich*, Vol. 4, pp. 191-193.

[63] Feuchtwanger, E. and Mayer-Gross, W. 1938. Hirnverletzung und Schizophrenie", *Schweiz Arch. Neurol. Psychiat.*, vol. 41, pp. 17-99.

[64] Liberman, J.I. 1964. The Influence of Head Injury in Peacetime on the Development of Schizophrenia (A Statistical Investigation), *Zh. Nuvropat. Psikhiat.*, vol. 64, pp. 1369-1373.

[65] Lobova, L.P. 1960. The Role of Trauma in the Development of Schizophrenia, *Zh. Nevropat. Psikhiat.*, vol. 64, pp. 1187-1192.

[66] Hillbom, E. 1960. After Effects of Brain Injuries, *Acta. Psychiat. Nerol. Scand.*, vol. 142, pp. ??.

[67] Davison and Bagley, *op. cit.*, pp. 113-184.

[68] *Ibid.*

[69] Davison, K. 1983. Schizophrenia-like Psychoses Associated with Organic Cerebral Disorders: A Review, *Psychiatric Developments*, vol. 1, pp. 1-34.

[70] Wilcox, J.A. and Nasrallah, H.A. 1987. Childhood Head Trauma and Psychosis, *Psychiatry Research*, vol. 21, pp. 303-306.

[71] Schultz, B. 1932. Zur Erbpathologie der Schizophrenie, *Ztschr. Neurol. Psychiat.*, vol. 143, pp. 175-293.

[72] Hillbom, *op. cit.*, pp. 36-47.

[73] Davison and Bagley, *op. cit.*, pp. 113-184.

[74] Hoffer, personal communication, 1991.

[75] Foster, *op. cit.*, pp. 135-140.

[76] Foster, H.D. 1986. *Reducing Cancer Mortality: A Geographic Perspective.* Western Geography Series, Vol. 23. Victoria: Department of Geography, University of Victoria.

[77] Foster, *op. cit.*, pp. 11-12.

[78] Foster, *op. cit.*, pp. 135-140.

[79] Templer and Veleber, *op. cit.*, pp. 74-76.

[80] Reichelt, K.L., Sagedal, E., Landmark, J. et al. 1990. The Effect of Gluten-free Diet on Urinary Peptide Excretion and Clinical State in Schizophrenia, *Journal of Orthomolecular Medicine*, vol. 5(4), pp. 223-232.

[81] Dohan, F.C. 1966. Cereals and Schizophrenia. Data and Hypothesis, *Acta Psychiatrica Scandinavia*, vol. 42, pp. 125-152.

[82] Dohan, F.C. 1969. Is Celiac Disease a Clue to the Pathogenesis of Schizophrenia?, *Mental Hygiene*, vol. 53, pp. 525-530.

[83] Dohan, F.C. and Grasberger, J.C. 1973. Relapsed Schizophrenia: Earlier Discharge from the Hospital after Cereal-free, Milk-free Diet, *American Journal of Psychiatry*, vol. 130, pp. 685-688.

[84] Singh, M.M. and Kay, S.R. 1976. Wheat Gluten as a Pathogenic Factor in Schizophrenia, *Science*, vol. 191, pp. 401-402.

85 Rice, J.R., Ham, C.H. and Gore, W.E. 1978. Another Look at Gluten in Schizophrenia, *American Journal of Psychiatry*, vol. 135, pp. 1417-1418.

86 Vlissides, Venulet et al., *op. cit.*, pp. 447-452.

87 Jansson, N., Kristjansson, E. and Nilsson, L. 1984. Schizophren Psykosbild Avklingag Naer Pasienten Gavs Glutenfri Kost, *Laekartidnigen*, vol. 81, pp. 148-149.

88 Dohan, F.C. 1980. Hypothesis: Genes and Neuroactive Peptides from Food as Cause of Schizophrenia, in Costa, E and Trabucchi, M. (eds.), *Neural Peptides and Neuronal Communication*. New York: Raven Press, pp. 535-538.

89 Dohan, *op.cit.*, pp. 535-538.

90 Reference not available at this time.

91 Hallert, C., Astrom, J. and Sedvall, G. 1982. Psychic Disturbances in Adult Coeliac Disease III. Reduced Central Monoamine Metabolism and Signs of Depression, *Scandinavian Journal of Gastroenterology*, vol. 17, pp. 25-28.

92 Ashbenazi, A., Drasilowsky, D., Levin, P. et al. 1979. Immunological Reaction of Psychotic Patients to Fraction of Gluten, *American Journal of Psychiatry*, vol. 136, pp. 1306-1309.

93 Dohan, F.C. 1978. Abnormal Behavior After Intracerebral Injection of Polypeptides from Wheat Gluten, *Journal of Biological Science*, vol. 13, pp. 73-82.

94 Dohan, *op. cit.*, pp. 535-538.

95 *Ibid.*

96 Graff, H. and Handford, A. 1961. Coeliac Syndrome in the Case Histories of Five Schizophrenics, *Psychiatry Quarterly*, vol. 35, pp. 306-313.

97 Dohan, *op. cit.*, pp. 535-538.

98 *Ibid.*

99 Dohan, F.C., Harper, E.H., Clark, M.N. et al. 1984. Is Schizophrenia Rare if Grain is Rare?, *Biological Psychiatry*, vol. 19, pp. 385-399.

100 Dohan and Grasberger, *op. cit.*, pp. 685-688.

101 Singh and Kay, *op. cit.*, pp. 401-402.

102 Potkin, S.G., Weinberger, D., Kleinman, J. et al. 1981. Wheat Gluten Challenge in Schizophrenic Patients, *American Journal of Psychiatry*, vol. 138, pp. 1208-1211.

103 Rice, Ham and Gore, *op. cit.*, pp. 1417-1418.

104 Vlissides, D.N., Venulet, A. and Jenner, F.A. 1986. A Double Blind Gluten-free/Gluten-load Controlled Trial in a Secure Ward Population, *British Journal of Psychiatry*, vol. 148, pp. 447-452.

[105] Dean, G., Hannify, L., Stevens, F. et al. 1975. Schizophrenic and Coeliac Exercise, *Journal of the Irish Medical Association*, vol. 68, pp. 545-546.

[106] Reference not available at this time.

[107] Foster, *op. cit.*

[108] Swank, R.L. and Pullen, M.H. 1977. *The Multiple Sclerosis Diet Book*. New York: Doubleday.

[109] Foster, *op. cit.*, pp. 11-12.

[110] Templer and Veleber, *op. cit.*, pp. 74-76.

[111] Dohan, *op. cit.*, pp. 535-538.

[112] Reference not available at this time.

[113] Templer, Hughey et al., *op. cit.*, pp. 125-127.

[114] Templer, Hintze et al., *op. cit.*

[115] Torrey, E.F. and Peterson, M.R. 1976. The Viral Hypothesis of Schizophrenia, *Schizophrenia Bulletin*, vol. 2(1), pp. 136-146.

[116] Fisman, M. 1975. The Brain Stem in Psychosis, *British Journal of Psychiatry*, vol. 16, pp. 414-422.

[117] Torrey, E.F., Peterson, M.R., Brannon, W.L. et al. 1978. Immunoglobulins and Viral Antibodies in Psychiatric Patients, *British Journal of Psychiatry*, vol. 132, pp. 342-348.

[118] Tyrrell, D.A.J., Crow, T.J., Parry, R.P. et al. 1979. Possible Virus in Schizophrenics and Some Neurological Disorders, *Lancet*, vol. 1, pp. 839-841.

[119] Crow, T.J., Johnstone, E.C., Owens, D.G.C. et al. 1979. Characteristics of Patients with Schizophrenia or Neurological Disorder and Virus-like Agent in Cerebrospinal Fluid, *Lancet*, vol. 1, pp. 842-844.

[120] Torrey, *op. cit.*

[121] King, D.J., Cooper, S.J., Earle, E.A.P et al. 1985. Serum and CSF Antibody Titres to Seven Common Viruses in Schizophrenic Patients, *British Journal of Psychiatry*, vol. 147, pp. 145-149.

[122] Bleuler, E. 1930. *Textbook of Psychiatry*. New York: MacMillan Company, pp. 163, 342-345.

[123] Reference not available at this time.

[124] Templer, Hintze et al., *op. cit.*

[125] Templer and Veleber, *op. cit.*, pp. 74-76.

[126] Reference not availabe at this time.

[127] Faris, R.E.L. and Dunham, H.W. 1939. *Mental Disorders in Urban Areas*. Chicago: University of Chicago Press.

[128] Hollingshead, A.B. and Redlich, F.C. 1958. *Social Class and Mental Illness*. New York: John Wiley.

[129] Kohn, M.L. 1973. Social Class and Schizophrenia: A Critical Review and a Reformulation, *Schizophrenia Bulletin*, vol. 7, pp. 60-79.

[130] Goldberg, E.M. and Morrison, S.L. 1963. Schizophrenia and Social Class, *British Journal of Psychiatry*, vol. 109, pp. 785-802.

[131] Templer and Cappelletty, *op.cit.*, pp. 255-260.

[132] Brenner, M.H. 1973. *Mental Illness and the Economy*. Cambridge: Harvard University Press.

[133] Torrey, *op. cit.*

[134] Reference not available at this time.

[135] Lambo, T.A. 1965. Schizophrenic and Borderline States, in Dereuck, A.V. and Porter, R. (eds.), *Transcultural Psychiatry*. Boston: Little Brown, pp. 62-75.

[136] Asai, T. 1964. The Contents of Delusions of Schizophrenic Patients in Japan: Comparison Between Periods From 1941-1961, *Transcultural Psychiatric Research Review*, vol. 1, pp. 27-28.

[137] Scott, E.H.M. 1967. A Study of the Content of Delusions and Hallucinations in 100 African Female Psychotics, *South African Medical Journal*, vol. 4, pp. 853-858.

[138] Schooler, C. and Caudill, W. 1964. Symptomatology in Japanese and American Schizophrenics, *Ethnology*, vol. 3, pp. 172-177.

[139] Leighton, A.H. 1965. Discussion, in Dereuck, A.V., and Porter, R. (eds.), *Transcultural Psychiatry*. Boston: Little Brown, 1965, pp. 75-83.

[140] Templer, D., Hartlage, L.C. and Cannon, W.G. 1992. *Preventable Brain Vulnerability and Brain Health*. New York: Springer Publishing Company.

[141] Drew, R.H., Templer, D.I., Schuyler, B.A., et al. 1986. Neuropsychological Deficits in Active Licensed Professional Boxers, *Journal of Clinical Psychology*, vol. 42, pp. 520-525.

[142] Abreau, F., Templer, D.I., Schuyler, B.A. and Hutchinson, H.T. 1990. Neuropsychological Assessment of Soccer Players, *Neuropsychology*, vol. 4, pp. 175-181.

[143] Templer et al., *op. cit.*

[144] Singer, R. Agricultural and Domestic Neurotoxic Substances, in *Preventable Brain Damage: Brain Vulnerability and Brain Health*. New York: Springer Publishing Company, in press.

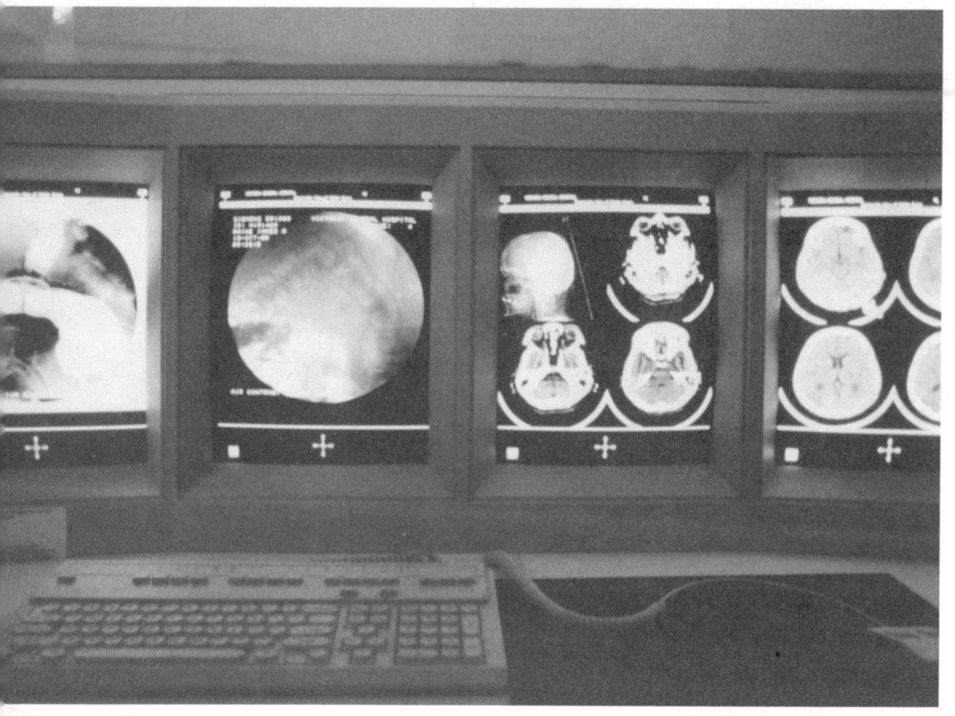

7 AMYOTROPHIC LATERAL SCLEROSIS: An Environmental Etiology?

Shane Snow

Department of Geography
University of Victoria

INTRODUCTION

Amyotrophic Lateral Sclerosis (ALS) is perhaps the most devastating of the neurological disorders affecting adults. It is characterized principally by the selective vulnerability of motor cells in the brain stem and in the anterior horns of the spinal cord. These gradually deteriorate, resulting in secondary atrophy of the muscle cells they enervate. As a consequence, the victim becomes progressively more paralyzed until death occurs.

Clinical symptoms of ALS and their rate of progression and severity vary. Symptoms are usually manifest sometime between the sixth and eighth decade of life, with 90 percent of patients dying within five years of diagnosis. However, there appears to be a more benign form of the disease, in which progression may arrest, without change, for periods of 10 years or longer.[1] The patient typically suffers impairment and eventual loss of mobility, dexterity and strength, the capacity to communicate, and the ability to eat and breathe. No available therapy currently reverses, arrests or slows the progression of the disease, which averages three years from diagnosis to mortality.[2,3] Approximately 10 percent of ALS cases are familial.[4] This inherited form of ALS has a typical onset at the age of 45.

ALS is not a common disease, having a prevalence of only three to four per 100,000 population in North America.[5] However, it has become widely known because of its association with celebrities, such as Lou Gehrig, the famed New York Yankees first baseman. The disease forced him to retire from the game in 1939. Indeed, it is commonly referred to as Lou Gehrig's Disease. The root cause of ALS's pathological process is unknown; that is it is idiopathic, although evidence is accumulating that supports an environmental interpretation of etiology.

SPATIAL AND GENDER VARIATIONS

Throughout the world, ALS generally has a relatively uniform incidence rate of approximately one per 100,000.[6,7] With the exception of clustering in the Western Pacific, there are no apparent differences in incidence rates according to hemisphere. It does, however, demonstrate considerable variations in its geographical distribution, if the western Pacific foci is included with the various community studies on incidence rates (Figures 1,7 and 2,7).

Clustering is apparent in Canada, as it is in the Western Pacific, but on a less dramatic scale. For instance, variations in incidence have been documented in southwestern Ontario counties, where rates range from 0.75 to 3.33 per 100,000 population.[12] These local and regional variations in incidence, and the existence of geographical disease foci, suggest that geographic environmental variables may play a role in the etiology of ALS.[13,14]

Community population surveys show a range of between 0.22 and 147 per 100,000 population (Figures 1,7 and 2,7), and several note progressive increases in incidence rates over time.[15,16,17] For example, research in Israel documented a rise in incidence, in both sexes, for the period 1959 to 1974, with the incidence being 66 percent greater in men than in women.[18] Research in the Rochester area of the United States also demonstrated that incidence rates increased from 2.1 for the period 1925 to 1954, to 2.65 per 100,000 population during the period 1970-1977.[19] Incidence rates also increased in Nova Scotia, Canada, from 1.5 per 100,000 population during the period 1974 to 1979 to 2.66 for the period 1981 to 1986.[20] Such apparent increases in ALS incidence may be a reflection of more accurate diagnosis in an aging population.[21,22] Alternatively, it may be an environmental association,[23] linked to increasing population exposures over time.

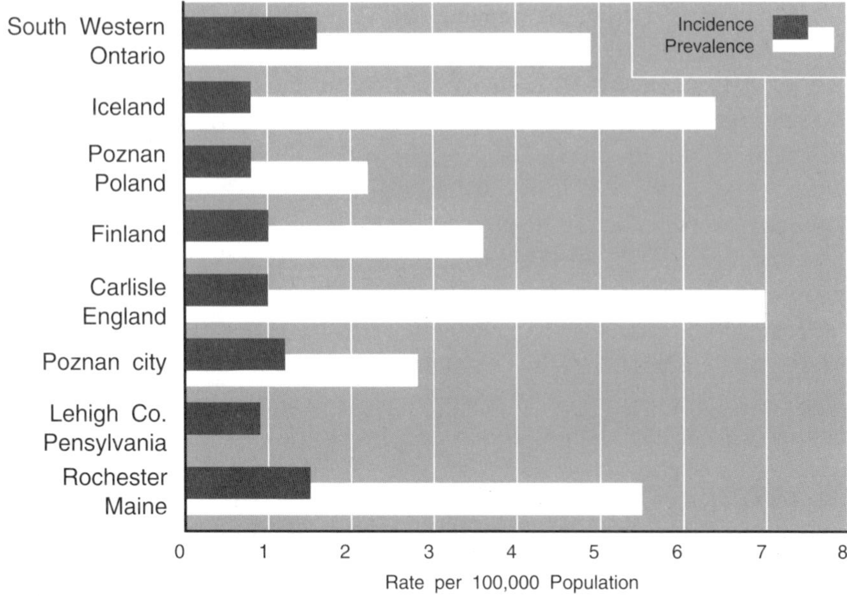

Figure 1,7 Community Population Surveys, Organized Geographically, Demonstrating Average Annual ALS Incidence Rates Per 100,000 Population (after A.J. Hudson et al., 1986,[8] and J.F. Kurtzke, 1982[9])

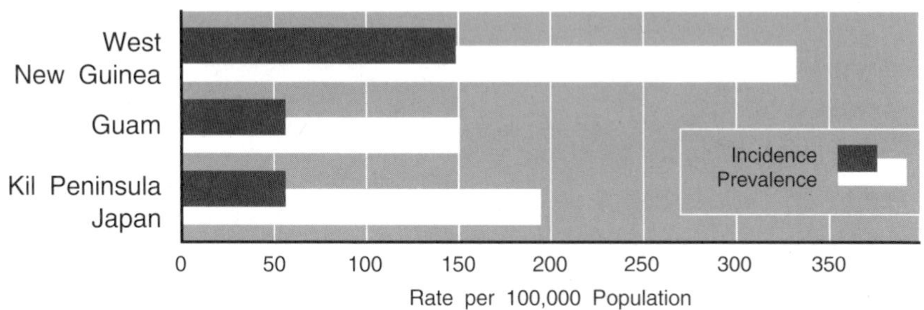

Figure 2,7 Crude Average Annual ALS Incidence and Prevalence Rates Per 100,000 Population, in Guam, West New Guinea, and the Kii Peninsula of Japan (after D.C. Gajdusek and A. Salazar, 1982,[10] and D.M. Reed et al., 1975[11])

Community studies in Canada, the United States and Finland have indicated that the risk of developing ALS increases with age, especially in the sixth through eighth decades, and the disease is more common in males (Figure 3,7).[24,25,26,27] Studies of international ALS death rates also demonstrate a male preponderance, and establish similarities in age predilection for both sexes. Few ALS-related deaths occur before 50 years of age, and an age specific death rate peak at age 70 seems widespread.[31] Globally, surveys of average annual ALS mortality have established rates that range between 0.2 and 1.5 per 100,000 population, and demonstrate a male predilection that varies between 1.5 and 2.9 to 1 (Figure 4,7). Taken in concert, studies of incidence and mortality, therefore, demonstrate a male ALS preponderance of approximately 2:1, and increased risk for both genders in the sixth to eighth decades of life.

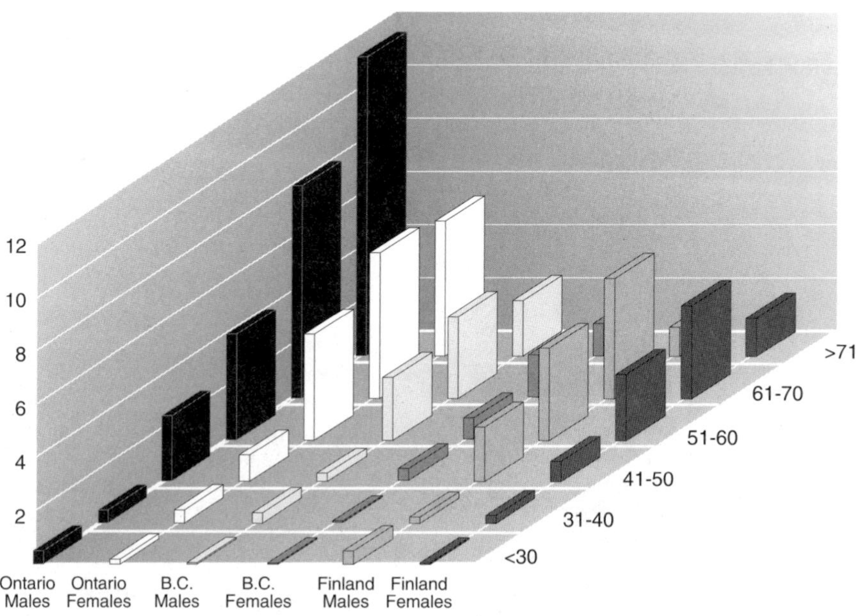

Figure 3,7 Age and Sex Specific Incidences of ALS Per 100,000 Population in Finland, British Columbia and Ontario (after A.A. Eisen, 1991,[28] M. Jokelainen, 1978,[29] and A.J. Hudson et al., 1986[30])

INFECTIOUS AGENTS

Considerable attention has been paid to the possible role of viruses in ALS, especially since it has been demonstrated that some chronic neurologic diseases, such as Creutzfeld-Jakob disease, are probably due to slow, or latent, viruses that are transmissible. Although virus-like particles have been observed in ALS patients,[34] efforts to recover an infectious agent from ALS victims have proven futile.[35]

ENVIRONMENTAL AND TEMPORAL DIMENSIONS

Foci of exceptionally high ALS risk have been identified in the Western Pacific. The foci include Guam and Rota Islands of the Marianas chain in Micronesia,[36] two villages on the Kii Peninsula of Honshu Island, Japan,[37] and a few specific lowland drainage basins in southern West New Guinea.[38]

Until recently, the crude ALS incidence rate for West New Guinea was 10 times that of Guam and the Kii Peninsula, and 100 times higher than the global norm. The spatial pattern of the disease in the Western Pacific foci is particularly interesting, because in West New Guinea, villages with high ALS risk are situated along rivers that originate on the coastal plain. Conversely, villages on drainage systems that originate in the Central Highlands do not have elevated ALS.[39] In the Kii Peninsula, the Japanese seem prone to ALS only when living in certain valleys; while members of the Chamorro linguistic group suffer from the disease only in particular villages, on two islands within the Marianas chain.

The possibility that environmental factors play a key role in the etiology of ALS, in Guam, is suggested by migration studies. Chamorros who migrate do not escape the disease, but suffer the same high incidence rates as island residents.[40] All Guam emigrants, who subsequently develop ALS, have spent their early years on Guam. However, ALS does not appear in Chamorros until up to 30 years after emigration. Typically, it develops some 18 years later,[41] suggesting an environmental latency period for ALS of up to three decades among long-term emigrants.

Studies of the Chamorros who immigrated to the United States have documented that, within 2 to 30 years, ALS developed at five times the typical incidence rate found in the rest of the U.S. population.[42] A similar excess in the incidence of ALS was also found among Filipinos who emigrated to Guam. Symptoms began to develop after a mean exposure time on the island of 17 years.[43]

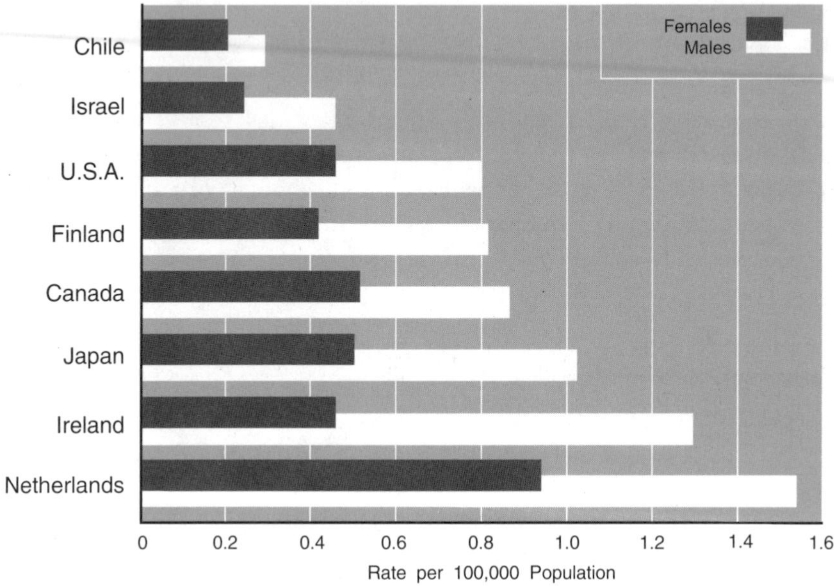

Figure 4,7 Mean Annual ALS Mortality Per 100,000 Population, Specific for Country and Gender (after L.T. Kurland et al., 1969,[32] and L. Olivares et al., 1972[33])

Although an environmental etiology has not been unequivocally proven for ALS, many lines of evidence tend to support this possibility. They include a steady decline in the annual incidence and mortality rates on Guam since World War II.[44] Such changes in disease patterns have coincided with alterations to the population's diet and water sources that have accompanied westernization. The involvement of environmental variables is also suggested by similarities in mean exposure times and latency periods for migrant ALS patients coming to, or leaving, Guam.[45] In West New Guinea, the lack of specific geographical barriers delimiting the affected from the unaffected suggest that genetic and communicable disease factors are of little significance.[46] It also indicates that etiology may not be dependent on focal infection, but may reflect mineral deficiency syndromes or excesses related to the geochemical environment.[47] For example, in the Western Pacific, all three geographically distinct high incidence foci are characterized by low concentrations of calcium and magnesium in their soil and water sources.[48]

The close covariation of Alzhemier's disease (AD), Parkinsonism with dementia (PD) and ALS, both spatially and temporally, in Guam, the Kii Peninsula of Japan and West New Guinea, also suggest that these illnesses may have one, or more risk factors in common.[49]

A number of environmental hypothesis have been postulated in attempts to explain ALS. Yase, in 1980, for example, has suggested that in both the Guam and Kii Peninsula foci, calcium and magnesium deficiencies, together with excesses of aluminum and manganese, are involved in the etiology of ALS.[50] Calne and associates, in 1986, proposed that motor neuron disease, Alzheimer's disease and Parkinson's disease may be due to subclinical environmentally induced damage that remains latent until age-related neuronal compromise and attrition occurs.[51] Indeed, many of the pathological features of ALS are similar to those of Parkinsonism with dementia, Alzheimer's and aging of the nervous system, supporting this possibility.[52] It has also been argued that, within the United States, statistical correlations indicate that soils very depleted in sodium are associated with a reduction in ALS mortality ($r = -0.46643$, $p = 0.0008$), and that soils deficient in iodine are linked to increased ALS death rates ($r = 0.38225$, $p = 0.0091$).[53]

Several sources have postulated that other elements may be involved in the pathogenesis of ALS. For example, concern has surfaced that ALS may be related to histories of increased metal exposures. In particular, lead and mercury have been posited, because exposures to these heavy metals have caused ALS-like symptoms.[54] It has also been suggested that heavy metals may injure motor neurons and cause their destruction.[55] Cadmium,[56] manganese,[57] selenium,[58] aluminum[59] and copper[60] have also been proposed as potential risk factors, because increased tissue levels have been found in ALS patients, or because increased exposure to these elements have suggested their possible involvement as etiological agents in the development of ALS.

As the preceding discussion illustrates, the etiology of ALS remains unclear. Allusion has been made to a viral etiology, a genetic abnormality, and an environmental etiology for ALS. Yet, despite considerable research, the ultimate cause or causes of ALS remains unclear. The investigation described in this chapter attempts to shed further light on this disease by using a gender and age matched case-control series from British Columbia, Canada, to examine exposures to environmental variables that may be involved in the etilogy of ALS.

CASE - CONTROL INVESTIGATIONS

A review of the literature demonstrates general agreement on the incidence, prevalence and mortality rates of ALS, its population characteristics, and its geographical distributions. However, ALS's cause(s) still remain a mystery, and relatively few case-control studies of potential risk factors have been conducted in attempts to clarify its etiology. If environmental variables, and possibly other risk factors, play roles in the etiology of ALS, they are likely to have affected ALS patients with greater frequency than controls.[61,62]

Kondo and Tsubaki, in 1981, suggested that a useful way of obtaining clues about the etiology of ALS was to use a case-control study to examine the history of victims' prior exposure to potential causal variables.[63] Such a retrospective approach samples prior exposure to a suspected risk factor and the presence, or absence, of a disease among individuals. With this approach, association is indicated if a significantly greater proportion of the diseased cases than controls have been exposed to the suspected risk factor. In such retrospective comparisons, the frequency of prior exposures can be used to estimate the relative risk of a disease attributable to a suspected risk factor.[64] In this way, it is possible to infer which environmental variables, if any, may be antecedent risk factors predisposing to the disease.

The composition of a comparable control population is important, in that it should be similar to the disease population with respect to variables that might reasonably be expected to influence the disease.[65] To illustrate, ALS demonstrates a male excess and tends to increase with age, especially in the sixth to eighth decades of life. Therefore, in order to control for possible biases due to the influence of different age and sex distributions within the samples, in this study matching was employed to ensure compatibility in the case and control samples, with respect to the variables of age (five year age categories) and gender. Such comparability was essential since, had it not been achieved, possible differences in sample gender and age, and not the presence or absence of ALS, might have explained differences in observed exposure patterns.[66,67] Thus, sample matching was clearly appropriate, since it helps to reduce misinterpretation of results.

The use of sex and age matched cases and controls has been successfully employed in previous ALS studies.[68-76] It must be assumed, however, that such case-control groups were equivalent prior to exposure to any environmental variables that are being studied. This assumption, however, cannot be proved in this type of design and, therefore, competing alternative explanations must still be considered.

METHODS AND MATERIALS

This study was undertaken from March to November, 1988. Cases were ascertained through the ALS Society of British Columbia, Canada. Controls were acquired with the aid of members and support staff of the ALS Society of Victoria, British Columbia. Only cases diagnosed as having ALS were considered in the investigation. Similarly, only controls, free of known neurologic disease, were considered for matching purposes of age (+ five years) and gender.

Data was collected by a mailed, self-administered questionnaire. One hundred and five case questionnaires were distributed and 61 cases responded; 50 were used after screening for a diagnosis of ALS and completeness of age and gender data. There were 28 males and 22 females. One hundred control questionnaires were distributed, and 56 were returned; 50 were matched with the cases on the basis of age (+ five years) and sex (Figures 5,7 and 6,7). In total, 100 questionnaires were accepted for analysis, providing some 6,200 individual question responses.

Cases and controls completing the questionnaire responded to demographic questions focussing on age and gender, and to spatial and temporal queries about residential locations. Occupational histories were requested of exposures to the metals aluminum, cadmium, copper, lead, mercury, manganese and selenium, which had been implicated in the literature as possible environmental influences in the pathogenesis of ALS. Information was also collected on family histories of diseases that have been linked to iodine deficiency, specifically ALS, goitre, Alzheimer's, Parkinsons disease, cancer of the nervous system and cancer of the thyroid.[77] Data were also sought on travel to Guam, Mariana Islands, New Guinea and the Kii Peninsula of Southern Honshu, Japan, a dietary history of specific goitrogens, and foods that might improve or adversely affect muscle movement or coordination. Other dietary questions collected data on food allergies, milk, salt and saltwater fish consumption, and the use of aluminum cookware. Miscellaneous questions sought information on the use of anti-perspirants, cortisone, the pain killer dimethyl sulfoxide, dianabol (an anabolic steroid), and the consumption of vegetables fertilized with milorganite; all of which have been implicated as possible influences in the pathogenesis of ALS.

Case and control samples were divided into four parts to examine the relation between exposure to environmental factors and ALS. The divisions resulted from classifying each member of the population on both variables of disease or non-disease and exposure or non-exposure. At the five percent level of significance, chi-square analysis based on discordant pairs, or chi-square analysis corrected for continuity (if expected frequencies were five or less), was used for four way contingency tables, to determine if a significantly greater proportion of ALS cases were previously exposed to an environmental factor than controls who were free from the disease. The frequencies of exposures in the diseased and non-diseased groups were also used to estimate the relative risk of ALS, attributable to the suspect determinant(s). To do this, the relative risk associated with the characteristic under study, and exact 95 percent confidence intervals were constructed. McNemar's test for matched pairs[78] was also used in the analysis of occupational histories. Analysis was conducted using the SAS software system for data analysis.

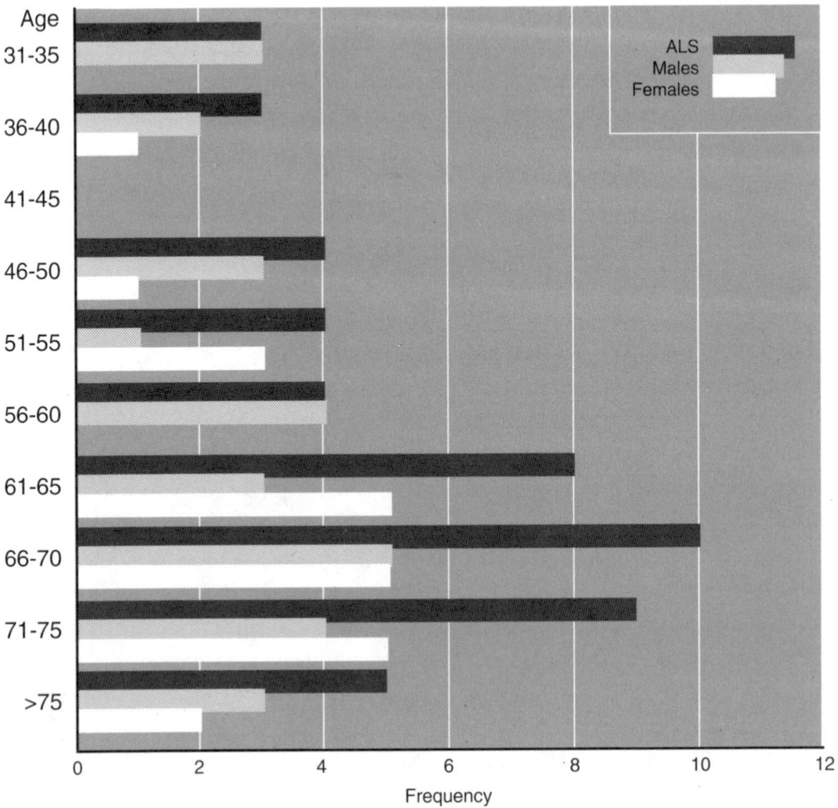

Figure 5,7 ALS Case Sample for British Columbia, Canada (sample size = 50)

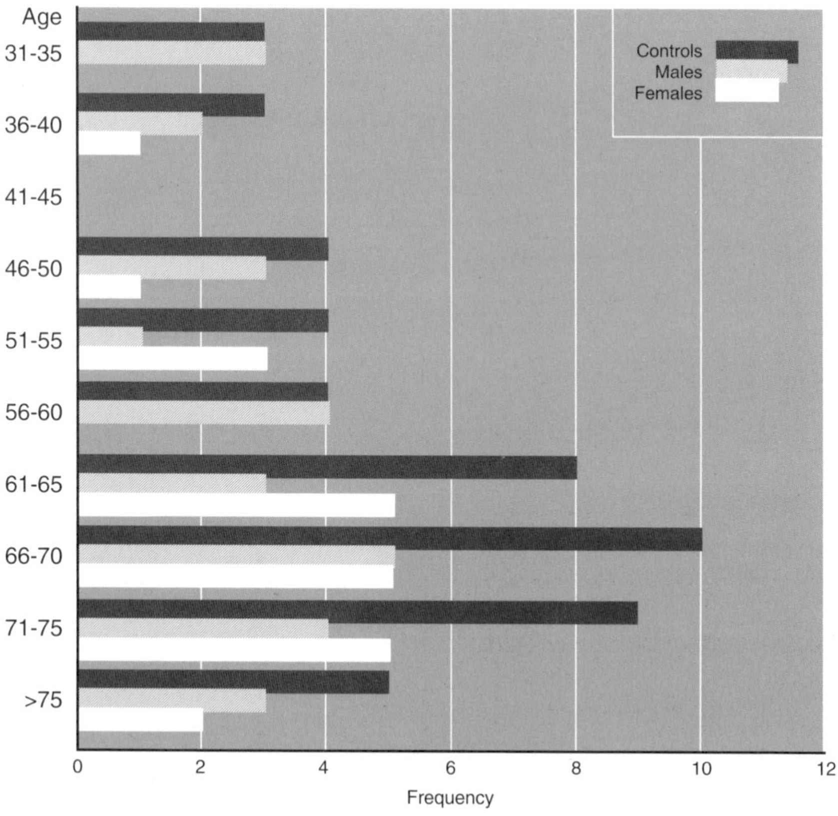

Figure 6,7 Control Sample for British Columbia, Canada (sample size = 50)

RESULTS

Occupational Metal Exposure

Table 1,7 lists the frequencies of reported exposures among cases and controls. Compared to controls, ALS cases more often reported occupational exposure to mercury, lead, copper, aluminum, selenium, cadmium and manganese. Among these, copper exposure ($p = 0.044$, rr = 2.708), lead exposure ($p = 0.043$, rr = 2.740), mercury exposure ($p = 0.045$, rr = 4.731) and aluminum exposure ($p = 0.007$, rr = 3.580) were statistically significant at the five percent level. No significant differences were found between cases and controls for occupational exposures to selenium ($p = 0.380$), cadmium ($p = 0.482$), or manganese ($p = 0.269$).

Table 1,7 Survey Results of 50 ALS Cases and 50 Age and Sex Matched Controls in British Columbia, Canada

Environmental Variable	*Number of ALS Cases N=50*	*%*	*Number of Controls N=50*	*%*	*Continuity Adjusted Chi-Square*	*Probability*	*Chi-Square*	*Probability*	*Phi*	*Relative Risk*	*95% Confidence Interval*
Occupational Exposures *											
Mercury	9	25.71	3	6.82	4.036	0.045	5.403	0.020	0.262	4.731	1.171 - 19.107
Lead	15	36.59	8	17.39	3.179	0.075	4.107	0.043	0.217	2.740	1.016 - 7.394
Copper	15	35.71	8	17.02	3.128	0.077	4.044	0.044	0.213	2.708	1.008 - 7.278
Aluminum	19	45.24	9	18.75	6.149	0.013	7.333	0.007	0.285	3.580	1.390 - 9.217
Selenium	2	6.06	0	0	0.769	0.380	2.554	0.110	0.186	6.587	0.305 - 142.114
Cadmium	3	8.57	1	2.38	0.494	0.482	1.486	0.223	0.139	3.844	0.382 - 38.720
Manganese	4	11.11	1	2.38	1.222	0.269	2.463	0.117	0.178	5.125	0.546 - 48.119
Diseases **											
Multiple Sclerosis	3	6.00	2	4.00	0.000	1.000	0.211	0.646	0.046	1.532	0.245 - 9.587
Goitre	11	22.00	6	12.00	1.134	0.287	1.772	0.183	0.133	2.068	0.700 - 6.116
Alzheimer's Disease	4	8.00	4	8.00	0.000	1.000	0.000	1.000	0.000	1.000	0.263 - 4.241
Parkinson's Disease	7	14.00	2	4.00	1.954	0.162	3.053	0.081	0.175	3.907	0.770 - 19.831
Cancer of the nervous system	3	6.00	0	0.00	1.375	0.241	3.093	0.079	0.176	7.442	0.374 - 147.925
Cancer of the thyroid	3	6.00	1	2.00	0.260	0.610	1.042	0.307	0.102	3.128	0.314 - 31.142
All diseases (Pooled)	31	62.00	15	30.00	9.058	0.003	10.306	0.001	0.321	3.807	1.657 - 8.747

Table 1,7 continued

Environmental Variable	*Number of ALS Cases N=50*	*%*	*Number of Controls N=50*	*%*	*Continuity Adjusted Chi-Square*	*Probability*	*Chi-Square*	*Probability*	*Phi*	*Relative Risk*	*95% Confidence Interval*
Diet											
Food allergies	4	8.51	7	14.00	0.283	0.595	0.726	0.394	-0.087	0.571	0.156 - 2.095
Peanuts	15	30.00	32	64.00	10.277	0.001	11.602	0.001	-0.341	0.241	0.104 - 0.556
Bananas	39	78.00	43	86.00	0.610	0.435	1.084	0.298	-0.104	0.577	0.204 - 1.636
Oranges	25	50.00	44	88.00	15.147	0.000	16.877	0.000	-0.411	0.136	0.049 - 0.377
Cabbage	29	58.00	36	72.00	1.582	0.208	2.154	0.142	-0.147	0.537	0.233 - 1.237
Onions	40	81.63	47	94.00	2.487	0.115	3.553	0.059	-0.189	0.284	0.072 - 1.120
Saltwater fish	47	95.92	48	96.00	0.000	1.000	0.000	0.984	-0.002	0.979	0.132 - 7.241
Milk	34	68.00	38	76.00	0.446	0.504	0.794	0.373	-0.089	0.671	0.278 - 1.618
Milk mean daily	15	44.12	8	21.05	3.394	0.054	4.391	0.040	0.247	2.961	1.068 - 8.208
Salt	37	74.00	40	80.00	0.226	0.635	0.508	0.476	-0.071	0.712	0.279 - 1.818
Foods which adversely affect muscle movement/ coordination	1	2.00	0	0.00	0.000	1.000	1.010	0.315	0.101	3.061	0.122 - 76.949
Foods which improve muscle movement/ coordination	3	6.00	1	2.00	0.260	0.610	1.042	0.307	0.102	3.128	0.314 - 31.142

Table 1,7 continued

Environmental Variable	*Number of ALS Cases N=50*	*%*	*Number of Controls N=50*	*%*	*Continuity Adjusted Chi-Square*	*Probability*	*Chi-Square*	*Probability*	*Phi*	*Relative Risk*	*95% Confidence Interval*
Travel History											
Guam	0	0.00	0	0.00	0.000	1.000	0.000	1.000	0.000	0.000	0.000 - 0.000
Mariana Islands	0	0.00	0	0.00	0.000	1.000	0.000	1.000	0.000	0.000	0.000 - 0.000
New Guinea	0	0.00	0	0.00	0.000	1.000	0.000	1.000	0.000	0.000	0.000 - 0.000
Kii Peninsula, Japan	1	2.00	0	0.00	0.000	1.000	1.010	0.315	0.101	3.061	0.122 - 76.949
Miscellaneous											
Aluminum cookware	37	80.43	33	70.21	0.814	0.367	1.305	0.253	0.118	1.744	0.668 - 4.555
Anti-perspirant	33	66.00	27	54.00	1.042	0.307	1.500	0.221	0.122	1.654	0.738 - 3.707
Painkiller Dimethyl Sulfoxide	1	2.08	0	0.00	0.000	1.000	0.990	0.320	0.102	3.000	0.119 - 75.519
Cortisone	16	32.65	22	44.90	1.075	0.300	1.547	0.214	-0.126	0.595	0.262 - 1.352
Anabolic Steroid/ Dianabol	0	0.00	0	0.00	0.000	1.000	0.000	1.000	0.000	0.000	0.000 - 0.000
Milorganite fertilizer	5	33.00	1	2.86	6.575	0.010	9.235	0.002	0.430	17.000	1.774 - 162.887

* A single respondent can have more than one type of exposure

** Frequency counts of cases and controls with afflicted relations, not number of relations

Continuity adjusted chi-square probabilities of <0.10, in occupational metal exposures to aluminum, copper, lead and mercury, suggested evidence of heterogeneity and led to matched pair analysis. However, calculation of gender and age matched pairs showed no association between cases and controls for occupational metal exposures to copper (p>0.05), mercury (p>0.05), aluminum (p>0.05) or lead (p>0.05) (Table 2,7).

Occupational exposures among males and females, for these metals, were not equal. Among males, 15 ALS patients reported occupational metal exposure to aluminum, 14 to copper, 14 to lead, and 7 to mercury. In comparison, among female ALS patients, 4 reported occupational metal exposure to aluminum, 1 to copper, 1 to lead and 2 to mercury. Compared to female cases, the male population reported increased frequencies of occupational exposures to these metals. As a consequence, matched male case-control pairs were calculated (Table 3,7). No association was found between male cases and controls for these variables. However, for each of these metals, the relative risk estimates for the matched analysis of male cases and controls were increased when compared to the matched ALS case-control pairs for British Columbia, Canada (Tables 2,7 and 3,7).

Table 2,7 Occupational Exposures of Matched ALS Cases and Controls in British Columbia, Canada

Occupational Exposure	*Number of Cases*	*Number of Controls*	*Chi-Square*	*Relative Risk*
Copper	8	5	0.692	1.600
Mercury	5	3	0.125	1.667
Aluminum	11	5	1.563	2.200
Lead	9	5	0.643	1.800

Table 3,7 Occupational exposures of Matched Male ALS Cases and Controls in British Columbia, Canada

Occupational Exposure	*Number of Cases*	*Number of Controls*	*Chi-Square*	*Relative Risk*
Copper	5	3	0.125*	1.667
Mercury	4	2	0.167*	2.000
Aluminum	9	4	1.231	2.250
Lead	6	3	0.444*	2.000

*chi-square test not valid as number of controls are not greater than 3.

Diseases

It has been argued that, because of genetic differences in their ability to utilize iodine, individual families vary in their susceptibility to certain diseases. These are goitre, multiple sclerosis, Parkinsonism, Alzheimer's disease, ALS and cancer of the thyroid and nervous system. If this hypothesis is correct, individuals suffering from ALS should have family disease histories that display an unusual number of instances of these diseases.[79] Data collected during the questionnaire survey were used to examine this possibility (Table 1,7). There were no cases or controls with concomitant disease associations. The number of ALS cases with a relative who had suffered from one of these diseases was elevated, particularily for goitre (11 cases and 6 controls) and Parkinson's disease (7 cases and 2 controls). Elevated levels were also found for cancer of the nervous system (3 cases and no controls), cancer of the thyroid (3 cases and 1 control) and multiple sclerosis (3 cases and 2 controls). For Alzheimer's disease, the frequencies were reported equally (4 cases and 4 controls). No significant difference was found between cases and controls for multiple sclerosis ($p = 1.000$), goitre ($p = 0.183$), Alzheimer's ($p = 1.000$), Parkinson's ($p = 0.162$), cancer of the nervous system ($p = 0.241$) or cancer of the thyroid ($p = 0.610$).

Combining disease categories showed that 31 ALS cases and 15 controls reported immediate family members, or blood relatives, diagnosed with these disorders ($p = 0.001$) (Table 1,7). The relative risk of ALS, within the samples, was 3.807 times as great in people whose relations were afflicted with these diseases as in people whose relations were not afflicted with these diseases.

Table 4,7 gives reported frequencies of disease in immediate family members (parents, siblings or children) and in non-immediate family members (blood relatives). More ALS cases than controls had disease afflicted immediate family (22 cases and 10 controls, $p = 0.01$). As a consequence, the relative risk of ALS, within the samples, was 3.143 times greater in people with diseased immediate family relations, than in people whose immediate family relations did not have these diseases. More cases than controls also had disease afflicted non-immediate family blood relatives (9 cases and 5 controls), however no significant association was found for this group ($p = 0.249$) (Table 4,7).

Four cases (8 percent) also reported relatives with ALS. These data suggest that familial ALS existed in the ALS sample from British Columbia, Canada.

Table 4,7 Reported Frequency of Disease in Family and Relatives of 50 ALS Cases and 50 Age and Gender Matched Controls in British Columbia

Disease	*Immediate Family** *(parents, siblings, children)*		*Non-Immediate Family** *(blood relatives)*	
	Cases	Controls	Cases	Controls
Multiple Sclerosis	2	0	1	2
Goitre	9	4	2	2
Alzheimer's Disease	3	3	1	1
Parkinson's Disease	5	2	2	0
Cancer of the Nervous System	1	0	2	0
Cancer of the Thyroid	2	1	1	0
Total	**22**	**10**	**9**	**5**

	Immediate Family	Non-Immediate Family
Chi-Square	= 6.618	= 1.3289
Probability	= 0.0010	= 0.249
Relative Risk	= 3.143	= 1.976

* Frequency counts of cases and controls with afflicted relatives, not number of relatives.

Travel

ALS is particularly common in certain foci in the Western Pacific. However, the survey demonstrated that no ALS patients, or controls, had ever lived in the Kii Peninsula of Japan, Guam, the Mariana Islands or New Guinea. Similarly, no patients or controls reported a travel history that included Guam, the Mariana Islands or New Guinea. Only one ALS case recorded travel to the Kii Peninsula, whereas no controls had travelled to this area ($p = 1.000$) (Table 1,7).

Diet

No difference was found between cases and controls for food allergies ($p = 0.595$), nor for dietary goitrogen consumption of onions ($p = 0.059$) or cabbage ($p = 0.142$). A significant difference was found for peanuts ($p = 0.001$), however, with 15 cases reporting diets containing peanuts, compared to 32 controls. No differences were found between groups for consuming saltwater fish ($p = 0.984$), nor for salt usage ($p = 0.476$), both dietary indicators of iodine intake (Table 1,7).

No significant difference was found between cases and controls for daily milk consumption (p = 0.373). However, ALS patients drank significantly more than the average daily ALS case and control milk ingestion of 297.6 grams (p = 0.040) (Table 1,7).

No significant differences were found between cases and controls for use of cortisone (p = 0.214), dimethyl sulfoxide (p = 1.000), dianabol (p = 1.000), anti-perspirants (p = 0.221), or for using aluminum cookware (p = 0.253). However, 80 percent of the case sample population had used aluminum cookware prior to a diagnosis of ALS. A significant difference was found between the groups for eating vegetables fertilized with Milorganite (p = 0.010) (Table 1,7).

INTERPRETATION

Considerations

Three assumptions had to be made to determine whether prior exposures to specific environmental variables were operating with greater frequency in the ALS sample population of British Columbia, Canada, than in unaffected peers. Firstly, it was assumed that the ALS sample cases had been reliably diagnosed. Secondly, it was accepted that the prior exposure frequencies in the sample drawn from the ALS Society of British Columbia, Canada, for each age and gender, is a fair and accurate representation of the total ALS population within British Columbia, Canada. Thirdly, it was assumed that prior exposures in the sample of controls, for each age and gender category, is representative of the non-ALS population in British Columbia, Canada. As little is known of the latency period of ALS, no attempt was made to select controls matched for a specific time period in the cases past, such as a neighborhood or work experience.

It is unavoidable that some subjects will be unaware of their exposures to particular environmental variables, and although case and control samples were post matched by gender and age to ensure comparability, it is possible that for some exposures, differences in age and/or gender distributions within the case-control samples may have influenced differences observed in exposure patterns.

Spouses, or friends, were asked to complete questionnaires on behalf of ALS patients if the latter were too incapacitated to do so. This process might bias results, because of incomplete knowledge of the victim's past history. It is also possible that patients with ALS may selectively recall events that occurred prior to disease onset more fully than do unaffected controls. To illustrate, some subjects with ALS may have had previous knowledge that particular environmental variables may be involved in the etiology of ALS. In their introspective search for explanations to the cause of their disease, therefore, they may assume that similar environmental variables may be responsible for their symptoms and may report their histories of exposures prior to onset more diligently than would controls, making positive answers more spuriously frequent in the case sample.[80] Thus, for some variables, errors of Type I (falsely rejecting the null hypothesis) may exist in the data.

For some variables examined, the effective sample size was fairly small. This weakens the reliability of the data used in some comparisons, and may lead to overestimates of associations. Although the continuity adjusted chi-square test was used to compensate for small frequencies, the potential for bias caused by the latter may still exist in this study.

IODINE DEFICIENCY

Iodine

Kurland and associates, in 1973, produced a series of monographs on the geographic distribution of diseases within the United States.[81] The disorders discussed included Parkinson's disease, multiple sclerosis, ALS and other motor neuron diseases. Multiple sclerosis and Parkinson's disease, by state of residence at death, displayed increased mortality in the northern latitudes of the United States. This suggested that one or more environmental variables may be responsible for their geographical distribution patterns.[82] The average annual age-adjusted ALS mortality rate for the United States (1959-1961), was determined to be 0.5 deaths per 100,000 population,[83] and appeared to demonstrate elevated mortality in the northern latitudes (Figure 7,7). To illustrate, 79 percent of the states with an above average morality rate were located at or above 40 degrees latitude ($p = 0.001$, $rr = 12.188$) (Figure 8,7). This suggests that one or more environmental risk factors may be influencing the distribution of ALS mortality. It has been argued that soils deficient in iodine in the United States were associated with increased ALS mortality ($r = 0.38225$, $p = 0.0091$).[86] Interestingly, the geographical availability of soil iodine in the United States tends to decrease with increasing latitude.[87] If increased ALS mortality is linked, at least in part, with iodine deficiency, then this element deserves investigation.

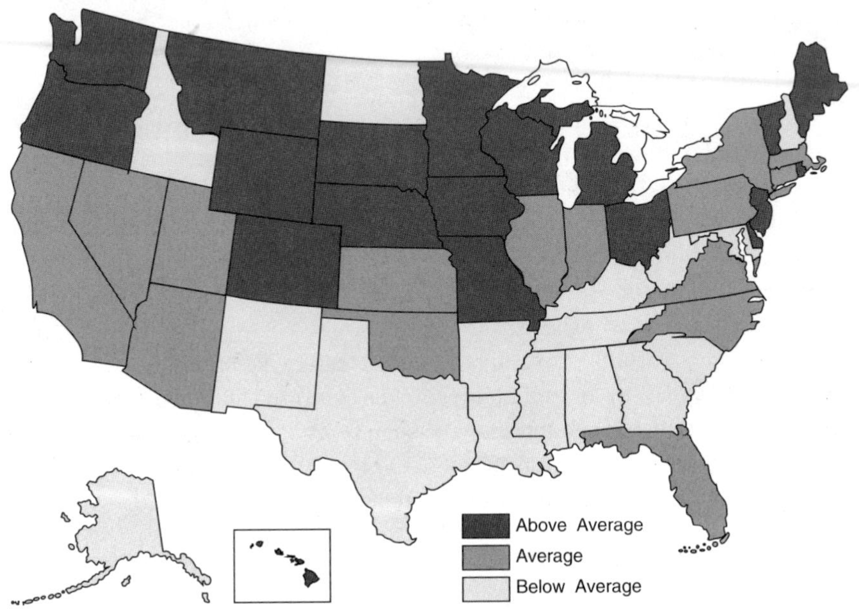

Figure 7,7 Mean Annual Age-Adjusted Mortality Rates Per 100,000 Population for the United States (1959-1961), Demonstrating an Above Average Mortality Rate Distribution in the Northern Latitudes (after L.T. Kurland et al., 1973[84])

In the human body, iodine is concentrated in the adrenal cortex, the ovaries and, in particular, the thyroid glands. A deficiency during pregnancy can result in abnormal development of the fetus, and, in severe cases, may result in cretinism.[88] Iodine deficiencies are also related to goitre and thyroid disorders. Iodine is required by the thyroid glands, which produce thyroxin, an iodine-containing hormone produced in normal amounts only with an adequate supply of iodine. Thyroxin has a profound effect upon both growth and health throughout life, and inadequate supplies are known to impair both mental and physical development.[89] A partial or severe deficiency of iodine causes goitre, an enlargement of the thyroid glands generally considered a response which occurs in an effort to utilize a limited supply of iodine more efficiently.[90]

Rat studies have also suggested a possible role for iodine in neurologic diseases, where thyroid hormones have been associated with the growth of axons and dendrites, and a thyroid deficiency has been linked to reduced myelin formation.[91] Though it is difficult to extrapolate from animals to humans, it is at least reasonable to assume that iodine imbalances, which affect thyroid deficiencies in experimental animals, may have similar effects on humans.[92]

It is evidenced in the literature that the geographical distribution of some diseases, including ALS, may be associated with the availability of iodine. For example, when U.S. troops were examined for goitre during World War I, a greater prevalence was found in those from states such as Montana, Michigan, Wisconsin and Washington, located in the northern U.S.[93] Furthermore, Figure 7,7 shows that these 'goitre states' all suffered above average mean annual age-adjusted ALS mortality rates.[94] In these high goitre, elevated ALS states, the soils are generally of glacial origin, and are deficient in iodine.[95] Indeed, it is interesting to note that many of the world's endemic goitre areas are found in regions that were recently glaciated.[96] The incidence of thyroid disorders also generally increases northwards in the U.S., probably because soils are deficient in iodine.[97] In general, young soils of post-glacial origin contain less iodine than older preglacial soils.

In the U.S., therefore, old soils in the north that probably had adequate time to accumulate supplies of iodine, were removed by Pleistocene ice sheets. To illustrate, the maximum ice advance, during the Wisconsin glaciation, extended from New England, westward through Ohio, Iowa and South Dakota. As a consequence, these areas were covered by extensive tills and glaciofluvial deposits. Further south, winds blowing from the Pleistocene ice-sheets often laid down thick loess deposits. Soils derived from such aeolian deposits are of the highest quality, but buried older iodine rich soils.[98]

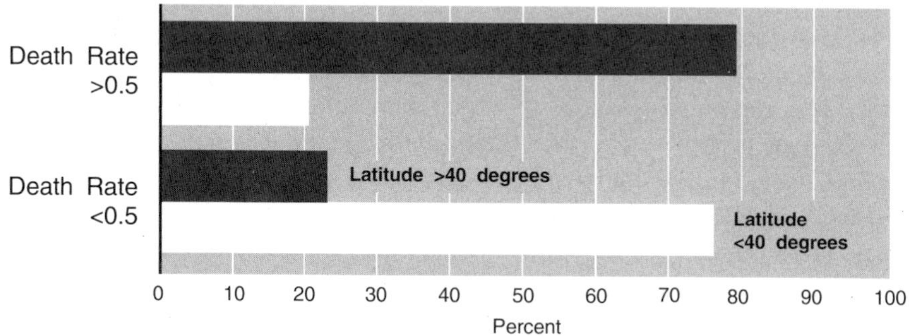

Figure 8,7 Latitude and ALS Mortality Rates for the United States (1959-1961) (after L.T. Kurland et al., 1973[85])

If ALS death rates are related, at least in part, to iodine deficiencies, then the geographic distribution of ALS mortality in the U.S. should indicate above average death rates in the northern U.S., on soils derived from glacial deposits. Some evidence to support this hypothesis can be drawn from the ALS mortality map of the U.S. (Figure 7,7). Indeed, northern states that were covered by the Wisconsin ice-sheet have above average ALS mortality rates, possibly because younger soils, formed on the glacio fluvial sediments and tills, are still iodine deficient.[99] However, states such as Nebraska and Missouri, which were beyond the terminus of the Wisconsin ice-sheet, also demonstrate above average ALS mortality rates due, perhaps, to loess deposition which buried older iodine rich soils. Indeed, as can be seen from the ALS mortality map (Figure 7,7), the terminal moraines of the Wisconsin ice-sheet, and loess deposition to the south, appear to separate above average ALS mortality rates from those that are average or below average.

Iodine is a component of igneous rocks, which have an average content of 0.3 ppm.[100] Natural weathering processes, which release much of the iodine content of igneous rocks, transport released iodine to the oceans, where it accumulates (up to 380 ppm) in marine sediments, such as shales and clays, or is returned to the continents through evaporation and precipitation.[101] Beeson, in 1958, calculated that 0.7 micrograms of iodine, per acre, per year is precipitated onto the continental interiors of the U.S., whereas between 22 and 50 micrograms, per acre, per year, fell onto the Atlantic coastal plain.[102] In the United States, therefore, adequate soil iodine concentrations, which are sufficient to maintain human health, are found along the Atlantic seaboard, and in areas which in recent geological times formed the ocean beds, such as parts of Kansas, Utah, western Texas and New Mexico.[103] It can be seen from Figure 7,7 that Kansas and Utah had average ALS mortality rates of 0.5 per 100,000 population, and New Mexico, Texas and states along the Atlantic seaboard and the Atlantic coastal plain experienced below average ALS mortality rates.

It is possible, therefore, that the distribution of ALS mortality in the U.S. may be associated with iodine deficiencies, but any attempt to explain such an association on a regional basis can only be speculative. There are many compounding factors that may dictate iodine's availability, and ultimately its association, if any, to ALS. For instance, individualized responses and susceptibility to variations in deficiencies of iodine, through pathways such as geographical residence, proximity to the ocean, nature of bedrock geology, dietary iodine consumption and the use of goitrogens in the diet, may directly or indirectly affect ALS. Matovinovic, in 1983, reported that many hard water areas are goitrous, since iodine tends to be insoluble in water enriched with calcium and magnesium.[104] In England and Wales, motor neuron disease mortality increases in the south,[105] and perhaps this may, in part, be related to the use of iodine deficient hard water derived from calcareous aquifers.

Disease Associations

In 1982, Tyler postulated that the frequent presence of neurological diseases in relatives of ALS cases suggested either a neuronal vulnerability with a genetic basis, or the possibility of a common environmental exposure.[106] It has also been postulated that individuals suffering from ALS should have family disease histories that display an unusual number of instances of goitre, multiple sclerosis, Parkinsonism, Alzheimer's disease, ALS and cancer of the thyroid and nervous system, because of genetic differences in their ability to utilize iodine.[107] A common environmental factor has also been suggested to explain the close covariation of endemic ALS, Alzheimer's disease and Parkinson's with dementia, within the Western Pacific.[108]

In the present study, pooling Alzheimer's (AD) and Parkinson's (PD) diseases and ALS demonstrated a significant difference between cases and controls for relations afflicted with these disorders (15 cases and 6 controls, p = 0.027) (Table 5,7). These results also demonstrate that the relative risk of ALS attributable to relations with these diseases was 3.143 times the risk of relatives without these diseases. However, it is unclear whether Parkinson's disease, Alzheimer's disease and ALS have similar etiologies. Reports of coincident occurrences of these diseases in families are not uncommon, but whether there are common biological, genetic or environmental mechanisms remains uncertain.[109] However, the possible role that iodine may play in both Parkinson's and Alzheimer's disease appears to warrant further study.[110]

In the present survey, a significant link was found between the cases and controls for relatives afflicted with multiple sclerosis, goitre, Alzheimer's disease, Parkinson's disease, cancer of the central nervous system and cancer of the thyroid (p = 0.001, rr = 3.807) (Table 1,7). The number of disease afflicted immediate family members was also found to be significantly higher among cases than controls (p = 0.010, rr = 3.143) (Table 4,7). The data suggest that the risk of having ALS is increased when relatives are afflicted with those diseases that Foster, in 1987, considered to be linked to iodine deficiency.[111] Clearly, this evidence is not conclusive, and any attempt to explain this link can only be speculative. However, it is suggestive that a family history of these diseases may lead to an increased risk of ALS (p = 0.001, rr = 3.807).

Table 5,7 Influence of Alzheimer's and Parkinson's Disease and ALS Among Cases and Controls With Relations Afflicted With These Disorders

Observed	*Cases*	*Controls*	*Total*
AD, PD and ALS	15	6	21
Non-diseased Relations	35	44	79
Total	50	50	100

Statistic	*Value*	*Probability*
Chi-Square	4.882	0.027
Continuity Adjusted Chi-Square	3.858	0.050
Phi	0.221	

Estimate of the Relative Risk

Type of Study	*Value*	*Sample Size*
Retrospective	3.143	100

Diet

If iodine deficiency plays a role in the etiology of ALS, then one might anticipate that ALS patients may have eaten fewer foods rich in this trace element than controls. Data was collected on this aspect. Although there are many possible dietary indicators of iodine, the present survey considered only the consumption of salt water fish, and the use of iodized salt.

Saltwater fish is considered a major source of iodine.[112] However, no difference was found between cases and controls for the consumption of salt water fish ($p = 0.984$). Approximately 96 percent of the cases ate salt water fish on a regular basis, suggesting that the ALS sample population might not have been iodine deficient (Table 1,7).

Iodized salt contains the amount of iodine that occurs naturally in unrefined ocean salt. In countries where the iodinization of salt is compulsory, such as Austria and Switzerland, goitre has all but disappeared.[113] In the questionnaire survey no difference was found between the case and control groups in their use of iodized salt ($p = 0.476$). Although there was a minimal reduction in the frequency of iodized salt usage in the case population (6 percent) compared to the control group, it did not establish that the ALS cases used significantly less iodized salt (Table 1,7). This also suggests that the ALS sample population may not have been iodine deficient.

If iodine deficiency is a factor promoting ALS in susceptible individuals, then it may be influenced by the consumption of goitrogens which may block the bodies ability to utilize iodine.[114] Therefore, the consumption of goitrogens, such as peanuts, onions and cabbage, were compared between the case and control samples. Approximately 82 percent of the cases had diets that normally included onions, 50 percent ate cabbage, and 30 percent consumed peanuts, however no significant differences were found between groups in the intake of these goitrogens (Table 1,7). Gajdusek, in 1982, reported that cassava, a well known goitrogen, was a staple food for many Guamanians,[115] and this would almost certainly interfere with iodine utilization in a population which demonstrates an increased risk for Alzheimer's, Parkinson's disease and ALS.

Iodine Overview

This study identifies several research areas that might be further explored. Iodine is lost continuously through perspiration, exhalation and in the urine.[116] Urinary iodine levels, therefore, might be investigated and compared among ALS cases and controls to determine whether there are any significant differences. Increased iodine requirements at different stages of development also deserve consideration, as they may be related, in part, to the lag-time, or increased incidence, that has been associated with ALS. For instance, iodine requirements are increased during pregnancy and lactation, early childhood, puberty and menopause.[117] Since the incidence of ALS is reported to approach unity in men and postmenopausal women,[118] perhaps iodine deficiencies in some menopausal women may be associated, in part, with increased incidence and mortality rates of ALS seen in populations in their sixth through eighth decades of life. It is also interesting to speculate that a deficiency in iodine during the different stages of development may predispose some individuals to the later development of ALS. Further examination of these possibilities appears to be warranted.

It was also noted that in South West New Guinea, endemic ALS, endemic goitre and cretinism had analogous spatial patterns.[119] This is of interest as it further suggests that iodine may be involved in the etiology of ALS in West New Guinea. These researchers also noted that the soils of southern West New Guinea were rich in iron and bauxite, and had been leached for millennia by heavy rainfall. Soft water, low in levels of calcium and magnesium, should favor iodine solubility, and elevation of iodine levels often occur as soils age, whereas other elements are leached out. Therefore, research directed at analysis of soil and drinking water for iodine levels in the Western Pacific foci may shed further light on iodine's role, if any, in the etiology of ALS.

Occupational Exposures

The occupational work environment can be directly related to acquiring many diseases.[120] Many neurotoxic elements can either mimic or cause neurological disorders, as illustrated by ALS-like syndromes caused by exposure to elemental mercury, and by outbreaks of Minamanta disease in Japan, which resulted from exposure to organic mercurial compounds.[121] Moreover, numerous examples of neuropathy have been recorded following occupational exposures to lead compounds.[122] The neurotoxic potential of elemental exposures is of concern because of the possibility that neurological damage may be related to occupational exposures.[123,124] As already described, the literature links the patheogenesis of ALS to occupational exposure to elements such as lead and mercury, aluminum, copper, cadmium, manganese and selenium. If any of these elements play a role in the pathogenesis of ALS, then exposure to them would have been expected to occur more frequently in ALS patients than in matched controls. This possibility was explored during the questionnaire survey (Table 1,7).

Lead has been identified as both an endemic environmental neurotoxin, and as a possible pathogenic agent in ALS.[125] Investigations have demonstrated greater lead concentrations in the nerve, muscle, tissue, plasma and cerebro spinal fluid of ALS patients, when compared to controls. However, the relevance of these findings to the pathogenesis of ALS has been questioned.[126] It is unclear whether increased lead levels are a cause, or a consequence, of ALS. What is clear is that more people are exposed to lead than develop ALS, suggesting that ALS is not merely a variety of conventional lead intoxication. The human nervous system is continually being burdened by naturally occurring, as well as manufactured, lead compounds, the metabolism, biologic effect and toxicity of which are as yet unclear.[127]

Although exposure to lead is encountered in the personal histories of many ALS patients, and provides indirect evidence that prior exposure may be associated with the disease, there is no unequivocal, conclusive proof of a causal association. If lead is indeed a pathogenic factor in ALS, then a vulnerability to lead can be postulated where there may be heightened sensitivity to lead neurotoxicity from occupational exposures, by absorption or metabolism in susceptible individuals. Lead is known to interfere with calcium homeostasis, and absorption and toxicity may involve interchanges with elements such as calcium, iron and zinc.[128] These interactions suggest that certain elements, in combination with lead, may be responsible for ALS. It may be possible, for example, that lead can potentiate other etiological agents, or vise versa. Clearly, further research on the possible pathological role of lead in ALS is warranted.

Mercury is distributed throughout the lithosphere, hydrosphere, atmosphere and biosphere, by processes such as the leaching of sediments and ore bodies and the venting of volcanic gases, and man induced activities. The latter include the mining of ores, combustion of fossil fuels, and several thousand industrial and other consumer uses of this metal. Despite mercury's ubiquitous occurrence in the modern world, which facilitates its availability for absorption or ingestion, it is toxic to many species. Exposure to this element, in its inorganic and organic forms, can cause neurotoxicity in man.[129] Occupational exposure to mercury has also been associated with ALS-like syndromes in which the manifestations are mainly peripheral.[130] As with lead, mercury toxicity is intriguing in that it may also precipitate the expression of ALS, from a latent form to an active state. Conversely, a similar argument to that postulated for lead can be put forward with regard to mercury exposure. It is possible that mercury exposure is an incidental feature, representing an epiphenomenon in the pathogenesis of ALS.

The role of copper is less understood. Although it is considered to be related to neurological diseases in humans, further research is required to determine if copper plays a role in the etiology of ALS. It remains equally unclear whether there is a neurotoxic role for aluminum in the pathogenesis of ALS. While research has found increased levels of aluminum in the tissues of ALS patients,[131] it must be emphasized that this may also be a consequence, rather than a cause of the disease (epiphenomenon).

Occupational Overview

The questionnaire survey and review of the literature suggests that ALS patients have been burdened by manufactured and naturally occurring neurotoxic elements. Exposure to such neurotoxins, either singularly or in combination, may predispose the development of ALS. Many occupational pursuits contribute to this metal burden, and may help to explain, in part, the preponderance of ALS in males. It may also help to account for the differences in the relative risk estimates of specific occupational exposures that are associated with ALS. Occupational exposures demonstrated an increased risk in the paired male case population, when compared to the paired case-control population (Tables 2,7 and 3,7). However, this preponderance cannot be explained solely on the basis of occupational exposure. The disparity among males and females suggests that it is not likely that occupational exposure would be a primary etiological factor in the development of the disease. The findings of statistically significant associations in the ALS cases from British Columbia, Canada, and in a number of other series, heightens the possibility that the etiology of ALS may be related to metal exposures. These findings are interesting, and probably warrant further investigation, the salient point being that heavy metals and trace elements may be involved, either singularly or in combination, in the development of the disease.

Milk and Calcium

Investigators have suggested that increased milk consumption may predispose to the development of ALS.[132] It has also been suggested that this may be a compensation mechanism, to increase bone mineralization, when these were inadequately storing calcium.[133] The present study shows that ALS patients consumed significantly more than the average daily ALS case and control milk ingestion of 297.6 grams ($p = 0.040$, $rr = 2.961$) (Table 1,7). To illustrate, the average daily ALS patient's milk consumption was 311.8 grams, compared to the control's 283.5 grams. ALS cases and controls, therefore, were subdivided into two categories according to the quantity of milk ingested: those who consumed more than the average daily case and control milk ingestion of 297.6 grams of milk per day, and those who drank 297.6 grams or less per day. Forty-four percent of the ALS patients recorded consuming more than 297.6 grams of milk a day compared with 21 percent of the controls (15 cases and 8 controls). The data indicate that increased milk consumption may be a risk factor in ALS ($rr = 2.961$), and further suggests that calcium deficiency may play a role in this disease.

Milorganite, Dimethyl Sulfoxide, Dianabol and Cortisone

Recent press reports have identified an apparent cluster of ALS among former San Francisco 49er football players, perhaps linked to the use of the fertilizer Milorganite on the practice field.[134] It has been argued that this cluster may also be due to the antecedent use of the pain killer dimethyl sulfoxide (DMSO), dianabol (an anabolic steroid) or cortisone. Questions about the use of these substances were included, therefore, in the questionnaire. Reported exposures among ALS cases, however, were lower than among controls for cortisone (16 cases and 22 controls, $p = 0.214$). No cases or controls had used anabolic steroids ($p = 1.000$), and only one case and no controls had used the painkiller dimethyl sulfoxide ($p = 1.000$) (Table 1,7).

The use of Milorganite fertilizer, however, was reported more often by ALS cases than controls (5 cases and 1 control, $p = 0.010$). Indeed, the relative risk associated with the use of this fertilizer was 17 times the unit risk for non-exposure. It is possible that this association may be an artifact of selective recall bias. However, it should be noted that this fertilizer contains the metals mercury and cadmium. A significant association found in the survey for prior occupational exposures to mercury ($p = 0.045$) further suggests that this metal may be involved in the etiology of ALS.

Sodium

In a case-control study of ALS, Hanisch and associates, in 1976, noted birthplace differences between controls and ALS cases, with more ALS cases being born in the western U.S. and fewer in the eastern states.[135] Foster, in 1987, documented that soils very deficient in sodium in the U.S. were associated with reduced ALS mortality.[136] Indeed, sodium enriched soils within the U.S. follow a westward trend, as does elevated ALS mortality. To illustrate, Shacklette and associates, in 1971, noted that sodium enriched soils are generally found in the central and western states.[137] The ALS monograph (Figure 7,7) also appeared to demonstrate elevated mortality in the western longitudes ($p>0.05$), and a reduced mortality in the eastern longitudes of the U.S. (Figure 9,7). These studies suggest that ALS is more likely to be found in the western U.S., which has sodium enriched soils. An increase in sodium, therefore, may be a risk factor in predisposing to the development of ALS.

Potassium and sodium are thought to be intimately involved in nerve and muscular functions.[139] Interestingly, treatment of multiple sclerosis through dietary modification demonstrated that particular foods, such as peanuts, oranges and bananas, adversely affected some patients.[140] This implies that food allergies may play a role in multiple sclerosis. Analysis of oranges, bananas and peanuts shows that they contain high concentrations of both sodium and potassium.[141] Since both elements are involved in nerve conduction processes, it is interesting to speculate that they may have been responsible for producing the adverse muscle movement and coordination noted with some multiple sclerosis patients. If increased sodium levels promote ALS, as has been suggested, it was considered of

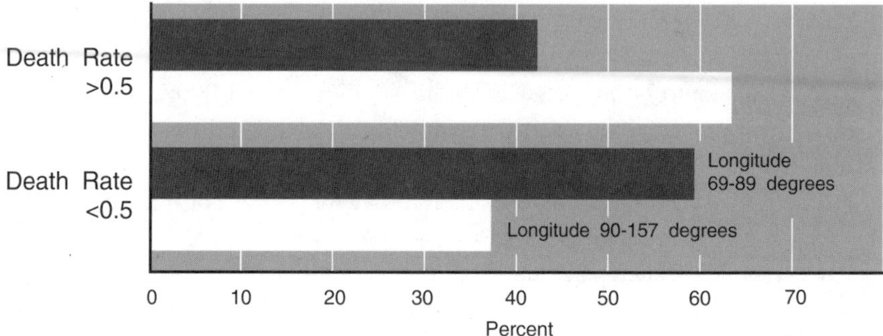

Figure 9,7 Longitude and ALS Mortality Rates for the United States (1959-1961) (after L.T. Kurland et al., 1973[138])

interest to know whether the foods which adversely affected multiple sclerosis patients had a similar effect on the ALS patients.

Therefore, queries about foods that might improve, or adversely affect, muscle movement or coordination in ALS patients was sought. Interestingly, only one reported that bananas had an adverse effect. In contrast, two patients reported that bananas improved their muscle movement and coordination, and one recorded that peanuts and oranges helped muscle movement. These results suggest the possibility that in three ALS cases, an increase of potassium and sodium may have improved nerve conduction.

Food Allergies

Research has shown that the symptoms of certain diseases disappear when specific foods are eliminated from the diet.[142] It has been argued further that in susceptible individuals, allergies to foods have caused the misdiagnosis of various diseases.[143]

No difference was found between cases and controls for food allergies ($p = 0.283$) (Table 1,7). However, two cases reported being allergic to peanuts, and a similar number identified food allergies to either mangos, cocoa, eggs or milk. Although again speculative, the two cases reporting allergies to peanuts and the one case reporting adverse muscle reactions to bananas, suggests that the elimination or addition of particular food groups may help alleviate the symptoms in some ALS cases.

CONCLUSIONS

This research suggests there is some evidence to support an environmental hypothesis of etiology for ALS. The study identified several risk factors that may be associated with ALS. The data indicate an increase, above the expected number, of ALS patients reporting occupational exposures to lead ($p = 0.043$, $rr = 2.740$), mercury ($p = 0.045$, $rr = 4.731$), copper ($p = 0.044$, $rr = 2.708$) and aluminum ($p = 0.007$, $rr = 3.580$) (Table 1,7). However, whether the development of ALS occurs as a consequence of increased occupational exposure to metals is unclear. It has been recognized for many years that human nervous system damage can be linked with exposures to toxic elements. In particular, exposure to toxic elements has frequently been linked to motor neuron disease in both humans and animals, and heavy metals, such as mercury and lead, have been implicated in some cases of ALS. The effects of these metals have been extensively studied in experimental settings and are known to produce neurotoxicity in the peripheral and central nervous systems. However, in rodent models, they have not produced the selective degeneration of motor neurons characteristic of ALS, suggesting that if these heavy metals have a relationship to ALS, it is not a simple or direct one.

An aspect of elemental toxicity arising from environmental excesses is that not all people with prolonged exposure manifest overt toxic effects. To illustrate, ALS is a frequent disease among individuals with a history of exposure to lead, however plumblism is often not overt at the time of ALS diagnosis. In addition, relatively few individuals with a history of lead exposure actually develop ALS. Indeed, if either copper, lead, mercury or aluminum are pathogenic factors in ALS, then a vulnerability to these metals can be postulated in susceptible individuals where there may be heightened sensitivity to metal neurotoxicity from occupational exposures. Such susceptibility may be caused by viral insult, or by genetic factors which could influence the absorption, metabolism and vulnerability to specific exposures. It may also be possible that exposure to these metals, either singularly or in combination, may predispose to the development of ALS. Alternatively, any one, or a combination of these metals may potentiate other etiological events, which may promote or modify the expression of toxicity, and ultimately be responsible for ALS, or the reverse may be true, with other events increasing the damage inflicted by such metals.

The significance of occupational exposure may rely on such factors as duration and concentration. Alternatively, the influence of exposures may be age-specific and not cumulative, so that a more complete history of exposures, prior to ALS onset, may provide further clues to the etiology of the disease.

Although there is no conclusive evidence of a causal association between ALS and metal exposures, there is every reason to explore the possibility of links between ALS and exogenous exposure to elements. Ultimately, ALS may be treatable, and if not, its rate of progression may be reduced by reducing exposures to elements that demonstrate heightened risks in ALS populations. Certainly, the statistically significant associations found between ALS and various metals in British Columbia, Canada, and in a number of other surveys, increase the possibility that ALS may be related to metal exposures.

The current research also shows that certain diseases, specifically multiple sclerosis, goitre, Alzheimer's disease, Parkinson's disease, cancer of the nervous system and cancer of the thyroid, are unusually common in both immediate family members ($p = 0.001$, $rr = 3.143$) and in relatives ($p = 0.010$, $rr = 3.807$) of ALS patients, when compared to controls. These data suggest that the risk of having ALS increases when relatives are afflicted with those disorders that are also considered to be linked to iodine deficiencies.[144] The evidence is not conclusive, but suggests that a family history of these diseases may lead to an increased risk of ALS.

Experimental studies have also suggested a possible role for iodine in neurologic diseases. In rats, thyroid hormones have been linked to the growth of axons and dendrites, and a thyroid deficiency has been associated with reduced myelin formation.[145] Iodine imbalances may have a similar effect on humans.[146] The hypothesis that iodine deficiency may be a risk factor contributing to the etiology of ALS has advantages. It might help to explain, in part, why the degeneration of motor neurons and the myelin covering that insulates the nerve cells, occurs in this disease.

There are many factors that may dictate iodine's availability, and ultimately its association, if any, to ALS. Any attempt, therefore, to explain an association between iodine and ALS can only be speculative. However, a question that requires resolution is whether increasing iodine in the diet may aid those afflicted with ALS. This question warrants research, because of the possibility that dietary adjustments may prove to be beneficial.

The present data also show that ALS patients consumed significantly more than the average daily case and control milk ingestion of 297.6 grams ($p = 0.040$, $rr = 2.961$), indicating that increased milk consumption may be a risk factor in ALS. This is of interest because milk is a major source of calcium,[147] and it is hypothesized that disordered calcium metabolism may contribute to the pathogenesis of ALS. Supporting this possibility are skeletal abnormalities, hypercalcemia and hypocalcemia suffered by many ALS patients.[148] A dietary calcium deficiency may result in increased parathyroid activity and the release of calcium from the bones. Indeed, elevated levels of parathyroid hormone have been observed in some ALS patients, and parathyroid abnormalities have been associated with atrophy of the muscles.[149] It has been suggested that increased milk consumption may be a compensatory mechanism, in an attempt to increase bone mineralization, when the latter were inadequately storing calcium.[150] Alternatively, susceptibility to ALS may be contingent upon the injury of cellular components associated with calcium metabolism. Calcium, stored in the tissues from dietary exposures, may later be removed, causing ALS to manifest itself, but the possible role that calcium may play in the etiology of ALS remains obscure.

In the present survey, an increase in exposure to Milorganite fertilized vegetables was reported more often by ALS cases than controls ($p = 0.010$, $rr = 17.000$). Milorganite fertilizer contains the metals mercury and cadmium, and if metals predispose to ALS, then a vulnerability to this fertilizer may be significant.

Difficulties in design of this study may have lead to false associations (errors of Type I and Type II), including the possibility that the samples do not constitute representative samples of ALS patients and their unaffected peers. Biases may also have been introduced into the retrospective inquiry because the exposures being studied may be remote in time, or having the disease may have influenced the recollection of exposures being studied. Also, with a large number of environmental variables under analysis and complete subjection of control upon statistical criteria of significance, an occasional false lead may have occurred. Nonetheless, a number of significant differences between cases and controls suggested the effect of more than random events, and enabled causal inferences to be drawn between some suspected environmental risk factors and the disease. Although significant associations may suggest a causal inference, they do not imply a biological cause-and-effect relationship. The causal significance of any association is a matter of judgement that goes beyond any statement of statistical probability.[151] Biologic plausibility and replication strengthen the presumption of causal relation. However, proof can be provided only by experimental demonstration of the association.

The goal of this study has been to identify environmental variables that may play a role in the etiology of ALS, in the belief that such information could perhaps generate specific hypotheses that can be tested more rigorously in subsequent investigations. Clearly, the statistical association of environmental exposures with ALS should not be equated with cause and effect, but considered as possible clues to the etiology of the disease. The major significance of the current study, therefore, lies in relation to the accumulating evidence that links ALS to the heavy metals lead and mercury, milorganite fertilizer, and increased milk consumption.

In light of the association of some exposures with ALS, it is reasonable to conclude that environmental variables might play a role in its etiology. Conceivably, a variety of etiological factors could share in the capacity to induce ALS. In view of the paucity of possible risk factors in ALS, the role that environmental variables may play in the etiology of this disease deserves further investigation.

The present research raises more questions than it answers. However, a case-control study is a screening method applied in an attempt to identify variables that can then be evaluated more specifically in subsequent investigations. Indeed, this type of research is considered useful for the evaluation of environmental exposures which might influence the pathogenesis of ALS. However, to establish bias-free answers, a prospective study designed to eliminate potential biases, and sampling of a larger population is requisite.

In conclusion, it seems appropriate to quote Kinnear Wilson (1907):

> We do not assert that a prima facie case has been made out for the toxic origin of Amyotrophic Lateral Sclerosis, but we cannot avoid the conclusion that the facts suggest the probability of such an origin.[152]

It is hoped that the data and discussion presented here are of sufficient interest to encourage further examination of the possibility that ALS is an environmental illness.

REFERENCES

[1] Kurtzke, J.F. 1982. Human Motor Neuron Diseases. Rowland, L.P. ed. *Advances in Neurology* 36, pp. 281-302.

[2] Cendrowski, W., Wender, M. and Owsianowski, M. 1970. Analyse epidemiologique de la sclerose laterale amyotrophique sur le territore de la Grande-Pologne. *Acta Neurologica Scandavia* 46, pp. 609-617.

[3] Juergens, S.M., Kurland, L.T., Okazaki, H. and Mulder, D.W. 1980. ALS in Rochester, Minnesota 1925-1977. *Neurology* 30, pp. 463-470.

[4] Emery, A.E. and Holloway, S. 1982. Familial Motor Neuron Diseases. Rowland, L.P. ed. *Advances in Neurology* 36, pp. 139-147.

[5] Kurtzke, *op. cit.*

[6] Juergens et al., *op. cit.*

[7] Kurtzke, *op. cit.*, p. 289.

[8] Hudson, A.J., Davenport, A. and Harder, W.J. 1986. The Incidence of Amyotrophic Lateral Sclerosis in Southwestern Ontario, Canada. *Neurology* 36, pp. 1524-1528.

[9] Kurtzke, *op. cit.*

[10] Gajdusek, D.C. and Salazar, A. 1982. Amyotrophic Lateral Sclerosis and Parkinsonian Syndromes in High Incidence Among the Auyu and Jakai People of West New Guinea. *Neurology* 32, pp. 107-126.

[11] Reed, D.M. and Brody, J.A. 1975. Amyotrophic Lateral Sclerosis and Parkinsonism-dementia on Guam, 1945-1972: Descriptive Epidemiology. *American Journal of Epidemiology* 101, pp. 287-301.

[12] Hudson, et al., op. cit., p. 1525.

[13] Gajdusek and Salazar, op. cit.

[14] Hudson, et al., op. cit., p. 1524.

[15] Juergens, et al., op. cit.

[16] Kahana, E., Alter, S. and Feldman, S. 1976. Amyotrophic Lateral Sclerosis: A Population Study. *Journal of Neurology* 212, pp. 205-213.

[17] Murray, T.I., Cameron, J., Hefferman, L.P., MacDonald, H.N. and King, S.R. 1987. Amyotrophic Lateral Sclerosis in Nova Scotia, in Cosi, V., Kato, A.C., Parlette, W., Pinelli, P. and Poloni, M. (eds.) *Amyotrophic Lateral Sclerosis.* Phenum Publishing, pp. 345-349.

[18] Kahana et al., *op. cit.*

[19] Juergens et al., *op. cit.*

[20] Murray, et al., *op. cit.*, p. 345.

[21] Durrleman, S. and Alperovitch, A. 1989. Increasing Trend of ALS in France and Elsewhere: Are the Changes Real? *Neurology* 39, pp. 768-773.

[22] Kurtzke, *op. cit.*, p. 285.

[23] Durrleman and Alperovitch, *op. cit.*, p. 771.

[24] Hudson, et al., *op. cit.*, p. 1524.

[25] Jokelainen, M. 1978. Amyotrophic Lateral Sclerosis in Finland. *Advancements in Experiments Medical Biology* 209, pp. 341-344.

[26] Juergens et al., *op. cit.*

[27] Murray et al., *op. cit.*, pp. 345-346.

[28] Eisen, A.A.. 1991. Age-specific Annual Incidence Rates per 100,000 of ALS in B.C. 1981-1986 (unpublished). Permission to use granted February 20, 1991.

[29] Jokelainen, *op. cit.*

[30] Hudson et al., *op. cit.*

[31] Kurtzke, *op. cit.*, p. 286.

[32] Kurland, L.T., Choi, N.W. and Sayre, G.P. 1969. Implications of Incidence and Geographic Patterns on the Classification of Amyotrophic Lateral Sclerosis, in Norris, F.H. Jr. and Kurland, L.T. (eds.) *Motor Neuron Diseases, Contemporary Neurology Symposia II*. New York: Grune and Stratton Inc., pp. 28-50.

[33] Olivares, L., San Esteban, E. and Alter, M. 1972. Mexican Resistance to Amyotrophic Lateral Sclerosis. *Archives of Neurology* 27, pp. 397-402.

[34] Oshiro, L.D., Cremer, N.E., Norris, F.H. Jr. and Lennette, E.H. 1976. Viruslike Particles in Muscle from a Patient with Amyotrophic Lateral Sclerosis. *Neurology* 26, p. 57.

[35] Gibbs, C.M. 1982. Attempts to Transmit Human Amyotrophic Lateral Sclerosis. *Pathogenesis of Motor Neuron Disease*. New York: Raven Press.

[36] Reed and Brody, *op. cit.*

[37] Kondo, K. 1979. Population Dynamics of Motor Neuron Disease, in Tsubaki, T. and Toyokura, Y. (eds.) *Amyotrophic Lateral Sclerosis*. Tokyo: University of Tokyo Press, pp. 61-103.

[38] Gajdusek and Salazar, *op. cit.*

[39] Gajdusek, D.C. 1982. Foci of Motor Neuron Disease in High Incidence in Isolated Populations of East Asia and the Western Pacific, in Rowland, L.P. (ed.) *Advances in Neurology* 36, pp. 363-393.

[40] Torres, J., Iriarte, L.L. and Kurland, L.T. 1957. Amyotrophic Lateral Sclerosis among Guamanians in California. *California Medicine* 86(4), pp. 385-388.

[41] Garruto, R.M., Gajdusek, D.C. and Chen, K.M. 1980. Amyotrophic Lateral Sclerosis among Chamorro Migrants from Guam. *Annals of Neurology* 8(6), pp. 612-618.

[42] Ibid.

[43] Ibid.

[44] Ibid.

[45] Reed and Brody, *op. cit.*

[46] Gajdusek and Salazar, *op. cit.*

[47] Ibid.

[48] Ibid.

[49] Gajdusek and Salazar, *op. cit.*

[50] Yase, Y. 1980. The Role of Aluminum in CNS Degeneration with Interaction of Calcium. *Neurotoxicology* 1, pp. 101-109.

[51] Calne, D.B., Elisen, A., McGeer, E. and Spencer, P. 1986. Alzheimer's Disease, Parkinson's Disease, and Motoneuron Disease: Abiotrophic Interaction Between Aging and Environment? *Lancet.* 11, pp. 1067-1070.

[52] Ibid.

[53] Foster, H.D. 1987. Disease Family Trees: The Possible Roles of Iodine in Goitre, Cretinism, Multiple Sclerosis, Amyotrophic Lateral Sclerosis, Alzheimer's and Parkinson's Disease and Cancers of the Thyroid, Nervous System and Skin. *Medical Hypotheses* 24, pp. 249-263.

[54] Kurtzke, J.F. and Beebe, G.W. 1980. Epidemiology of Amyotrophic Lateral Sclerosis: A Case-control Comparison Based on ALS Deaths. *Neurology* 30, pp. 453-462.

[55] Campbell, A.M.G., Williams, E.R. and Barltrop, D. 1970. Motor Neuron Disease and Exposure to Lead. *Journal of Neurology, Neurosurgery and Psychiatry* 33, pp. 877-885.

[56] Felmus, M.T., Patten, B.M. and Swanke, L. 1976. Antecedent Events in ALS. *Neurology* 26, pp. 167-172.

[57] Roelofs-Iverson, R.A., Mulder, D.W., Elverback, L.R., Kurland, L.T. and Molgaard, C.A. 1984. ALS and Heavy Metals: A Pilot Case-control Study. *Neurology* 34, pp. 393-395.

[58] Gresham, L.S., Molgaard, C.A., Golbeck, A.L. and Smith, R. 1986. ALS and Occupational Heavy Metal Exposure: A Case-control Study. *Neuroepidemiology* 5, pp. 29-38.

[59] Pierce-Ruhland, R.A., Patten, B.M. "Analytical cross-sectional case controlled study of antecedent events in motor neuron disease." International Congress on Neuromuscular Diseases. Montreal 1978; Abstract no. 337.

[60] Felmus et al., *op. cit.*, p. 171.

[61] Ibid.

[62] Gawel, M., Zaiwalla, Z. and Rose, F.C. 1983. Antecedent Events in Motor Neuron Disease. *Journal of Neurology, Neurosurgery and Psychiatry* 46, pp. 1041-1043.

[63] Kondo, K. and Tsubaki, T. 1981. Case-control Studies of Motor Neuron Diseases. *Archives of Neurology* 38, pp. 220-226.

[64] Friedman, G.D. 1980. *Primer of Epidemiology*. New York: Oxford University Press, pp. 103-119.

[65] Fox, J.P., Hall, C.E. and Elveback, L.R. 1970. *Epidemiology: Man and Disease*. New York: Macmillian Company, 339 pp.

[66] Cornfield, J. and Haenszel, W. 1960. Some Aspects of Retrospective Studies. *Journal of Chronic Diseases* 11, pp. 523-534.

[67] Fox et al., *op. cit.*, pp. 294-295.

[68] Deapen, D.M. and Henderson, B.E. 1986. A Case-control Study of Amyotrophic Lateral Sclerosis. *American Journal of Epidemology* 123(5), pp. 790-799.

[69] den Hartog Jager, W.A., Hanler, P.W., Anisink, B.J. and Vermeulen, M.B. 1987. Results of a Questionnaire in 100 ALS Patients and 100 Control Cases. *Clinical Neurology and Neurosurgery* 89(1), pp. 37-41.

[70] Felmus et al., *op. cit.*

[71] Forbes, H.N. and Padia, L.A. 1989. Toxic and Pet Exposure in ALS. *Archives of Neurology* 46, p. 945.

[72] Gawel et al., *op. cit.*

[73] Gresham et al., *op. cit.*

[74] Hanisch, R., Dworsky, R.H. and Henderson, B.E. 1976. A Search for Clues to the Cause of Amyotrophic Lateral Sclerosis. *Archives of Neurology* 33, pp. 456-457.

[75] Kondo and Tsubaki, *op. cit.*

[76] Pierce-Ruhland, R. and Patten, B.M. 1981. Repeat Study of Antecedent Events in Motor Neuron Disease. *Annals of Clinical Research* 13, pp. 102-107.

[77] Foster, 1987, *op. cit.*

[78] Lilienfeld, A.M. 1976. *Foundations of Epidemiology*. New York: Oxford University Press.

[79] Foster, 1987, *op. cit.*

[80] Kondo and Tsubaki, *op. cit.*, p. 224.

[81] Kurland, L.T., Kurtzke, J.F., Goldberg, I.D. and Choi, N.W. 1973. Amyotrophic Lateral Sclerosis and Other Motor Neuron Diseases, in Kurland, L.T., Kurtzke, J.F. and Goldberg, I.D. (eds.) *Epidemiology of Neurologic*

and Sense Organ Disorders. Cambridge, MA: Harvard University Press, pp. 108-127, 350-354.

[82] Kurtzke, *op. cit.*, p. 291.

[83] Kurland et al., 1973, *op. cit.*, p. 353.

[84] Kurland et al., 1973, *op. cit.*, pp. 350-354

[85] Ibid.

[86] Foster, 1987, *op. cit.*, p. 253.

[87] Shacklette, H.T., Hamilton, J.C., Boerngen, J.G. and Bowles, J.M. 1971. Elemental Composition of Surficial Materials in the Conterminous United States. Statistical Studies in Field Geochemistry. *Geological Survey Professional Paper 574-D*, Washington, DC.

[88] Davis, A. 1954. *Let's Eat Right To Keep Fit*. New York: Harcourt and Brace, pp. 192-202.

[89] Ibid., p. 193.

[90] Ibid.

[91] Balazs, R. n.d. cited by Foster, H.D. 1987. Disease Family Trees: The Possible Roles of Iodine in Goitre, Cretinism, Multiple Sclerosis, Amyotrophic Lateral Sclerosis, Alzheimer's and Parkinson's Disease and Cancers of the Thyroid, Nervous System and Skin. *Medical Hypotheses* 24, pp. 249-263.

[92] Foster, 1987, *op. cit.*, p. 259.

[93] Pendergrast, W.J., Milmore, B.K. and Marcus, S.C. 1961. Thyroid Cancer and Thyrotoxicosis in the United States: Their Relation to Endemic Goitre. *Journal of Chronic Diseases* 13, pp. 22-38.

[94] Kurland et al., 1973, *op. cit.*, p. 353.

[95] Foster, 1987, *op. cit.*, p. 252.

[96] Foster, H.D. 1988. Reducing the Incidence of Multiple Sclerosis. *Environments*. 19(3), pp. 14-34.

[97] Shacklette et al., *op. cit.*

[98] Foster, 1988, *op. cit.*, p. 30.

[99] Goldschmidt, V.W. 1954. *Geochemistry*. Oxford: Claredon Press, p. 615.

[100] Fleischer, M. n.d. cited by Keller, E.A. 1976. *Environmental Geology*. Columbus, Ohio: Charles E. Merill, p. 329.

[101] Foster, 1988, *op. cit.*, p. 29.

[102] Beeson, K.C. 1958. The Relation of Soils to the Micronutrient Element Content of Plants and to Animal Nutrition, in Lamb, L.A., Bentley, O.G. and Beattie, J.M. (eds.) *In Trace Elements*. New York: Academic Press, pp. 67-69.

[103]Davis, *op. cit.*, p.196.

[104]Matovinovic, J. 1983. Endemic Goitre and Cretinism at the Dawn of the Third Millennium. *Annual Review of Nutrition* 3, pp. 341-412.

[105]Martyn, C.N., Barkler, D.J. and Osmond, C. 1988. Motoneuron Disease and Past Poliomyelitis in England and Wales. *Lancet* 9, pp. 319-322.

[106]Tyler, R.T. 1982. Nonfamilial Amyotrophy with Dementia or Multisystem Degeneration and Other Neurological Disorders, in Rowland, L.P. (ed.) *Advances in Neurology* 36, pp. 173-180.

[107]Foster, 1987, *op. cit.*, pp. 250-252.

[108]Gajdusek and Salazar, *op. cit.*

[109]Ibid.

[110]Katzman, R. 1986. Alzheimer's Disease. *New England Journal of Medicine* 314(15), pp. 964-973.

[111]Foster, 1987, *op. cit.*

[112]Fredericks, C. 1976. *New and Complete Nutrition Handbook.* California: Majour Books, 272 pp.

[113]Davis, *op. cit.*, p. 197.

[114]Matovinovic, *op. cit.*

[115]Gajdusek, *op. cit.*, p. 367.

[116]Davis, *op. cit.*, p. 202.

[117]Ibid., p. 197.

[118]Kurtzke, *op. cit.*

[119]Gajdusek and Salazar, *op. cit.*

[120]Fox et al., *op. cit.*, p. 87.

[121]Demays, A., Reader, S.W. and Taylor, M.C. 1979. *Mercury.* Environment Canada, Inland Waters Directorate, Water Quality Branch, Ottawa.

[122]Jacobs, J.M. 1979. Perikaryal Neurotoxins, in Aguayo, A.J. and Karpati, G. (eds.) *Current Topics in Nerve and Muscle Research.* Amsterdam: Excerpta Medica, pp. 299-308.

[123]Conradi, S., Ronnevi, L.O. and Norris, F.H. 1982. Motor Neuron Disease and Toxic Metals, in Rowland, L.P. (ed.) *Advances in Neurology* 36, pp. 201-231.

[124]Yanagihara, R.T. 1982. Heavy Metals and Essential Minerals in Motor Neuron Disease, in Rowland, L.P. (ed.) *Advances in Neurology* 36, pp. 223-247.

[125]Spencer, P.S. and Schaumburg, H.H. 1982. The Pathogenesis of Motor Neuron Disease; Perspectives from Neurotoxicology, in Rowland, L.P. (ed.) *Advances in Neurology* 36, pp. 249-266.

[126]Ibid.

[127]Gresham et al., *op. cit.*, p. 35.

[128]Conradi et al., *op. cit.*

[129]Ibid.

[130]Barber, T.E. 1978. Inorganic Mercury Intoxication Reminiscent of Amyotrophic Lateral Sclerosis. *Journal of Occupational Medicine* 20, pp. 667-669.

[131]Yoshimasu, F., Yasui, M. and Yase, Y. 1980. Comparative Study of Analytical Results on Guam PD, Japanese ALS and Alzheimer's Disease Cases. *Folia Psychiatrica et Neurologica Japonica* 34, pp. 75-82.

[132]Felmus et al., *op. cit.*, p. 172.

[133]Pierce-Ruhland and Patten, 1981, *op. cit.*, p. 106.

[134]Petit, C. San Francisco Chronicle. April 13, 1989.

[135]Hanisch et al., *op. cit.*, p. 456.

[136]Foster, 1987, *op. cit.*, p. 255.

[137]Shacklette et al., *op. cit.*

[138]Kurland et al., 1973, *op. cit.*, pp. 350-354

[139]Kirschmann, J.D. and Dunne, L.J. 1984. *Nutrition Almanac*. New York: McGraw-Hill.

[140]Mandell, M. and Mandele, F.G. 1981. *Dr. Mandell's Allergy-Free Cook Book*. New York: Simon and Schuster, pp. 15-25.

[141]Kirschmann and Dunne, *op. cit.*

[142]Finn, R. and Cohen, H.N. 1978. 'Food Allergy': Fact or Fiction? *Lancet* 2, pp. 426-428.

[143]Randolph, T. 1962. *Human Ecology and Susceptibility to the Chemical Environment*. Springfield, Illinois.

[144]Foster, 1987, *op. cit.*

[145]Balazs, *op. cit.*

[146]Foster, 1987, *op. cit.*, p. 259.

[147]Fredericks, *op. cit.*, p. 231.

[148]Yanagihara, *op. cit.*, p. 239.

[149]Felmus et al., *op. cit.*, p. 172.

[150]Pierce-Ruhland and Patten, 1981, *op. cit.*, p. 106.

[151]Fox et al., *op. cit.*, p. 309.

[152]Wilson, K. 1907 cited in Spencer, P.S. and Schuamburg, H.H. 1982. The Pathogenesis of Motor Neuron Disease; Perspectives from Neurotoxicology, in Rowland, L.P. (ed.) *Advances in Neurology* 36, pp. 249-266.

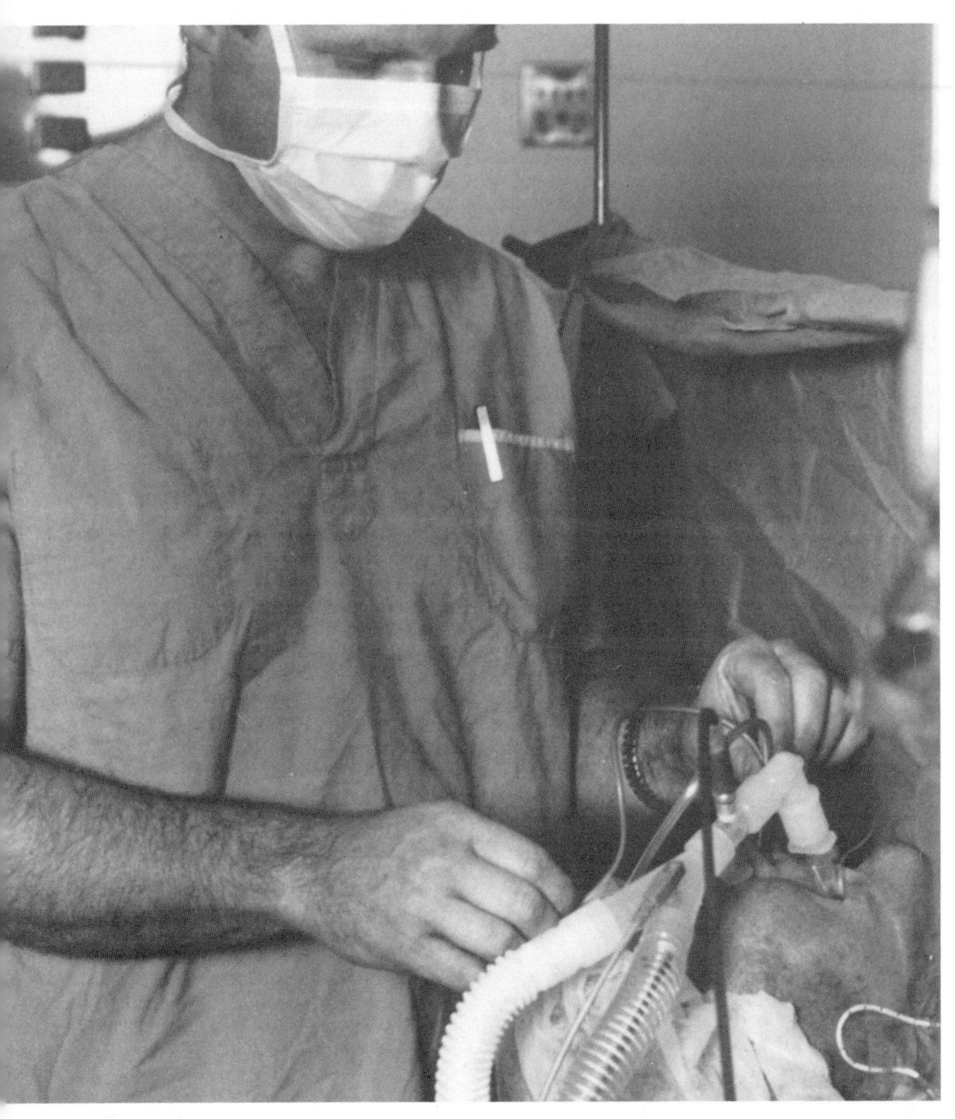

HEALTH AND HOMELESSNESS

Michael Dear
Lois Takahashi

Department of Geography and
School of Urban & Regional Planning
University of Southern California, Los Angeles

INTRODUCTION

As the crisis of homelessness in the U.S. deepens, scholars and service providers have turned their attention to a new set of concerns. During the 1980s, a major effort went toward estimating the numbers of homeless people, and providing a comprehensive account of the causes of homelessness. The expectation seemed to be that such documentation would provide the necessary ammunition for solving the crisis. Many solutions were devised, but they were implemented only in a piecemeal manner, often by private sector or voluntary agencies cajoled by the forces of welfare state restructuring. No concerted attack on homelessness was ever made by any level of government, federal, State, or local. Few public agencies *acted* to eradicate homelessness.

At the opening of a new decade, the problems of homelessness persist, and show no signs of disappearing. Consequently, researchers have turned their attention to a different set of issues, namely, the lives of the homeless. This work includes documentation of the physical and mental health status of the population; ethnographic studies of coping on the streets; and policy and program evaluation. The unspoken assumption behind this current round of work appears to be that since we are likely to have homeless people in our midst for some considerable time into the future, how can we alleviate the worst depredations of their existence?

This chapter draws on these sources to argue that the problems of illness and health among the homeless are best understood in a *contextual* manner. That is to say, population health is highly contingent upon environmental factors, including local climatic conditions; the prevalence of violence; the availability of drugs, problems of poor nutrition, and the absence of friends and family. The argument is simply that a comprehensive knowledge of this broad range of "environmental" conditions facilitates illness prevention, health promotion and (ultimately) exit from the condition of homelessness.

In order to advance this viewpoint, this chapter first briefly reviews the existing literature pertaining to the health of the homeless population in the U.S. Next, it focusses on the complex web of structural, institutional and human factors that operate at different scales through time and space to cause homelessness. (This section is less concerned with abstract theory and more with demonstrating the specific etiology of homelessness.) Finally, it shows, in a preliminary way, how an understanding of context and geographical theory helps to determine appropriate interventions on behalf of the homeless.

In presenting the case for a contingent, contextual social theory of illness and health, the authors are fully aware of the extensive arguments that have already been made in favor of a holistic approach to understanding population health. Indeed, there are few who would today question the notion that individual well-being is linked to the vagaries of time and place; or that geography plays an important role in access to health care. However, such contextual factors are too-often forgotten in practice. And even when they are acknowledged, there remains the problem of determining exactly what personal, social, political and economic conditions have led to the particular health crisis under review. In the absence of information on these contextual determinants, the specification of appropriate and timely interventions remains problematic.

HEALTH STATUS OF THE HOMELESS

Homeless people experience illness and injury to a much greater extent than the homed, or domiciled, population.[1] Three sets of health problems are relevant to our concerns. (1) Those problems that *cause*, or at least contribute to homelessness. These include alcoholism, AIDS and mental disabilities, as well as employment-related injuries that may lead to job loss and, eventually, to homelessness.[2] (2) Health problems also *result* from homelessness, such as hypothermia, skin disorders, and trauma-related events associated with living conditions (e.g. the prevalence of rape and assault). And (3), there are problems in which homelessness is a *complicating* factor in providing care. This category involves most injuries and illnesses, and refers to the difficulties of following the most basic medical advice (for example, bed rest), as well as more complex prescriptions (such as maintaining special dietary restrictions). In many instances, therefore, we may say that homeless people suffer the illnesses of place, situation and circumstances.

The most comprehensive account to date of the prevalance of acute, chronic and infectious diseases has been provided by the U.S. Institute of Medicine (see appendix). The following account draws extensively on their findings.

General Health Problems

Homeless people are at high risk of traumatic injuries, largely because they are frequent victims of violent crimes, and because their primitive living conditions limit access to household remedies and moreover expose them to unusual risk.[3] Skin disorders are prevalent, largely because of inadequate opportunities to shower or bathe. Respiratory diseases are also common (Table 1,8). Tuberculosis has recently become a major health

Table 1,8 Rates of Occurrence (percentages) of Acute Physical Disorders

Diagnosis	*Homeless*	*Domiciled*	*Diagnosis*	*Homeless*	*Domiciled*
INF	4.9	0.1	TRAUMA		
NUTDEF	1.9	0.1	ANY	23.4	NA
OBESE	2.3	2.7	FX	4.5	2.2
MINURI	33.2	6.7	SPR	7.1	3.1
SERRI	3.4	1.0	BRU	5.6	1.0
MINSKIN	13.9	5.0	LAC	8.6	1.2
SERSKIN	4.2	0.9	ABR	2.2	0.4
			BURN	1.1	0.2

Homeless: N = 11,886 Domiciled: N = 28,878

SOURCE: Institute of Medicine, 1988. *(see key pp. 189-191)*

issue among the homeless. This has been associated with exposure, poor diet, acoholism, substance abuse, AIDS and other conditions leading to decreased resistance in the host (Table 2,8). Rates of chronic illness are much higher than in the domiciled population (Table 3,8). Homelessness makes the long-term dietary and pharmacological management (of conditions such as hypertension) extremely difficult; and individual compliance with recommended treatment regimens is also problematic.

Table 2,8 Rates of Occurrence (percentages) of Infectious and Communicable Diseases

Diagnosis	*Homeless*	*Domiciled*	*Diagnosis*	*Homeless*	*Domiciled*
AIDS/ARC	0.2	NA	SEXUALLY TRANSMITTED DISEASES		
			VDUNS	0.7	0.6
TUBERCULOSIS			SYPH	0.2	0.1
TB	0.5	0.1	GONN	0.8	0.1
PROTB	4.5	NA	ANYSTD	1.6	NA
ANYTB	4.9	NA			
			OTHER		
			INFPAR	0.3	0.7
			ANYPH	17.4	NA

Homeless: N = 11,886 Domiciled: N = 28,878

SOURCE: Institute of Medicine, 1988. *(see key pp. 189-191)*

Other health problems that occur with great regularity among the homeless are related to teeth (due to poor oral hygiene and lack of dental care); feet (usually infections, corns, etc. resulting from ill-fitting shoes); and environmental exposure (e.g. hypothermia, hyperthermia and frostbite). In addition, specific subgroups of the homeless have their own particular health problems. Dependent children seem to exhibit upper respiratory and ear infections, and many need standard immunizations.[4] Homeless people who are older (e.g. between 50 and 78 years) appear to have problems more similar to geriatric people in the domiciled population rather than to younger homeless individuals.[5] Veterans appear to have more physical problems than the homeless population as a whole.[6]

Table 3,8 Rates of Occurrence (percentage) of Chronic Physical Disorders

Diagnosis	*Homeless*	*Domiciled*	*Diagnosis*	*Homeless*	*Domiciled*
ANYCHRO	41.0	24.9	COPD	4.7	3.2
			GI	13.9	5.6
CANC	0.7	3.5	TEETH	9.3	0.3
ENDO	2.2	1.6	LIVER	1.3	0.3
DIAB	2.4	2.7			
ANEMIA	2.2	0.9	GENURI	6.6	2.9
			MALEGU	1.9	3.2
NEURO	8.3	1.8	FEMGU	15.6	7.3
SEIZ	3.6	0.1	PREG	11.4	0.5
EYE	7.5	5.5	PVD	13.1	0.9
EAR	5.1	1.6	ARTHR	4.2	3.7
CARDIAC	6.6	6.2	OTHMS	6.0	5.8
HTN	14.2	8.0			
CVA	0.3	0.7			

Homeless: N = 11,886 Domiciled: N = 28,878

SOURCE: Institute of Medicine, 1988.

Explanation of Abbreviations used in Tables 1,8-3,8

Acute Disorders

INF	Infestational ailments (e.g. pediculosis, scabies, worms)
NUTDEF	Nutritional deficiencies (e.g. malnutrition, vitamin deficiencies)
OBESE	Obesity
MINURI	Minor upper respiratory infections (common colds and related symptoms)
SERRI	Serious respiratory infections not classified elsewhere (e.g. pneumonia, influenza, pleurisy)
MINSKIN	Minor skin ailments (e.g. sunburn, contact dermatitis, psoriasis, corns, calluses)
SERSKIN	Serious skin disorders (e.g. carbuncles, cellulitis, impetigo, abscesses)
TRAUMA	Injuries
ANY	Any trauma
FX	Fractures
SPR	Sprains and strains
BRU	Bruises, contusions

LAC	Lacerations, wounds
ABR	Superficial abrasions
BURN	Burns of all severity

Infectious and Communicable Disorders

AIDS/ARC	Acquired immune deficiency syndrome, AIDS-related complex
TB	Active tuberculosis infection, any site
PROTB	Prophylactic anti-TB therapeutic regimen
ANYTB	Either TB or PROTB or both
VDUNS	Unspecified venereal disease disease, herpes
SYPH	Syphilis
GONN	Gonnorhea
ANYSTD	VDUNS, SYPH or GONN, or any combination
INFPAR	Infectious and parasitic diseases (e.g. septicemia, amebiasis, diphtheria, tetanus)
ANYPH	AIDS, ANYTB, ANYSTD, INFPAR, SERURI, INF or SERSKIN

Chronic Disorders

ANYCHRO	Any chronic physical disorder
CANC	Cancer, any site
ENDO	Endocrinological disorders (e.g. goiter, thyroid, pancreas disease)
DIAB	Diabetes mellitus
ANEMIA	Anemia and related disorders of the blood
NEURO	Neurological disorders, not including seizures (e.g. Parkinson's disease, multiple sclerosis, migraine headaches, neuritis, neuropathies)
SEIZ	Seizure disorders (including epilepsy)
EYE	Disorders of the eyes (e.g. cataracts, glaucoma, decrease vision)
EAR	Disorders of the ears (e.g. otitis, deafness, cerumen impaction)
CARDIAC	Heart and circulatory disorders, not including hypertension and cerebrovascular accidents
HTN	Hypertension
CVA	Cerebrovascular accidents/stroke
COPD	Chronic obstructive pulmonary disease
GI	Gastrointestinal disorders (e.g. ulcers, gastritis, hernias)
TEETH	Dentition problems (predominantly caries)
LIVER	Liver diseases (e.g. cirrhosis, hepatitis, ascites, enlarged liver or spleen)
GENURI	General genitourinary problems common to either sex (e.g. kidney, bladder problems, incontinence)
MALEGU	Genitourinary problems found among men (e.g. penile disorders, testicular dysfunction, male infertility)
FEMGU	Genitourinary problems found among women (e.g. ovarian dysfunction, genital prolapse, menstrual disorders)

PREG	Pregnancies
PVD	Peripheral vascular diseases
ARTHR	Arthritis and related problems
OTHMS	All musculoskeletal disorders other than arthritis

Mental Disability

As a rule of thumb, it is commonly estimated that one-third of all homeless people are mentally disabled.[7] The proportion of mentally disabled people in the homeless population as a whole ranges from 25 to 50 percent.[8] The U.S. Department of Housing and Urban Development (HUD) reports that, among the sheltered homeless, the proportion of those with mental disabilities increased between 1984 and 1988.[9] Major mental disabilities (principally schizophrenia and affective disorders) when untreated are often causal factors inducing homelessness.[10] Other psychiatric difficulties -- personality disorders, anxiety and phobic disorders, dementia -- may impair an individual's ability to cope with the challenges of everyday life, and thus contribute to homelessness. Homelessness may be a consequence of illness, but life on the streets can also cause or exacerbate emotional difficulties.[11] Without the resources that are common to the rest of the population, homeless people are vulnerable to demoralization, which can exacerbate mental disabilities.[12] In one Chicago study, the homeless had significantly more mental health problems than a group of domiciled extremely poor with similar demographic characteristics.[13]

In one startling study of mental disability on Skid Row in Los Angeles, homeless men (when compared with domiciled men in the same catchment area) were given a current diagnosis of schizophrenia 38 times more frequently, major affective disorders 4 times more frequently, antisocial behavior disorders 13 times more frequently, dementia 3 times more frequently, and substance abuse disorders 3 times more frequently. In sum, a total of 83 percent of the homeless sample (N=379) was found to be suffering from some kind of mental disability or substance abuse.[14] These dimensions of the problem have been corroborated in other cities. However, it has also been observed that street-based populations are at greater risk than shelter-based populations; and that homeless women with families have markedly different psychiatric profiles than individual single women. Homeless women, moreover, demonstrate almost twice the rate of mental illness as that of homeless men.[15]

Most studies confirm that the central problems in aiding the homeless mentally disabled are the lack of (a) treatment facilities and (b) adequate housing.[16]

Substance Abuse: Alcohol and Drugs

The single most frequently observed disorder among the homeless is alcoholism (with the exception of homeless women with children). Between 25 and 40 percent of homeless men have serious alcohol problems.[17] This compares to a rule of thumb estimate that approximately 10 percent of the population as a whole has a significant drinking problem.[18] Alcohol abuse is related to a higher incidence of disease among homeless people. Heavy alcohol consumption can interact and excerbate other physical health problems common to the homeless, and moreover, many types of medication are not recommended in the presence of alcohol.[19] Alcohol is also a problem for a small percentage of homeless children. In the National Health Care for the Homeless Program, for example, about 8 percent of teenage girls and more than 10 percent of teenage boys had problems with alcohol.[20] Alcoholism is just as likely to be cited as a cause, as well as a consequence of homelessness.

Illicit drug use is more common than alcoholism among the younger, "new" homeless population (who are also characterized by high rates of sexually transmitted diseases, and pregnancy). Estimates of the rates of drug abuse vary from 3 to 48 percent of the homeless population.[21] Drug abuse, like heavy alcohol consumption, is also related to a higher incidence of other diseases among the homeless. These include liver disease, cardiac disease, peripheral venous stasis disease, as well as chronic disorders such as diabetes, and diseases of the liver and genitourinary tract. Most worrying, of course, are the problems of HIV infection and AIDS. Prevalence rates are notoriously unreliable, but one estimate puts the rate of HIV infection for one sample of homeless people at 185/100,000, compared with 144/100,000 in the general population.

There is growing concern about homeless people with dual or multiple diagnoses.[22] One estimate in three California counties is that from 54 to 93 percent of homeless people who abuse drugs or alcohol also have some kind of severe mental disorder.[23] Much comorbidity data point to correlations among drug abuse, alcoholism and mental disability.[24] Such dual-diagnosed homeless people tend to be underserved by county agencies because of program funding restrictions, weak institutional connections among the various service agencies, lack of specific treatment and programs for dually-diagnosed people and lack of staff experience.[25] Primary medical care facilities, however, would provide an excellent opportunity for delivering mental health services to homeless people given the resolution of such problems.[26] In addition to the institutional problems with treating dual-diagnosed homeless individuals, however, people with multiple diagnoses tend to be more difficult to entice into treatment.

UNDERSTANDING HOMELESSNESS

The process of homelessness has three levels of explanation (Figure 1,8). First, a set of structural (or contextual) factors operate on the national and state/provincial levels over the long term. Those factors relate especially to underlying changes in the economy and in the patterns of welfare provision. One important effect of those changes has been to increase the demand for temporary shelter. Second, there are a number of components on the supply side that contribute to the increase in homelessness. Those components have combined to drastically reduce the amount of affordable accommodation available to people in marginal economic circumstances. Third, our framework focusses on the individual to account for particular adverse events that propel people into homelessness. From a geographical viewpoint, interest lies in exactly how these processes are localized (or concretized) in particular places.

The Structural Context of Homelessness

Since the 1960s, two major national trends have contributed to an increase in the population of "proto-homeless", defined as those living in marginal economic and housing circumstances. The trends are: the reduction in public expenditures on welfare and other service-related programs, and in particular the development of the deinstitutionalization movement; and the trend toward deindustrialization and its concomitant unemployment and poverty which are associated with deep-seated changes in the restructuring of the economy.

Deinstitutionalization in the United States received a federal seal of approval in 1963 with the passage of the Community Mental Health Centers Act. That legislation cleared the way for moving psychiatric patients out of mental hospitals and, according to the plan, into community-based treatment and service settings. Other deinstitutionalized groups included the mentally retarded, the physically disabled, the dependent elderly, and probationers and parolees.[27] Unfortunately, the government subsequently allocated a woefully inadequate amount for community-based programs. As a consequence, the deinstitutionalized tended to drift toward inner-city neighborhoods where cheap rental accommodation and most health and welfare services existed.[28] The dearth of adequate community-based shelter and service facilities even in inner-cities has caused the deinstitutionalized to become a major component of today's homeless population.[29]

Recent reductions in social expenditures have exacerbated the plight of these vulnerable, welfare-dependent populations.[30] Significant shifts in

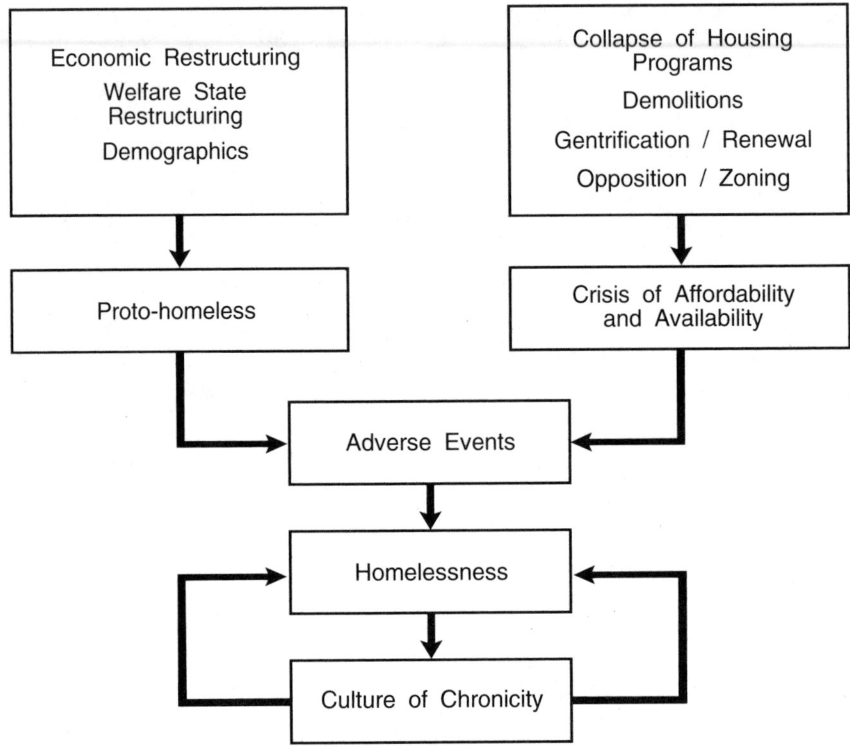

Figure 1,8 Pathways to Homelessness

welfare state budgets began in the 1970s and accelerated under the Reagan administration. Many federal programs fell to state and local governments; and the private, voluntary, and nonprofit sectors had to replace other federal efforts, a trend usually referred to as "privatization". The reorganization of federal spending has pushed millions of people who depend upon social services and welfare checks to the brink of poverty. It has caused many to become homeless.

The victims of *deindustrialization* have joined the deinstitutionalized in the streets and on the sidewalks. Deindustrialization refers to the declining fortunes of the manufacturing sector in general and particularly to large-scale plant closures in the traditional centers of production in the "snowbelt" cities of the northeast.[31] The process has accelerated through the 1980s as a result of economic recession, fluctuations in the value of the dollar, and the decline of union membership and influence. Such factors raised unemployment levels and created the highest rate of official poverty since the early 1960s. By 1982, 15 percent (34.4 million) of the nation's population was living below the poverty line, an increase of 40 percent since 1978.[32]

The economic recovery in the late 1980s did not fundamentally change prospects for the poor. While expansion in service sector employment has somewhat offset the decline of the manufacturing sector, many of the new jobs being created are low-wage, low-skilled and part-time. The economic security these jobs offer is tenuous and such trends have seriously hurt many workers and their families.

Simple *demographic trends* are exaggerating the demand for low-cost housing and social assistance programs. The "graying" of our populations has been well-documented. Recent figures suggest that the Bureau of Census has grossly underestimated in its predictions: (1) the number of people over 65 will be 87 million by 2040 (20 million more than Census Bureau estimates); and (2) the number over 85 will be 24 million (more than double the Census Bureau figure). Medical and psychiatric advances imply that physically and mentally disabled populations will be living longer. Many of these populations are not able to afford preventative health care now, and are thus likely to surface later with more chronic illnesses.

There has also been a phenomenal rise in the number of single-person households and single-parent families. Between 1970 and 1990, the number of people living alone in the U.S. has increased by 112 percent. Additional single-person households translate to increasing demand for relatively small housing units. Growing numbers of female-headed households all too often mean growing demands for low-cost, also relatively small units. Together, these trends in the numbers of dependent persons and demographics of housing demand suggest even tighter low-end housing markets and increased need for more, and increasingly specialized, human services (as seen in the growing demand for child care services).

In summary, the restructuring of the economy and welfare state infrastructure as well as significant demographic change have caused a massive increase in the demand for low-cost housing and welfare services. These are people living on the margins of the housing market, on the edge of homelessness. We shall refer to them as the *proto-homeless*.

The Diminishing Supply of Affordable Housing

There has been a substantial decline in the number of housing units that low-income people and those in need of shelter assistance can afford.

Those losses have resulted primarily from downtown urban renewal, gentrification, abandonment, and from suburban land use controls. The *elimination and reduction of federal low-income housing programs* has also dramatically curtailed the supply of affordable shelter. Construction of low income and assisted housing has essentially stopped.[33] Currently, the net change in the publicly assisted housing stock is negative; more units are being demolished or released from subsidy requirements than are being constructed.[34]

The amount of housing available in the private sector rental stock is also diminishing rapidly. As more and more landlords abandon apartment buildings and houses rather than repair them, the housing supply for the poor is declining at an accelerating pace in some cities.[35] The growth of service-sector employment in central business districts has attracted white-collar professionals, many of whom prefer to live in accessible central city neighborhoods,[36] where they compete with poor, indigenous residents for private market housing.[37] The result is *gentrification* of that inner-city housing which has traditionally been the major source of low-income housing. At the same time, downtown service-sector expansion has created jobs for many low-waged ancilliary workers. This has increased the demand for low-cost shelter readily accessible to the downtown.[38]

The response to the pressures of gentrification and urban renewal has been the *demolition* of thousands of single-room occupancy (SRO) units or their conversion to condominiums. In Chicago, for example, 300 SRO units were lost to gentrification in 1981-1982, and a total of 18,000 units have been demolished or converted since 1973.[39] In New York City, city-sponsored *urban renewal* legislation encouraged the demolition of SROs and their replacement by luxury condominiums; more than 31,000 units were lost between 1975 and 1981. Chester Hartman has estimated the number of New York City SROs dropped from 170,000 in 1971 to 14,000 by the mid-80s.[40]

The geographical concentration (or "ghettoization") of the homeless in downtown areas has been accelerated by *exclusionary zoning practices and community opposition* to local siting of shelters and services for the homeless.[41] Many suburban jurisdictions have used zoning to limit the number and types of community-based service facilities and to restrict the development of subsidized housing projects. One of the most common zoning approaches is to require a conditional use permit for service facility siting; neighborhood opposition to the service or its clients is mobilized and often results in the denial of the use permit.[42]

As a consequence of these trends, there is a *chronic shortage of affordable and appropriate housing.* Two million households that could afford to buy a home in 1980 cannot do so today. Rents have risen at a rate 14 percent higher than prices generally. The result of rising rents and (for many) declining incomes is a dramatic increase in the amount of income people pay for rent. It is now typical for households to spend 30 to 50 percent of income on rent; the proportion reaches almost 60 percent for single-parent households. Only 28 percent of poor renter households live in public housing or receive federal housing subsidies, leaving 5.4 million poor households competing for the stock of private sector rental housing.[43] Since 1978, over 2 million low-cost units have been removed from the inventory. The "housing market" for many people consists of friends' couches, illegal garage conversions, all-night buses, hotbedding, cars, bicycles, rooftops, cardboard boxes and sidewalks.

Adverse Events That Precipitate Homelessness

Increased demand for and diminished supply of shelter together underlie much of the process of homelessness. But those problems do not explain the actual event(s) that may cause a person to become homeless. Many people experience adverse events in the housing market (eviction is one of them). For most people those occurences represent only a temporary setback. However, for those proto-homeless who already live in marginal economic and housing conditions, a single adverse event can be sufficient to act as a catalyst for the descent into homelessness.[44] The five most common immediate causes of homelessness that individuals report are: eviction, discharge from an institution, loss of a job, personal crisis (including divorce or domestic violence), and removal of monetary or nonmonetary welfare support.

The Culture of Chronicity

Once homeless, many people are caught in a vicious cycle of deteriorating circumstances -- a downward spiral that affects their mental, social and physical well-being. Unable to help themselves, refused aid, or given inappropriate assistance, their difficulties accumulate: families break up; health and appearance decline; and victimization (robbery, mugging) increases. Such circumstances threaten to create a new class, the "chronically" homeless, people for whom the experience of homelessness itself creates a new set of social and personal crises that tend to perpetuate the problem. These individuals inhabit what we term a "culture of chronicity."

What causes a person, once homeless, to remain so? The answer to that question lies in the pathology of everyday life on the streets. Evidence suggests that five factors determine whether or not an individual will escape homelessness:[45] experiences in temporary shelter; financial status; availability of assistance; personal status (including health); and street experience.

For instance, living conditions in temporary and emergency shelters are often so bad that many homeless prefer to avoid them.[46] Some shelters are centers of crime, including substance abuse and personal violence.[47] The simple experience of shelter may, therefore, seriously depress the morale of the newly-homeless.[48] The management and operation of shelters can also affect the lives of the homeless. Some shelters systematically exclude certain groups through eligibility requirements (for instance, refusing admission to women or the mentally disabled).[49] Others have obtrusive routines, including long and detailed intake procedures, or degrading and humiliating residence rules such as mandatory gynecological examinations for women.[50] Women are also vulnerable to victimization, such as rape and physical abuse, which later become associated with specific mental disabilities and substance abuse.[51]

With or without shelter, the homeless lead precarious daily lives. Public entitlement programs often present difficult bureaucracies to navigate, especially for the homeless person who is mentally disabled. Eligibility rules make obtaining entitlement benefits a difficult and time-consuming process.[52] Moreover, the financial resources most public programs offer are so low that the recipients can barely survive. For example, general relief payments in Los Angeles County in the late 1980s were approximately $250 per month. That allowed for three weeks' SRO accommodation, with nothing left for the fourth week's shelter or for food.[53] The homeless typically supplement their incomes by casual day labor, panhandling, and prostitution.[54] They rarely earn enough to rent permanent shelter.

The homeless depend not only on income and financial assistance, but also on other forms of support such as food programs, job search services, and clothing provision. However, access to those services is limited and frequently involves long waiting lists and intrusive procedures. For example, Los Angeles County's "60-day rule" allows welfare officers to suspend benefits for two months for some real or perceived breach of agency rules.[55] Those transgressions can include arriving late for an interview or failing to have the required number of job interviews in a month. Suspended claimants receive no welfare payments. For those who avoid suspension, the day is often spent standing in lines or moving between agencies in search of benefits.[56]

An individual's personal strength, both emotional and physical, is an important determinant of how well he or she will stand up to the rigors of life on the street. It is difficult to remain optimistic and healthy when cleanliness is an impossible goal, sleep a luxury, nutritious food scarce, and health care nonexistent. To sleep in the open in wintertime can cause death through hypothermia (even in southern California); in the summer, such exposure can cause sunburn or sunstroke. To stay in public shelters overnight often leaves lice infestations as well as empty pockets.[57]

Life on the street tends to exacerbate the experience of homelessness.[58] The homeless reap some benefits from gathering in the inner city, including access to services and social support from other homeless people.[59] But they also have to contend with life in degraded physical environments. Muggings are common, as is harassment by police and other street people. Many homeless report that their daytime activity is "moving or walking," largely for self-protection.[60]

In sum, the condition of homelessness appears to have a cumulative effect on its victims. Once on the street, physical and mental health problems rapidly surface, even among those with no previous history of such problems. As time passes, the dividing line between those with a history of mental disability and those with street-induced emotional problems becomes increasingly fuzzy. Many people find that the descent into "chronic" homelessness can be sickeningly quick. In a bizarre concession to the speed with which lives can unravel, one shelter in Long Beach, California, refused beds to people who had been homeless for over two weeks on the grounds that they were already beyond rehabilitation.[61] The primary impetus for chronic homelessness appears to be the failure of formal and informal support networks. The social dislocation associated with being homeless is exacerbated by a welfare system which is being stretched beyond its limits. As a consequence, the experience of homelessness itself perpetuates homelessness.

THE FUTURE OF HOMELESSNESS

The problem of homelessness is unlikely to diminish, at least in the immediate future. The difficulties and adjustments of economic restructuring and of welfare state reorganization continue relatively unabated. It took two decades to create the current crisis of homelessness; if we look ahead to future social conditions, there seems little hope that current policies will lead to quick solutions to the crisis.

Economic recovery and expansion may provide employment opportunities that could make it possible for some of the unemployed homeless to support themselves. However, industrial restructuring has led to the disappearance of many job categories that those people might have filled in the past.[62] Increasingly, new jobs, primarily in the service sector, tend to be either high-skill/high-wage positions or low-skill/low-wage posts. The result has been a growing inequality in income distribution.[63] Many of the homeless lack skills and training for high-wage jobs, and job training opportunities are scarce. A large proportion of the remaining jobs are low-wage, part-time, and/or casual, and do not provide benefits such as health insurance or sick leave. Between 1979 and 1984, 44 percent of the net new jobs created only paid poverty-level wages.[64] Even full-time work at the current minimum wage may not remove people from poverty status. Finally, we cannot forget that, although fully one-third of the homeless work on a regular basis, many others remain outside the labor market -- either permanently (e.g., the chronic mentally disabled), or temporarily (e.g., the single parent committed to child care).

Changes in the economic climate are therefore likely to have only a limited impact on homelessness. What future lies in the reorganized welfare state? Between 1982 and 1985, the Reagan administration cut federal programs targeted to the poor by $57 billion (adjusted for inflation).[65] Housing in particular has fallen relative to other federal spending priorities. Federal authorizations for housing were 7 percent of the total federal budget in 1978; by 1988, they amounted to only 0.7 percent.[66] There have also been major cuts in housing, job training, food and nutrition, social services and income maintenance. The federal government assisted an estimated one million fewer households in 1985 than in the pre-Reagan era as a result of cuts in subsidized housing; 300,000 more families were living in substandard housing; hundreds of thousands have been removed from job training programs, medical insurance plans, and disability rolls; and many others have lost Aid for Families with Dependent Children, foodstamps, and food nutrition benefits.[67]

The Reagan administration provided only extremely limited programs to the homeless. Until 1987, the administration's involvement was restricted to provision of emergency relief to meet immediate needs for food and shelter, through the Federal Emergency Management Agency, the

Department of Defense, the Department of Agriculture, and the Department of Housing and Urban Development. There was also some effort at coordination of federal efforts through a federal task force on the homeless, and administrative reforms to make existing services more readily available to the homeless. Although estimates of the amount of assistance provided to the homeless through the means-tested income maintenance and service programs are difficult to obtain, in 1987 only $250 million was specifically targeted for the homeless.[68] In late 1987, Congress passed the McKinney Act, which authorized approximately $442 million in 1987 and $616 million for 1988, for a variety of homeless assistance programs; the actual appropriations, however, came to $365 million for 1987, and $356 for 1988.[69] Also, a housing bill was ratified very late during Reagan's administration, authorizing $15 billion to augment housing and community development programs.[70]

These, and more recent, spending commitments are important but they will not be sufficient to bridge the growing gap between the demand for and supply of affordable housing. The National Association of Housing and Redevelopment Officials estimates that the low-income population will grow by more than five million and that nearly eight million additional low-cost housing units will be needed by the year 2000.[71] In addition, the government will decontrol 1.9 million currently-subsidized units over the next two decades.

Other forces, beyond the purview of the federal government, influence the provision of welfare and social services to the homeless. For instance, very few state and local governments have the financial resources -- or political will -- to address the problem comprehensively. Under Reagan's "new federalism", federal contributions to city budgets dropped by 64 percent (1980-90). Many urban governments have sought to shift responsibility to other tiers of government, through law suits and lobbying efforts.[72]

More and more homeless people are moving to older suburbs and outlying communities. Their visibility is mounting in tolerant liberal communities, racially-mixed suburbs, and lower income inner-ring localities, as well as in more conservative, affluent, single-family housing zones. That visibility, and the fiscal burdens associated with the homeless, have generated a backlash.[73] According to national and local advocacy groups, "1987 marked the beginning of a dangerous trend that places the aesthetic concerns of select groups of business and property owners above the life-or-death needs of the homeless".[74] Even traditionally accepting neighborhoods are starting to squeeze out the homeless. Their efforts include "anti-bum" ordinances, increased enforcement of vagrancy laws, park-watering policies designed to make public parks soggy and uninviting, and exclusion of public and private service agencies viewed as magnets that attract more homeless persons.[75]

Another trend, of vital importance to the mentally disabled homeless, is the growth of a "new asylum movement". This has been backed by many human services professionals who perceive the need to reestablish "comfortable, friendly asylums" for chronically ill homeless people.[76] Advocates of the movement come mainly from among psychiatric and penal workers. Their impetus seems to derive from a feeling of hopelessness in the face of the massive problems of the homeless population, as well as from a somewhat belated recognition of the fact that many chronically ill alcoholics, other substance abusers, and mentally disabled have already been institutionalized in prisons (in a process commonly termed "trans-institutionalization"). For example, one-eighth of California's prison population is classified as "severely mentally ill".[77] If the new asylum movement continues to gain impetus, the threat facing many of the homeless will be *reinstitutionalization*.[78] Reinstitutionalization represents a retreat to incarceration of the mentally disabled. It implies a collapse of the incipient community-based support network and a return of most (if not all) of the mentally disabled to some form of institution-based care, including those never previously institutionalized. Professional support for reinstitutionalization seems to be related to concern for the client, but also with re-establishing territorial claims over traditional client groups, many of whom were lost to other community-based helping professions and the legal system with the onset of deinstitutionalization.

GEOGRAPHICAL INTERVENTIONS

> "The most basic health problem of homeless people is the lack of a home; to condemn someone to homelessness is to visit him or her with a host of other evils. Ignoring the causes of homelessness leads to treating only symptoms and turns medical programs into costly but necessary stopgap measures. Attempts to address the health problems of homeless persons separately from their systemic causes is largely palliative."
>
> -- *Institute of Medicine Minority Report, 1988, p. S-3.*

There is no shortage of ideas about how to address the crisis of homelessness. Indeed, the general prescription for solving homelessness is already well-known; *emergency shelter* now; then *affordable housing*; *jobs* paying enough to rent or buy that housing; adequate *welfare payments* for those who inevitably will remain outside the labor market; and *services* as well as shelter for those whose disabilities prevent them from returning to the mainstream. As pointed out in the introductory remarks, the most important inhibiting factor so far has been the lack of political will.

In designing practical solutions, several important geographical considerations should be borne in mind. At the structural level, there is no way policy-makers can avoid the ramifications of deindustrialization and recession, particularly as these impact on local communities. Neither is it possible to solve homelessness without specific programs designed to address the shortage of affordable housing, which, again, is a highly place-specific deficit. Finally, the restructuring of the welfare state will have to be re-examined for its catastrophic impact on certain population groups in parts of our cities. Instead of pursuing these issues in the concluding remarks, the remainder of the chapter shall, instead, focus on the geographical dimensions of caring for those people who are already homeless.[79] These are principally linked to the fact that the distribution of homeless people and the services designed to assist them have been highly localized in very specific places.

The geographical distribution of homeless people is very uneven. The homeless are overwhelmingly concentrated in cities, mostly in downtown "skid rows" or other inner-city locales where human services and cheap rental accommodations are available. On a more macro-scale, homelessness is a greater problem where deindustrialization has hit hardest, and where deinstitutionalization has proceeded furthest. Geographical concentrations of homeless people also result from the migration of homeless people from service-poor environments, as well as professional referrals to service-rich areas.

The geographical clustering (ghettoization) of service opportunities and homeless people is beneficial and functional for both. Service operators in the "ghetto" can interact more closely with each other by dint of proximity, and thus serve their clients' multiple needs more effectively. Clients benefit from direct physical proximity to a wide range of service opportunities; this aids consumer choice and facilitates access to health and welfare services. Clients' social networks are also promoted by physical proximity. The "ghetto" is a positive asset in that it promotes interaction and mutual support between those in need. The spatial juxtaposition of the clients' informal street networks and the formal networks of service providers permits ready contact between the two. However, specific program outreach is usually needed to actually affect the link between them.

The positive aspects of the "ghetto" of clients and service facilities may be replicated elsewhere in the city to create a network of what we term "service hubs" that will provide the homeless with choice in housing and residential location. Service hubs rely on the spatial juxtaposition of affordable housing, jobs, and service opportunities, as well as access to public transportation. A proliferation of service hubs would also diminish the deleterious effects of ghettoization (e.g., stigmatization, and victimization).

Decentralized service opportunites might also assist in relieving the pressures of neighborhood resistance to human service facilities. Community opposition to the homeless and the services designed to assist them is unevenly distributed throughout most urban areas. Typically, suburban areas tend to close ranks to exclude unwanted populations; inner city locations tend to be more tolerant or accepting, or on the other hand, may not be able to organize political opposition. As services to assist homeless people spread throughout metropolitan jurisdictions, efforts will be needed to emphasize community obligations to care for the homeless as much as community rights to self-determination. Use of the service hub concept would promote a "fair-share" approach to assisting homeless people, and help to overcome the proximity-induced NIMBY (not-in-my-back-yard) syndrome.

Outreach to deliver services to homeless people *in situ* is of cardinal importance in ensuring accessible services. Monolithic hospital-based programs remote from the centers of demand are not likely to be utilized. Needless to say, accessibility is not solely a geographical concept. In 1988, approximately 38 million U.S. inhabitants were without health insurance; the equivalent figure in 1977 was 25 million. The principal factor behind this increase related to changes in the labor market, especially the decline in unionized manufacturing industries with good insurance benefits, and the growth of non-unionized, minimum-wage service industries, which often offer only part-time or casual labor and minimum benefits. Aside from the short-term implications (the inability to access the health care system), the other, longer-term effect of this situation is to leave unattended many minor ailments that are likely to surface in the future as chronic conditions.

In terms of geographical access, the factors limiting availability of health care to the homeless include:

(a) *On the supply side*

1. geographical maldistribution of hospital and other service facilities;
2. lack of transportation to those facilities that are available;
3. a chronic undersupply of certain services (for the mentally disabled and homeless veterans in particular);
4. variations in availability and eligibility of MEDICAID (the health insurance scheme for low income individuals); and
5. intrusive bureaucracies which inhibit (often deliberately) the availability of services.

(b) *On the demand side*

6. the complex daily lives of many homeless persons, who spend much of each day waiting in line, and who have to be in different places at different times (for food, shelter, work, counselling, etc.);
7. the difficulties of comprehensively addressing the multiple diagnoses of many homeless people;
8. the acute social disaffiliation experienced by many homeless people;
9. the distrust of, and passive resistance offered to health and welfare services because of negative prior experiences (this seems particularly true of ex-patients of psychiatric services, veterans, and women); and
10. cultural differences between consumers and providers (usually related to race, ethnicity or language).

CONCLUSION

The crisis of homelessness has complex origins. At least in the United States, its roots can be traced back more than two decades. A crisis that was 20 years in the making is hardly likely to disappear overnight. Moreover, intervention in any single sector (housing, health care, etc.), or at any one "level" of the problem (structures, institutions, agency), is unlikely to eliminate homelessness. This is the prinicipal message from this analysis of homelessness. It should be clear that multiple solutions are required, operating at many different levels and sectors. There is no "silver bullet" for homelessness.

Only a brave and/or foolish person would predict that homelessness will not be with us in the year 2001. Recession and war have placed great stresses on resources, and society's ability/willingness to respond to the needy. Crises in the U.S. financial system (especially the Savings & Loan industry's collapse and the costs of the associated bail-out) not only erode trust and confidence, but also deleteriously affect the everyday lives of most individuals. Complaints over the ethical standards of elected representatives (including notorious scandals at the U.S. Department of Housing and Urban Development) further undermine faith in effective public action. In the meantime, social divisiveness increases in a country that is becoming re-polarized into a society of haves and have-nots.

ACKNOWLEDGEMENT

This research is supported by a grant from the U.S. National Science Foundation.

REFERENCES

[1] Gelberg, L., Linn, L.S., Usatine, R.P. and Smith, M.H. 1990. Health, Homelessness, and Poverty, *Archives of Internal Medicine* 150 (November), pp. 2325-2330.

[2] For a review, see Shinn, M., Burke, P.D. and Bedford, S. (eds.) 1990. *Homelessness: Abstracts of the Psychological and Behavioral Literature, 1967-1990.* Washington, D.C.: American Psychological Association.

[3] Rossi, P.H. 1989. *Down and Out in America: The Origins of Homelessness.* Chicago: University of Chicago Press. The prevalence of physical health problems among the homeless also depends on the sampling technique used; see, for example, Gelberg, L. and Linn, L.S. 1989. Assessing the Physical Health of Homeless Adults, *The Journal of the American Medical Association* 262(14), pp. 1973-1979.

[4] Wright, J.D. 1989. *Address Unknown: The Homeless in America.* New York: Aldine de Gruyter. Wright's data come from the National Health Care for the Homeless (HCH) Program which was funded by the Robert Wood Johnson Foundation.

[5] Gelberg, L., Linn, L.S. and Mayer-Oakes, S.A. 1990. Differences in Health Status Between Older and Younger Homeless Adults, *Journal of the American Geriatric Society* 38(11), pp. 1220-1229.

[6] Robertson, M.J. 1987. Homeless Veterans: An Emerging Problem?, in Bingham, R.D., Green, R.E. and White, S.B. (eds.) *The Homeless in Contemporary Society.* Newbury Park, California: Sage Publications, pp. 64-81; See Dear, M. and Takahashi, L. "Homeless Veterans in Los Angeles." Working Paper #31, Los Angeles Homelessness Project, Department of Geography, University of Southern California, December 1990, for review of literature on characteristics and problems of homeless veterans.

[7] See Wilson, J.C. and Kouzi, A.C. 1990. A Social-Psychiatric Perspective on Homelessness: Results from a Pittsburgh Study, in Momeni, J. (ed.), *Homelessness in the United States.* New York: Praeger, pp. 95-110, for brief review of the literature; also U.S. Department of Housing and Urban Development, Office of Policy Development and Research, "A Report on the 1988 National Survey of Shelters for the Homeless". Washington, D.C.: U.S. Department of Housing and Urban Development, March 1989. See Morrissey, J.P. and Dennis, D.L. (eds.) "Homelessness and Mental Illness: Toward the Next Generation of Research Studies". Proceedings of a NIMH-Sponsored Conference, Bethesda, Maryland, January, 1990 for shortcomings of existing research and suggestions for addressing these shortcomings; Koegel, P., Burnam, M.A. and Farr, R.K. 1988. The Prevalence of Specific Psychiatric Disorders Among Homeless Individuals in the Inner City of Los Angeles, *Archives of Gen. Psychiatry* 45 (December), pp. 1085-1092.

[8] Vernez, G. et al. 1988. Review of California's Program for the Homeless Mentally Disabled. Prepared for the California Department of Mental Health. Santa Monica, California: The RAND Corporation, February, 1988.

[9] U.S. Department of Housing and Urban Development, 1989, *op. cit.*

[10] Lamb, H.R. and Lamb, D.M. 1990. Factors Contributing to Homelessness Among the Chronically and Severely Mentally Ill, *Hospital and Community Psychiatry* 41(3), pp. 301-305.

[11] Stefl, M.E. 1987. The New Homeless: A National Perspective, in Bingham, R.D., Green, R.E. and White, S.B. (eds.) *op. cit.*, pp. 46-63.

[12] Rossi, *op. cit.*

[13] *Ibid.*

[14] Koegel, P., Burnam, M.A. and Farr, R.K., 1988, *op. cit.*

[15] Wright, *op. cit.*; Vernez et al., *op. cit.*

[16] Rossi, *op. cit.*

[17] Stefl, *op. cit.*; also U.S. Department of Housing and Urban Development, 1989, *op. cit.*

[18] Wright, *op. cit.*

[19] Wright, J.D. 1987. The National Health Care for the Homeless Program, in Bingham, R.D., Green, R.E. and White, S.B. (eds.), *The Homeless in Contemporary Society*. Newbury Park, California: Sage Publications, pp. 150-169.

[20] Wright, *op. cit.*

[21] Milburn, N.G. 1990. Drug Abuse Among Homeless People, in Momeni, J. (ed.), *Homelessness in the United States*. New York: Praeger, pp. 60-80; U.S. Department of Housing and Urban Development, 1989, reports an estimate of 25 percent of sheltered homeless having abused drugs.

[22] Rossi, 1989, *op. cit.* and Milburn, *op. cit.*

[23] Vernez et al., 1988, *op. cit.*

[24] Wright, 1989, *op. cit.*

[25] Vernez et al., 1988, *op. cit.*

[26] Linn, L.S., Gelberg, L. and Leake, B. 1990. Substance Abuse and Mental Health Status of Homeless and Domiciled Low-Income Users of a Medical Clinic, *Hospital and Community Psychology* 41(3), pp. 306-310.

[27] Dear, M.J. and Wolch, J.R. 1987. *Landscapes of Despair: From Deinstitutionalization to Homelessness*. Princeton, New Jersey: Princeton University Press; Lerman, P. 1982. *Deinstitutionalization and the Welfare State*. New Brunswick, New Jersey: Rutgers University Press.

[28] Wolpert, J., Dear, M. and Crawford, R. 1975. Satellite Mental Health Facilities, *Annals of the Association of American Geographers*, Vol. 65, pp. 24-35; Dear, M.J. 1977. Psychiatric Patients in the Inner City, *Annals of the Association of American Geographers* 67, pp. 588-594; Wolch, J. 1980. Residential Location of the Service-Dependent Poor, *Annals of the Association of American Geographers* 70, pp. 330-341.

[29] Lamb, R. 1984. *The Homeless Mentally Ill*. Washington, D.C.: American Psychiatric Association; Torrey, E.F. 1988. *Nowhere to Go: The Tragic Odyssey of the Homeless Mentally Ill*. New York: Harper & Row.

[30] Dear and Wolch, 1987, *op. cit.*; Wolch, J.R. and Akita, A. 1988. The Federal Response to Homelessness and Its Implications for American Cities, *Urban Geography* 10(1), pp. 62-85.

[31] Bluestone, B. and Harrison, B. 1982. *The Deindustrialization of America*. New York: Basic Books.

[32] Danzinger, S. and Feaster, D. 1985. Income Transfers and Poverty in the 1980s, in Quigley, J.M. and Rubinfeld, D.L. (eds.), *American Domestic Priorities*. Berkeley, Calif.: University of California Press, pp. 89-117.

[33] *Newsweek*, "Homeless in America", 2 January 1984, CIII, pp. 20-29. See also: City of Chicago, Department of Planning, *Housing Needs of Chicago's Single, Low-Income Renters*. Chicago: City of Chicago Department of Planning, 1985; Baxter, E. and Hopper, K. 1982. The New Mendicancy: Homeless in New York City, *American Journal of Orthopsychiatry* 52, pp. 392-408.

[34] Herbers, J., "Outlook for Sheltering the Poor Growing Even Bleaker," *New York Times*, 8 March 1987.

[35] Sternlieb, G. et al. 1980. *America's Housing*. New Burnswick, New Jersey: Center for Urban Policy Research; Dowall, D. 1985. *The Suburban Squeeze*. Berkeley, California: University of California Press; Wolch, J. and Gabriel, S. 1985. Dismantling the Community-Based Human Services System, *Journal of the American Planning Association* 51, pp. 24-35; Palmer, J. and Sawhill, I. 1984. *The Reagan Record: An Assessment of America's Changing Doestic Priorities*. Washington, D.C.: The Urban Institute; Wolpert, J. and Seley, J. 1986. Urban Neighborhoods as a National Resource: Irreversible Decisions and Their Equity Spillovers, *Geographical Analysis* 18, pp. 81-93.

[36] Soja, E., Morales, R. and Wolff, G. 1983. Urban Restructuring: An Analysis of Social and Spatial Change in Los Angeles, *Economic Geography* 59, pp. 195-230.

[37] Lipton, G. 1977. Evidence of Central City Revival, *Journal of the American Institute of Planners* 43, pp. 136-147; Noyelle, T. 1983. The Rise of Advanced Services, *Journal of the American Planning Association* 49, pp. 280-290.

[38] Sassen-Koob, S. 1984. The New Labor Demand in Global Cities, in Smith, M.P. (ed.), *Cities in Transformation*. Beverly Hills, California: Sage Publications, pp. 139-171.

[39] Fustero, S. 1984. Home on the Street, *Psychology Today* 18 (February), pp. 56-63; City of Chicago, 1985, *op. cit.*

[40] Holden, C. 1986. Homelessness: Experts Differ on Root Causes, *Science* 232 (2 May), pp. 569-570.

[41] Dear, M.J. and Taylor, S.M. 1982. *Not On Our Street*. London: Pion; Wolch and Gabriel, 1987, *op. cit.*

[42] Dear and Wolch, 1987, *op. cit.*

[43] Brown, H.J. and Apgar, W. 1988. *The State of the Nation's Housing*. Cambridge, Massachusetts: Harvard Joint Center for Housing Studies.

[44] McChesney, K. 1986. New Findings on Homeless Families, Working Paper, Social Science Research Institute, University of Southern California. Los Angeles: University of Southern California; Sullivan, P. and

Damrosch, S. 1987. Homeless Women and Children, in Bingham, R., Green, R. and White, S. (eds.), *The Homeless in Contemporary Society*. Newbury Park, California: Sage Publications, pp. 82-98; Farr, R., Koegel, P. and Burnam, A. 1986. A Study of Homelessness and Mental Illness in the Skid Row Area of Los Angeles. Los Angeles: County of Los Angeles Department of Mental Health; Baxter and Hopper, 1982, *op. cit.*

[45] See for example, Hope, M. and Young, J. 1986. *The Faces of Homelessness*. Lexington, MA: Lexington Books.

[46] Baxter and Hopper, 1982, *op. cit.*; *New York Times*, "Fear of Shelters Sends Many to Street", 5 March 1989.

[47] Grunberg, J. and Eagle, P.F. 1990. Shelterization: How the Homeless Adapt to Shelter Living, *Hospital and Community Psychiatry* 41(5), pp. 521-525.

[48] Coleman, J. 1986. Diary of a Homeless Man, in Erickson, J. and Wilhelm, C. (eds.), *Housing the Homeless*. New Brunswick, New Jersey: Center for Urban Policy Research, pp. 37-53.

[49] Koegel, P. "Ethnographic Perspectives on Homeless and Homeless Mentally Ill Women". Proceedings of a two-day workshop sponsored by the Division of Education and Service Systems Liaison National Institute of Mental Health. Washington, D.C.: U.S. Department of Health and Human Services, October 30-31, 1986.

[50] Redburn, F.S. and Buss, T. 1986. *Responding to America's Homeless*. New York: Praeger.

[51] D'Ercole, A. and Struening, E. 1990. Victimization Among Homeless Women: Implications for Service Delivery, *Journal of Community Psychology* 18(April), pp. 141-152.

[52] Vernez et al., 1988, *op. cit.*

[53] Dear and Wolch, 1987, *op. cit.*

[54] City of Chicago, 1985, *op. cit.*, pp. 33-55; Farr, Koegel and Burnam, 1986, *op. cit.*

[55] Dear and Wolch, 1987, *op. cit.*

[56] Rousseau, A.M. 1981. *Shopping Bag Ladies*. New York: Pilgrim Press.

[57] Baxter and Hopper, 1982, *op. cit.*

[58] Erickson, J. and Wilhelm, C. (eds.) 1986. *Housing the Homeless*. New Brunswick, New Jersey: Center for Urban Policy Research; Hope and Young, 1986, *op. cit.*

[59] See Rowe, S. and Wolch, J. 1990. Social Networks in Time and Space: Homeless Women in Skid Row, Los Angeles, *Annals of the Association of American Geographers* 80(2), pp. 184-204; See Levine, I.S., Lezak, A.D.

and Goldman, H.H. 1986. Community Support Systems for the Homeless Mentally Ill, in Bassuk, E.L. (ed)., *The Mental Health Needs of Homeless Persons*. San Francisco: Jossey-Bass, pp. 27-42, for a discussion of the federal Community Support Program, a demonstration program to set up local support networks.

[60] City of Chicago, 1985, *op. cit.*

[61] Dear and Wolch, 1987, *op. cit.*

[62] Roderick, K., "Homeless - Left Behind by Recovery", *Los Angeles Times*, 2 February 1985. For a good overview of the way global economic trends are influencing development in individual countries, see Henderson, J. and Castells, M. (eds.) 1987. *Global Restructuring and Territorial Development*. London: Sage Publications.

[63] Storper, M. and Scott, A.J. 1990. The Geographical Foundations and Social Regulation of Flexible Production Complexes, in Wolch, J. and Dear, M. (eds.), *The Power of Geography: How Territory Shapes Social Life*. London: Allen & Unwin, pp. 21-40.

[64] Bluestone, B. and Harrison, B., "The Grim Truth About the Job 'Miracle'", *New York Times*, 1 February 1987; Harrison, B. and Bluestone, B. 1988. *The Great U-Turn: Corporate Restructuring and the Polarization of America*. New York: Basic Books.

[65] Wolch and Akita, 1988, *op. cit.*

[66] Dear, M., "Our 'Third World' of Housing Have-nots Needs Action", *Los Angeles Times*, 8 February 1988.

[67] Wolch and Akita, 1988, *op. cit.*

[68] *Ibid.*

[69] *Ibid.*

[70] *Safety Network*, "McKinney Funds Slashed; Homeless Lose $250 Million", January 1988a; *Safety Network*, "Housing Bill Passed, Funds Community Programs", January 1988b.

[71] National Association of Housing and Redevelopment Officials. 1987. Keeping the Commitment: An Action Plan for Better Housing and Communities for All. Washington, D.C.: National Association of Housing and Redevelopment Officials.

[72] Dear and Wolch, 1987, *op. cit.*

[73] Muir, F., "Homeless Plight Made Worse by Growing Backlash, Advocates Report", *Los Angeles Times*, 22 December 1987.

[74] See, for example, National Coalition for the Homeless. 1987. Less than Zero: Backlash against Homeless People and the Programs that Serve Them. Washington, D.C.: National Coalition for the Homeless; and

National Campaign to End Hunger and Homelessness in America. 1986. A Survey of Attitudes Toward Hunger and Homelessness in America. Washington, D.C.: National Campaign to End Hunger and Homelessness in America.

[75] Winerip, M., "Applause of Help Homeless", *New York Times*, 26 February 1988.

[76] Bassuk, 1984, *op. cit.*

[77] Dear and Wolch, 1987, *op. cit.*

[78] Wolch, J., Nelson, C. and Rubalcaba, A. 1991. Back to Back Wards? Prospects for the Reinstitutionalization of the Mentally Disabled, in Smith, C. and Giggs, J. (eds.), *Location and Stigma*. London: Allen and Unwin, pp. 264-284.

[79] More general considerations of health care for the homeless are to be found in Brickner, P.W., Scharer, L.K., Conanan, J.A., Savarese, M. and Scanlan, B.C. 1990. *Under the Safety Net: The Health and Social Welfare of the Homeless in the United States*. New York: Norton.

THE RHETORIC OF HEALTH PROMOTION AND THE REALITY OF VANCOUVER'S DOWNTOWN EASTSIDE: Breeding Cynicism

Michael Hayes

Department of Geography
Simon Fraser University

INTRODUCTION

Achieving Health for All: A Framework for Health Promotion,[1] released in November, 1986 by then Minister of Health and Welfare, the Honorable Mr. Jake Epp, was meant to signal a new era in social policy.[2] The heart of this new policy direction was health promotion, defined as "the process of enabling people to increase control over, and to improve, their health".[3]

In September, 1987, community leaders in Vancouver's Downtown Eastside neighborhood (DES) organized a workshop to help local residents become familiar with the principles of health promotion and strategies contained in the Epp Report (as *Achieving Health for All* is commonly called). The DES is among Canada's poorest inner city neighborhoods, and home to the type of population identified in the Epp Report as deserving of special attention to reduce the effects of material deprivation upon health status. The workshop was subsequently cited in *Health Promotion*,[4] a journal published by Health and Welfare Canada to popularize the concept, as an example of how public participation in health promotion can be achieved.

According to the Epp Report, health promotion "recognizes freedom of choice and emphasizes the role of individuals and communities in defining what health means to them".[5] The DES residents who participated in the workshop did just this, and identified the following concerns in their community:[6]

- insufficient services to people who "fall through the cracks" (for example, native people, deinstitutionalized individuals, street kids, the isolated, women with children, and people aged 35 to 55);
- inadequate planning, consultation and coordination of services;
- lack of recognition or support for the existing self-help system;
- a shortage of quality housing;
- inadequate treatment programs for the addicted; and
- the need for purposeful activity.

DES residents are still waiting for their concerns to be addressed. Despite the health promotion rhetoric about individual "empowerment", public participation, social justice and equity, healthy environments, healthy public policy and reducing inequalities in health, people in the neighborhood continue to fall through the cracks of a fragmented (and shrinking) social welfare "safety net". There is no evidence of any tangible improvement in living conditions of area residents, and little reason to hope for improvements in the near future. The promise of health promotion rings hollow for residents of the DES.

The purpose of this chapter is to examine the rhetoric of health promotion, to distill from the rhetoric its apparent policy implications, and to identify the reasons why health promotion, as a policy thrust, has failed to serve such populations as Vancouver's Downtown Eastside. The chapter begins by presenting the apparent aims of health promotion. This is followed by a description of the conditions in the DES. The chapter concludes with a critical evaluation of the health promotion rhetoric, identifying reasons why the potential it implies for the DES has not been realized.

HEALTH PROMOTION AND ITS POLICY IMPLICATIONS

In Canada, "health promotion" has gained currency as both a social policy goal and an organizing framework for social services. In the early 1980s Health and Welfare Canada established the Health Promotion Directorate, and since then has played a major role in nurturing the ideology of health promotion. Its most significant contributions in this regard are (i) *Achieving Health for All: A Framework for Health Promotion* (the Epp Report), the statement of official federal policy, and (ii) its co-sponsorship in November, 1986, of the international conference that produced the Ottawa Charter for Health Promotion,[7] the touchstone of the principles and objectives of health promotion unanimously endorsed by participants representing 44 countries.[8] It was at that conference that *Achieving Health for All* was officially released.

Health promotion as conceived in the Epp Report, the Ottawa Charter, publications sponsored by the WHO,[9] and related literature[10] is meant to represent an important shift in public policy conceptions of health and its determinants, away from biomedical orientations concerned with curing sickness and the biological basis of disease toward a much more holistic view of health. According to the Ottawa Charter:

> Health promotion is the process of enabling people to increase control over, and to improve, their health. To reach a state of complete physical, mental and social well-being, an individual or group must be able to identify and to realize aspirations, to satisfy needs, and to change or cope with the environment. Health is, therefore, seen as a resource for everyday life, not the objective of living. Health is a positive concept emphasizing social and personal resources, as well as physical capacities. Therefore, health promotion is not just the responsibility of the health sector, but goes beyond healthy life-styles to well-being.
>
> The fundamental conditions and resources for health are peace, shelter, education, food, income, a stable eco-system, sustainable resources, social justice and equity. Improvement in health requires a secure foundation in these basic prerequisites.[11]

Further, health promotion is meant to focus on achieving equity in health status between groups and ensuring equal opportunities and resources to enable all people to achieve their fullest potential. The prescription for action is contained in three words: advocate, enable and mediate. To ensure that individuals obtain access to the basic prerequisites for health, the Charter calls upon health professionals and social groups to act as advocates for the socially disadvantaged and as mediators in conflicts between interest groups. The purpose of health promotion activity is to enable people to achieve their fullest potential.

Five specific types of action are called for in the Charter. The first is to build healthy public policy. All levels of the state (federal, provincial and municipal institutions) are called upon to review all policies that have a bearing upon health regardless of the portfolios of specific ministries or agencies. Healthy public policy requires cooperation between various ministries or agencies within the same level of state and between levels to ensure that policies and practices do not have an adverse effect upon individual well-being. Given the all encompassing nature of the determinants of health, building healthy public policy implies that all public policies must be reviewed to establish their implications for the health of individual members of society.

The second type of action identified in the Charter is to create supportive environments. In effect, this means creating conditions, settings and relationships that foster mutual aid and mutual support. The third is to strengthen community action by empowering communities to take control of their own endeavors and destinies. Developing personal skills through providing information and education about health and its determinants, and by improving the life skills of individuals, is the fourth type of action. Finally, the Charter calls for the reorientation of health services away from curative medicine toward health education and toward the recognition of the roles that other forms of social welfare play in maintaining and securing individual well-being.

The Epp Report echoes many of the sentiments contained in the Ottawa Charter. In the words of the Report:

> Today, we are working with a concept which portrays health as a part of everyday living, an essential dimension of the quality of our lives. "Quality of life" in this context implies the opportunity to make choices and gain satisfaction from living. Health is thus envisaged as a resource which gives people the ability to manage and even to change their surroundings...It is a basic and dynamic force in our daily lives, influenced by our circumstances, our beliefs, our culture and our social, economic and physical environments.[12]

The framework for health promotion proposed by Mr. Epp (Figure 1,9) contains the same basic themes as those identified in the Charter presented in a somewhat different format. The Epp Report identifies three "health challenges" for social policy. The first is to reduce inequities in health status between socioeconomic groups. The second is to increase prevention of disease by increasing individual awareness and changing

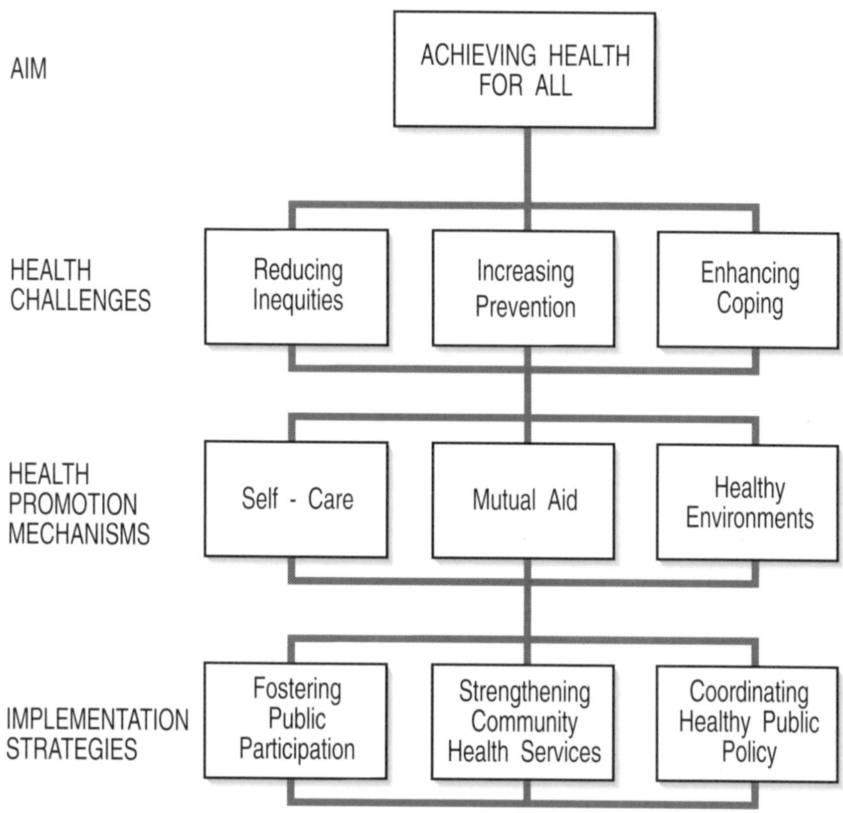

Figure 1,9 The Epp Report Framework for Health Promotion

behaviors that are detrimental to health, such as smoking, certain dietary practices, or driving while impaired. The third challenge is to enhance the individual's ability to cope with chronic conditions, disabilities and mental health problems.

Health promotion is the vehicle for addressing these challenges, and three mechanisms for increasing the uptake of health promotion activities are proposed: improving individual self-care, increasing mutual aid (social support networks), and creating healthy environments (i.e., conditions and surroundings conducive to health). To achieve these ends, three implementation strategies are described: fostering public participation, strengthening community health services, and coordinating healthy public policy.

The concept of "health promotion" described in the two documents discussed thus far is fundamentally different from the one outlined in the *American Journal of Health Promotion.* Here, health promotion is defined as "the science and art of helping people to change their lifestyle to move toward a state of optimal health".[13] This definition contains no reference to social justice or equity, or to social determinants of health. Responsibility for "lifestyle" is placed squarely upon the shoulders of the individual.

It is important to note that, by contrast, health promotion in the Epp Report and the Ottawa Charter recognizes broader social determinants of health that are beyond the control of the individual and beyond the capacity of the health care system to address:

> ... we cannot invite people to assume responsibility for their health and then turn around and fault them for illnesses and disabilities which are the outcome of wider social and economic circumstances. Such a 'blaming the victim' attitude is based on the unrealistic notion that the individual has ultimate and complete control over life and death.[14]

The Epp Report implicitly and explicitly acknowledge's the state's responsibility for ameliorating the negative health effects of these social determinants. The notion of healthy public policy carries with it explicit reference to the state's role in ensuring the well-being of members of society. Expressed concerns with issues of "social justice", "equity", and with reducing "inequities in the health of low- versus high-income groups", implicates the state because of its central role in the redistribution of public resources. The Epp Report identifies specific low-income groups with particularly poor health status: older people, the unemployed, welfare recipients, single women supporting children, and minorities such as natives and immigrants - persons who are commonly dependent upon the state for their daily maintenance. The assertion is that "social justice" in social policy would give special consideration to these groups.

Casting health as "a resource for daily living" or a "dynamic force in our daily lives" indicates a concern with the maintenance of individual well-being in every day living. This implies a redirection of resources from health care to social programs that improve well-being in daily life. Given the recognition of the influence of the environment upon health, and identification of shelter, income, education, etc. as prerequisites of health (or, in the Epp Report, as being fundamental dimensions of well-being), it is reasonable to assert that health promotion entails a greater policy commitment to housing, income maintenance and education.

The stated objective of health promotion, to help individuals manage, control or change those circumstances of their daily lives that threaten their physical well-being or ability to "realize aspirations", "make choices" or "gain satisfaction from living", implies that the purpose of health promotion is to do more than maintain the status quo. Persons who are not satisfied with their life situations, or those whose aspirations have yet to be realized, have not achieved their full health potential. They are to be encouraged and supported in their efforts to seek change. In this regard, the suggested role of social policy is to help to create opportunities for relatively disadvantaged persons to change their situations. There is also the suggestion that it is the individual to be empowered who will make the relevant choices, and that those choices will be respected and accepted by others.

These themes indicate the major policy implications of the health promotion rhetoric. They suggest a strong commitment on the part of the institutions and individuals who ascribe to the principles of health promotion to social policies that reduce disparities. Certainly, health promotion sets an ambitious agenda for social policy and social service provision in Canada. In the words of the Epp Report:

> The challenge here is...to integrate our thinking about health care and health promotion to a greater extent than we have done before. It is like drawing the appropriate linkages between economic and social policy spheres...the kind of thinking that produces healthy public policy.[15]

REALITIES IN THE DES: POPULATION, HEALTH AND SOCIAL CHARACTERISTICS

It is easy to understand why the Epp Report generated such enthusiasm among residents and service providers in Vancouver's Downtown Eastside neighborhood. The personal characteristics of area residents and elements of the physical, social and environmental conditions that obtain in the DES are typical of the low-income populations identified in the Epp Report as having inequitable access to health, a suspicion which is supported by the limited health status information that is available for this population. Also, there was (and still is) a strong desire to create a "healthy community" on the part of the majority of DES residents, many of whom have lived in the area for more than a decade. Moreover, the types of interventions sought by local residents (see introduction) are completely consistent with the ideology of health promotion.

Available health status measures for the DES illustrate the comparatively poor health status of this population. Tollefson[16] found the incidence of tuberculosis (76/100,000) in the DES to be 2.4 times higher than the Vancouver rate (31/100,000) and almost 7 times higher than the provincial rate (11/100,000). She reported that the average team caseload for the area mental health service (40/1,000) was almost 6 times greater than the Vancouver average (12-16/1,000). She also reported that the homicide rate for males in the DES (225/100,000) was about 9 times higher than the Vancouver rate (25/100,000), while for females the rate (35/100,000) was about 6 times greater than that of Vancouver (6/100,000). The risk of being a victim of crime in the DES is almost 5 times greater than in the rest of the city.

In a study of mortality disparities by census tract income quintiles for Vancouver and Victoria, Thomson[17] found a six year difference in life expectancy for all males in the highest and lowest quintiles, and an eight year difference when the analysis was restricted to males born in Canada. About half of the DES population was born in Canada. In recent housing surveys of persons living alone in single room occupancy (SRO) units and socially assisted housing conducted by the Downtown Eastside Resident's Association,[18] half of all respondents reported some physical disability. Serious disabilities were reported by about 15 percent of respondents.

These health status measures reflect the economic and social realities reported in the two DERA surveys. The average annual income of those surveyed in 1987 was about $5,300, less than half the estimated poverty line for Vancouver.[19] The annual income for about 90 percent of respondents in the 1990 survey fell below the National Council of Welfare (NCW) poverty line for single person households, and was only half the poverty line for about half the survey sample. About 90 percent of respondents in both surveys were unemployed and had not worked for some time. Their primary source of income was some form of maintenance program from the state (Guaranteed Annual Income (GAIN - welfare) 67.7 percent, Canada Pension Plan 11.5 percent, Handicapped Persons Income Assistance 6.2 percent, Old Age Pension 3.1 percent, unemployment insurance 1.0 percent of respondents in the 1990 survey). As of September, 1989, the monthly maximum GAIN allowance for employable persons was $500, for single parent families and single persons aged 60 to 64 years it was $550, and for handicapped people and persons over age 65 it was $694. The NCW poverty line in 1989 for a single person in an urban centre with 500,000 or more persons was $1,003.

Households that spend in excess of 30 percent of the household income on rent and are not able to obtain cheaper housing of comparable size elsewhere within the urban area are identified as being in "core need" of housing by the Canada Mortgage and Housing Corporation.[20] The average monthly rent for an SRO unit is set at or near the maximum GAIN shelter allowance of $300. This means that persons on GAIN spend between 45 and 60 percent of their incomes on rent.

Although rents in the DES are about $2.75 per square foot[21] - the highest per square foot in Vancouver - accommodation in the SROs is typically poor. About half of the lodgings have no cooking facilities, and most of those with facilities have a simple hot plate. Most units do not have private toilets, showers or baths. Residents have limited access to telephones, and few or no amenities in their rooms. Many SROs violate municipal building codes, and some owners defy orders to rectify violations. There is little security of tenure for SRO residents. Eviction and rent increases occur with 24 hours notice, despite recent changes in municipal bylaws governing these units which make this illegal. Many SROs do not allow guests in the rooms, or levy a charge for visitors.

Despite these poor conditions, there is considerable demand for housing in the DES. Between 1985 and 1990 the vacancy rate has plummeted from about 16 percent to under 4 percent in the SROs, and to under 2 percent overall in the DES.[22] This decrease reflects, in part, a shrinking housing supply in the area. Between 1985 and 1989, an estimated 1,150 SRO units (10 percent of all SRO units) were demolished, closed by the city or lost through gentrification and conversion.[23] Another 17 hotels (983 units) are known to have applied for or received development permits.[24] It also reflects an increased demand for low rent housing as a consequence of increasing rents and housing prices in Vancouver. Service providers in the area have noticed an increase in the number of single parents and families moving into the DES. Increasing numbers of persons deinstitutionalized from psychiatric and/or penal institutions are also competing for housing in the DES. The Strathcona Mental Health Team, responsible for servicing this population in the DES, estimates that it now takes about 16 weeks to obtain a boarding home placement in the area. In 1988 it took about three weeks to find a placement.[25]

Not everyone in search of accommodation in the DES can find it. As of December, 1989, there were 43 persons (all male) known to be without accommodation by the Vancouver Urban Core Workers Shelterless Committee.[26] Over the past three years the estimated number has fluctuated between 30 and 80 persons, and typically includes a small number of females. These estimates exclude persons sleeping in emergency shelters and persons staying with friends. The emergency shelters provide dormitory accommodation and persons who make use of them often spend nights sleeping out. A total of 199 beds are available in four facilities (Catholic Charities 80, Dunsmuir House 30, Lookout 63, and Triage 26). Member agencies of the Shelterless Committee providing emergency shelter reported a constant heavy demand in 1989, and Lookout, Triage and Dunsmuir House all turned people away in November and December. Lookout Emergency Shelter reports that 5 years ago 80 percent of its users were DES residents. Now only 45 percent are from the neighborhood.[27] The shelter suggests that the increased demand from out-of-area persons is tied to downsizing of Riverview Psychiatric Hospital in Coquitlam and Tranquille in Kamloops.

Recent increases in demand for affordable accommodation are reflected in the increased number of persons seeking assisted housing. The number of persons on DERA waiting lists for assisted housing almost doubled between 1987 and 1990 and the number of families increased by 80 percent.[28] According to DERA:

> DERA housing has amassed a wait list of 2772 "active" applicants of which 319 are families and 1286 seniors. Of these, 50 families and 448 seniors have listed single room occupancy residential hotels as their principal residence. These families and seniors, along with the "traditional" Downtown Eastside resident, are competing for a shrinking supply of affordable housing with an influx of deinstitutionalized mental patients and others from outside the area, who are flooding into the Downtown Eastside which is seen as the last refuge for affordable housing in Vancouver. The result of this competition will mean the ultimate displacement of many residents and the destabilization of the community.[29]

Despite the quality of housing many residents wish to remain in the DES. The 1987 DERA survey found that the average length of residence among respondents was greater than 10 years. About 43 percent of respondents in the 1990 survey had lived in the DES for 11 or more years, and a further 21.5 percent had lived there between 6 and 10 years. The locational advantages of living in the DES include proximity to shops and social services, friends and support networks, as well as cheaper rents. For a significant proportion of the population the SRO unit is "home", and the DES is their community. Thus, destabilization of the DES is deeply threatening to area residents, many of whom have already experienced dislocation from their homes during B.C.'s Expo '86. The proposed development of the Expo lands, to take place over the next 15 years, poses yet another threat to the social fabric of the DES.

In light of the preceding, it is easy to understand why residents of the DES would readily embrace the notion of health promotion and be such willing participants. Although it appears to offer considerable hope for change in the living conditions and health status of residents of the DES, health promotion as a policy initiative has not delivered. In the following section of this paper reasons for this failure are explored.

WHY HEALTH PROMOTION FAILS THE RESIDENTS OF THE DES

The most obvious failure of health promotion in general is that the spirit of its rhetoric has not been translated into action - the political will required to give meaning to notions of social justice and equity in communities like the DES simply does not exist.

Although the vision of health promotion contained in both the Epp Report and the Ottawa Charter may readily be interpreted as a call for expansion of the welfare state, the federal government obviously has no intention of increasing social welfare expenditures. Its agenda is just the opposite: to increase privatization and reduce the state's commitment to social programs.[30] Funding for social housing has been systematically reduced over the past decade, despite that in most urban areas of Canada shortages of affordable housing have reached crisis proportions.[31] Reduction in federal spending has not been fully offset by increases in spending at other levels of government, so there has been an overall reduction in the number of new socially assisted housing units being built. The consequences of reduced expenditure on social housing will have considerable impact in the DES. When Expo '86 was first proposed, 20 percent of the residential units to be constructed on the site after Expo were to be socially assisted units built by the provincial government. Now that the Expo lands have been sold to a private developer and construction of private market units has begun, there is great concern that the social housing component of this enormous redevelopment will not be realized. The fear is that if the space allocated for social housing cannot be developed at the same time as the private market development, the residents of that new private development will oppose the social housing when (and if) it is built. Moreover, the price of the private market accommodation may further increase rents for existing units throughout the area, driving long-time residents out of the neighborhood.

The significant proportion of DES residents identified above as dependent on state programs have incomes that are about half the poverty line for a city the size of Vancouver. Clearly the opportunity for these individuals to make choices is severely constrained by their state-controlled incomes, and choices become more and more limited as levels of social support are diminished by policy decisions. The value of family benefits has decreased by about three percent per year since the mid-1980s. Benefits paid through unemployment insurance have been reduced, and in 1989 the provincial government actually reduced welfare payments by $50 per month to provide an incentive for individuals to "get off welfare". (This decision was later reversed as a result of public outrage.) But perhaps the height of hypocrisy in light of the Epp Report came in February 1991, when the federal government passed the Transfer Payments Expenditures Act (Bill C69) cutting transfer payments to the provinces for health care and post-secondary education. The passing of this act flies in the face of the ideology of health promotion.

A second failing of health promotion is that it has little or no legitimacy as a social policy thrust, despite claims that health promotion is meant to signal a new era of social policy. The Ottawa Charter has no "official" policy status, and the influence of the Epp Report is limited to the Ministry of National Health and Welfare. The primary responsibility for social welfare policy rests with provincial governments, although municipal governments also may play important roles in providing social assistance. Furthermore, most of the federal programs that are implicated by health promotion - housing, employment and unemployment insurance, the quality of food, air and water, etc. - are outside the purview of Health and Welfare. Thus, while the Epp Report claims an intention to coordinate healthy public policy, its ability to do so is extremely limited.

Few of the concerns identified above by DES residents could, in fact, be met directly through Health and Welfare even if that ministry was truly committed to the vision of health promotion contained in its rhetoric. Most services required by persons identified as "falling through the cracks" (native people, deinstitutionalized individuals, street kids, women with children, and persons aged 35 to 55) would be provided by the provincial government, although Health and Welfare could supply financial support to the province to run specific programs, and could conceivably support directly mental health services and health services for Native Canadians and veterans. Improvement in the adequacy of planning, consultation and coordination of services, or in provision of programs for the addicted, support for self-help systems, or creating opportunities for purposeful activity are also primarily provincial (or municipal) concerns. The shortage of decent, affordable housing, perhaps the most pressing problem in the DES, could be eliminated by the provision of socially assisted housing, in which the federal government could play a central role. This, however, would not involve Health and Welfare. In effect, the Epp Report has encouraged residents of the DES to identify local health concerns in the pretext of taking some action to address these concerns while at the same time having no legitimate basis (and, therefore, no responsibility) for taking action.

Another, perhaps less obvious, failure is that the epistemological implications of health promotion have not been fully appreciated by its advocates. Health promotion programs or actions systematically avoid addressing those elements of social structure acknowledged as being important determinants of health. If health promotion is concerned with the **process** of enabling people to increase control over and improve their health, and the fundamental conditions and resources for health are peace, shelter, education, income, stability, social justice and equity, then clearly the epistemology of health promotion (i.e., its underlying theory of knowledge) - the conceptual glue that holds the notion of health promotion together and enables us to make sense of how health is to be promoted - must concern the processes that give rise to and distribute these fundamental conditions and resources: processes by which income is created and distributed and what it is about these processes of distribution that results in inequity and social injustice; of how shelter is created and distributed; of the nature of education and access to educational opportunity; and so on. By its own logic, therefore, health promotion calls for an analysis of the political economy of social reproduction, although it is quite certain that Mr. Epp did not have such a radical form of analysis in mind when he released *Achieving Health for All.*

Despite stated intentions about moving beyond lifestyle to broader issues of well-being, the primary focus of health promotion programs continues to be on modifying individual behavior. Consider the national programs identified in the Epp Report as examples of the health promotion strategy - *Dialogue on Drinking*, the *Breast-feeding Program*, *It's Just Your Nerves* (a program on women's use of alcohol and tranquillizers), *Time to Quit* and *Break Free* (smoking cessation program), and *Stay Real* (substance abuse prevention). In the absence of critical analysis of the underlying processes of distribution of income, education, housing, etc., health promotion will continue to focus on lifestyle and disease. This is not to say that programs directed toward smoking cessation, substance abuse or breast-feeding do not have their place, but these are insufficient in and of themselves to address the underlying prerequisite conditions for well-being.

A related aspect of the epistemology of health promotion that is under-appreciated in the rhetoric concerns the nature of situated human behavior. If health is part of everyday living, and the aim of health promotion is to improve quality of life, then the conditions that exist in specific places at specific times in the course of routinized behavior of specific individuals must be more fully understood.[32] As it stands, health promotion is somewhat insensitive to individual biography and geography. Daily routines take place in contexts which both enable and constrain, but in the health promotion rhetoric the enabling aspects of context are emphasized while the constraining aspects are played down. To illustrate, consider the "mutual aid" "mechanism" for nurturing the uptake of health promotion identified in the Epp framework. Mutual aid

> refers to people's efforts to deal with their health concerns by working together. It implies people helping each other, supporting each other emotionally, and sharing ideas, information and experiences. Frequently referred to as social support, mutual aid may arise in the context of the family, the neighborhood, the voluntary organization, or the self-help group.[33]

The assumptions here are i) that people live in supportive neighborhoods, with ready access to supportive families, and/or self-help groups, and so on, and ii) that they are secure in their locations (ie., they don't have to move if they choose not to). The concerns raised by DES residents indicate that not everyone enjoys living in a supportive community. Isolation, idleness and loneliness are major problems in the DES - thus the desire among residents for purposeful activity and for greater support for the existing self-help system. But even for those residents who do feel they live in a supportive community, changes in the local land market in the form of increased rents and demand for space, and in the local labor market in the form of economic restructuring (job loss and low wages), threaten to disrupt their social support networks.[34]

Health promotion is totalizing discourse - virtually every aspect of human behavior and experience relates in some way to individual well-being. Key words and phrases in this discourse are ambiguous, multi-dimensional terms - environment, community, health, empowerment, etc. In the Epp Report the concepts represented by each of the nine cells in the framework (fostering public participation, healthy environments, etc.) are not explicitly defined, thus what constitutes public participation or a healthy environment depends entirely upon who is providing the definition. Consequently, players of all political stripes and with diverse interests can legitimately appropriate the language of "health promotion", almost regardless of the exact nature of their activity. In becoming everything, the focus on social justice and equity is lost and health promotion becomes nothing of particular relevance to residents in the DES.

The net result of residents having "participated" in health promotion is that their expectations have been raised while the conditions of their daily lives have remained the same or have deteriorated. In this sense, both the Epp Report and the Ottawa Charter have, in fact, adversely influenced the health of area residents. The price paid for grand rhetoric about social justice and equity in the absence of firm political action to reduce social disparities is credibility. And at a time when Canadians have never before been so cynical in their attitudes toward politicians and the governing institutions of this country, it's a price the state can ill afford.

REFERENCES

1 Health and Welfare Canada. 1986. *Achieving Health for All: A Framework for Health Promotion.* Ottawa: Ministry of Supply and Services.

2 Epp, J. 1987. Address: The Honourable Jake Epp, *Health Promotion International* 1(4), pp. 413-417.

3 Health and Welfare Canada, 1986, *op. cit.*, p. 6.

4 Brooks, J., Chisholm, M., Cairns, R., Jones, J., Jones, D., Harris, S., Martin, S. and Stern, R. 1988. Taking it to the Street: A Vancouver Community Responds to Achieving Health for All. *Health Promotion* 27(1), pp. 2-6.

5 Health and Welfare Canada, *op. cit.*, p. 3.

6 Brooks et al., *op. cit.*, p. 5.

7 Government of Canada. 1986. *Ottawa Charter for Health Promotion.*

8 The other co-sponsors of the First International Conference on Health Promotion were the WHO and the Canadian Public Health Association (CPHA).

9 WHO Healthy Cities Project. 1988. *Promoting Health in the Urban Context.* WHO Healthy Cities Papers No. 1. Copenhagen: FADL. See also *The New Public Health in an Urban Context.* WHO Healthy Cities Papers No. 4. 1989. Copenhagen: FADL.

10 Rootman, I. 1988. Inequities in Health: Sources and Solutions. *Health Promotion* 26(3), pp. 2-8; Rootman, I. and Munson, P. 1990. Strategies to Achieve Health for All Canadians. *Canadian Geographer* 34(4), pp. 332-334.

[11] Government of Canada, *op. cit.*

[12] Health and Welfare Canada, *op. cit.*, p. 3.

[13] O'Donnell, M. 1986. Definition of Health Promotion. *American Journal of Health Promotion* 1(1), p. 4.

[14] Health and Welfare Canada, *op. cit.*, p. 12.

[15] Health and Welfare Canada, *op. cit.*, p. 12.

[16] Tollefson, J. 1990. *A Preliminary Study of Selected Morbidity and Mortality Indicators in Census Tracts 57, 58, 59.01 and 59.02, Vancouver, British Columbia.* Vancouver: Vancouver Health Department.

[17] Thomson, M. 1990. Association Between Mortality and Poverty. *B.C. Medical Journal* 32, pp. 337-338.

[18] Downtown Eastside Resident's Association. *Downtown Eastside Housing and Residents Survey (1987-88)*; and *Downtown Eastside Housing and Residents Survey (1990).*

[19] National Council of Welfare. 1989. *Poverty Line 1989.* Ottawa: Ministry of Supply and Services.

[20] Murray, A. 1990. Homelessness: The People, in Fallis, G. and Murray, A. (eds.) *Housing the Homeless and Poor.* Toronto: University of Toronto Press.

[21] Downtown Eastside Resident's Association, 1990, *op. cit.*

[22] Downtown Eastside Resident's Association, 1990, *op. cit.*

[23] Hulchanski, D.J. 1989. Low Rent Housing in Vancouver's Central Area: Policy and Program Options. *UBC Centre for Human Settlements*, Housing and Community Planning Occasional Paper, UBC, September, 1989.

[24] Downtown Eastside Resident's Association, 1990, *op. cit.*

[25] *Ibid.*

[26] Buckley, L.R., "Shelterless Committee Newsletter, January 1990", Vancouver: Urban Core Shelterless Committee.

[27] *Ibid.*

[28] Downtown Eastside Resident's Association, 1990, *op. cit.*

[29] *Ibid.*, p. iii.

[30] Mishra, R. 1990. The Collapse of the Welfare Consensus? The Welfare State in the 1980s, in Fallis, G. and Murray, A. (eds.) *Housing the Homeless and Poor.* Toronto: University of Toronto Press.

[31] Fallis, G. 1990. The Urban Housing Market, in Fallis, G. and Murray, A. (eds.) *Housing the Homeless and Poor.* Toronto: University of Toronto Press.

[32] See, for example: Wolch, J.R. and Dear, M.J. 1989. *The Power of Geography*. Boston: Unwin Hyman; and Gregory, D. and Urry, J. (eds.). 1985. *Social Relations and Spatial Structures*. London: Macmillan.

[33] Health and Welfare Canada, *op. cit.*, p. 7.

[34] Wolch, J.R., Dear, M.J. and Akita, A. 1988. Explaining Homelessness, *American Planning Association Journal*, Autumn 1988, pp. 443-53.

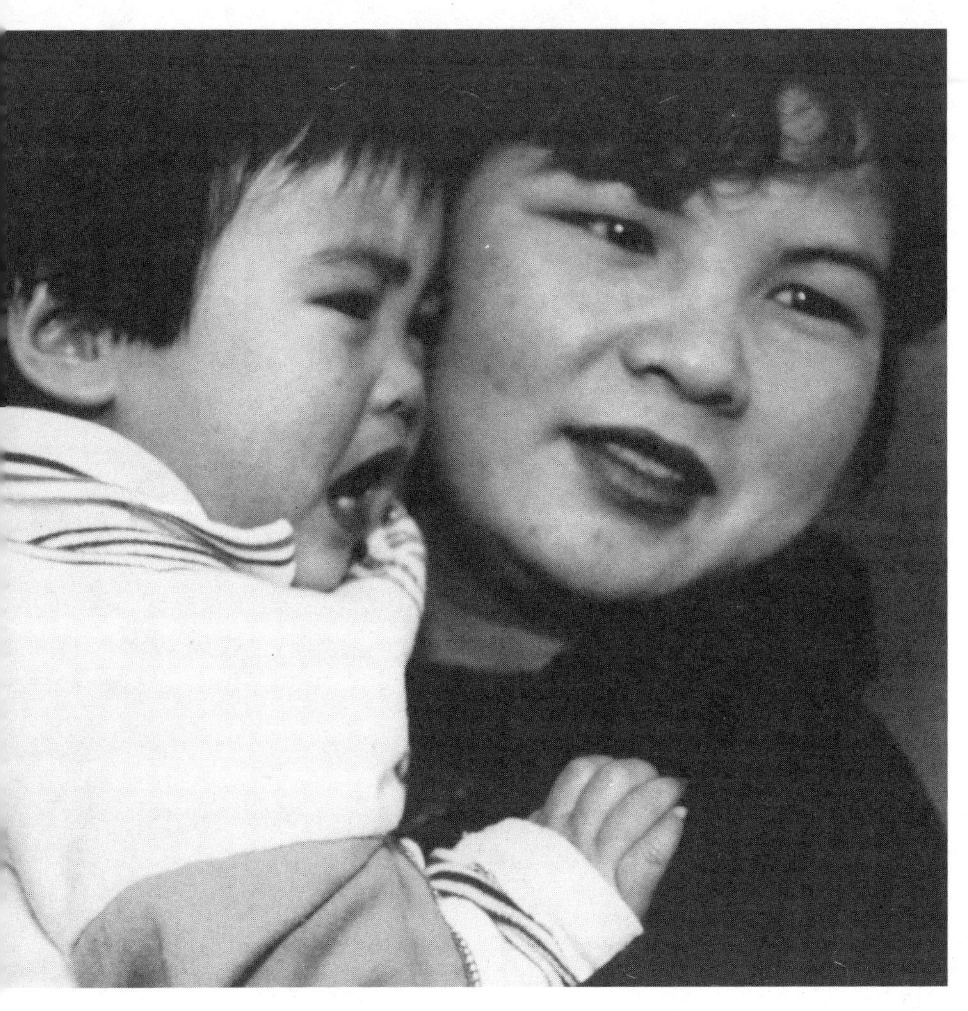

10

HEALTH AND HEALTH CARE EXPERIENCES OF THE IMMIGRANT WOMAN: Questions of Culture, Context and Gender

Isabel Dyck

School of Rehabilitation Medicine
University of British Columbia

INTRODUCTION

This chapter explores the potential contribution of a social and contextual approach to the investigation of health and health care experiences of immigrant, visible minority women living in western society. The issues here are broadly framed by a contemporary health promotion mandate, which adopts an holistic concept of health.[1] Such a concept of health emphasizes control by individuals over their own lives, but also understands health to be closely related to the environment and shaped by social, economic and cultural influences. It stresses the interplay of the individual and society. However, translation of the concept into practice and research has largely entailed a division between 'individualistic' and 'societal' or macro-level approaches, as in the tendency towards a 'lifestyle' as against a 'healthy community' division in health promotion work. The author maintains that this is a false division, that individual lifestyles, health experiences and coping strategies cannot be divorced from their context, and explores the possibilities of integrating what appear to be divergent approaches through considering the contextualization of human action and ideas in time and space. The chapter focuses specifically on the notion of culture, as this concerns health-related knowledge about visible minorities, as a vehicle to explicate this argument.

Culture needs to be recognized as a complex and dynamic concept, closely linked to the processes and structures embedded in society that shape individual and group behavior. Attention is then directed away from the individual to the contexts in which concepts of health and health-related experiences and action are generated and take place. This acts as a necessary corrective to work within a positivist paradigm that has focused on culture as an autonomous social characteristic, 'added on', as with other variables, in explanations of ethnic differences in health status and difficulties associated with health care use and provision. Instead, investigation of the complexity of processes and meanings surrounding experiences of health and health care is advocated; this enhances understanding of the significance of differentials in power relations, wrought within gender, class and racial divisions, that forge the circumstances of people's lives, including their health and responses to illness and disability.

The first section focuses on 'immigrant women' as it relates to the health promotion mandate. The next section discusses links between culture and health ideas and practices, and provides the conceptual framework of the argument presented in the chapter. The major section considers ways in which perceived links between culture and health and health care experiences have been approached at different contextual scales in research, and draws upon models in traditional medical anthropology and critical perspectives spanning recent work in geography, anthropology and sociology. An approach is proposed that attempts to integrate different levels of context in analyzing health and health care experiences, through transcending the common analytic dualism of 'individualistic' and 'societal' or macrolevel oriented research. Implications of this theoretical orientation for guiding health promotion related research are discussed in the final section.

HEALTH FOR ALL AND THE IMMIGRANT WOMAN

Health promotion concepts, following from the World Health Organization's vision of health for all, have attracted considerable comment in Canada since the publication in 1986 of "Achieving Health for All" and, following that, the document "Mental Health for Canadians", published in 1988.[2] Reducing inequities, increasing prevention and enhancing coping are the major health challenges identified within this health promotion mandate. And while the health promotion initiative is couched in terms of moral responsibility to the country's population, its anticipated cost-effectiveness is also recognized as a desirable and needed outcome.

Canada's health-care costs are of considerable concern to government, and the thrust toward community-based care is one attempt toward long-term cost-cutting by reducing reliance on medical practitioners, high-tech intervention and expensive institutional care. Indeed, a dominant emphasis in medicine around costly acute care has earned the health-care system the alternative conceptualizaton of an illness-care system, that reflects little of the disease prevention and health maintenance needs of contemporary society.[3] The federal government's call for the development of a knowledge base to guide policy and activity toward the goal of "health for all" is indicative of an intellectual move away from an illness-care approach to health problems. Researchers are urged to take explicit account of the environmental, social, economic and cultural complex in which an individual's ability to control his or her own health is embedded. Yet, although social and economic circumstances are recognized as important to people's health status and the 'community' is singled out as an appropriate focus for health intervention, an emphasis remains on individual responsibility through self-help and compliance with advocated healthy lifestyle components or medically prescribed treatment procedures. There is little evidence of any radical change occurring in conceptions of health, illness or health care delivery at the level of the institution.[4]

On the other hand, there is awareness of inadequacies in the effectiveness of current health care provision, particularly as this pertains to minority groups. Immigrant groups tend to make relatively less use of health services, to terminate treatment early, and to receive poorer quality health services.[5] Many immigrants also experience social and economic disadvantages that are understood to contribute to high risk for poor physical and mental health. Furthermore, immigrants who are elderly or adolescent, and women from traditional cultures, are at particularly high risk for emotional distress.[6]

Links between culture and health are also indicated in studies that report associations between culture or ethnicity and the experience of pain, incidence rates of certain diseases and psychiatric conditions and access to and use of health services.[7] These discoveries have stimulated interest in how culture may act as a barrier to health. The scope of this concern has increased with contemporary changes in Canadian society, such as Canada's multicultural mandate and profound changes in immigration patterns that have resulted in an increasingly heterogenous population. Between 1981 and 1986, Asian born people accounted for 43 percent of all those immigrating to Canada, and the growth in actual numbers of immigrants from Asia and the Pacific region has continued into the late 1980s with, for example, a percentage increase between the years 1985-6 and 1986-7 of nearly 40 percent. Indeed, numbers of those born in Asia make up an increasing proportion of those entering Canada in all three major categories of immigration, these being the family, humanitarian and business categories.[8] Those officially destined for the labor force and those classified as dependants have been of approximately equal numbers, and places of settlement indicate a clustering in the cities of Toronto, Montreal and Vancouver.

This author's interests in singling out immigrant women for discussion reflect both empirical problems surrounding health and health service usage and theoretical concerns over the intersections of gender and ethnicity in the analysis of immigrant women's health experiences. A growing literature points out that women's health experiences differ from those of men, as does their use of the health care system,[9] and that poor women from minority groups are likely to have serious health needs that will not be adequately met.[10]

A number of issues need to be taken up with regard to the categories used in understanding links between culture and immigrants' experiences of health and health care. A main interest in this author's research has been the question of how theoretical understandings of culture and usage of the term influence the interpretation of its meaning in different health-related situations and outcomes. While such terms as 'immigrant women' and 'culture', and indeed, health and illness, have "taken-for-granted" meanings, these need to be deconstructed and commonsense understandings challenged. A further strand of enquiry concerns the nature of the intersections of gender, class and 'culture' or race in health experiences, and how these may vary in particular places. Attention to social processes is essential in investigating such issues as they pertain to the interconnections between culture and health-related phenomena. What, in fact, is it about culture that is important to health, illness and health care experiences?

HEALTH IDEAS AND PRACTICES AS LOCAL, CULTURAL KNOWLEDGE

Culture

How culture is conceptualized is critical to how health problems of immigrants from visible minority groups are perceived and acted upon. The term 'culture' has been defined in a variety of ways, but common working definitions adopted in health care disciplines usually refer to it in terms of the ideas, beliefs and customary practices of an interacting group that are learnt and transmitted from generation to generation. Specific dress styles, religion, music, architecture, literature, food, childrearing practices, economic practices and everyday objects are all aspects of life that, when clustered together, form a pattern of 'cultural traits' by which people may identify themselves and be identified by others as a distinctive social group. A particular dimension of culture that is currently emphasized in anthropology is that of knowledge, so that culture may be defined, as by Spradley, as "The acquired knowledge people use to interpret experience and generate behavior."[11] Customs and artifacts are included in this understanding of culture, for it is knowledge that 'drives' or generates these. They are considered to be the manifestations of a world view, rooted in much common knowledge and understanding of how the world works, that is developed over time among people living together in a particular geographical area. Culture, in effect, consists of 'recipes for action' or 'recipes for living in the world' - the "taken-for-granted", commonsense knowledge for dealing with everyday life.

'Culture', however, is not autonomous or static, coming in a discrete bundle to be passed on to the next generation, but is a complex, dynamic phenomenon closely linked to the political, social and economic institutions of a society, themselves dynamic and changing over time.[12] Furthermore, while there are commonalities in the knowledge and practices of a society, broad cultural notions will also be mediated by local contexts, or settings for interaction. Culture in this way is local knowledge, its nature and acquisition closely tied to the social and material circumstances in which a person lives.[13] It is through the experiences of living in particular places and within localized sets of relationships that meaning is given to the ideas, beliefs and actions of everyday life, including those surrounding health and illness. Such a contextual understanding of culture opens up the possibility of a considerable variation in ideas and practices to exist, not only between societies but also within societies.

Health care ideas and practices, as with other types of knowledge, will not be homogenous but may be anticipated to vary within a society, fitting particular situations and open to different meanings in different places. Western professional medicine, with its biomedical disease categorization and application of associated techniques, is also cultural knowledge, and so represents only one view, although a powerful one, of the world. Indeed, rather than being objective, scientifically neutral and universal knowledge, biomedicine is understood in critical perspectives to be a social and moral practice, reflecting the sociocultural values and norms and dominant economic and political interests of a society.[14] Concepts of health, illness, disease and disability are thus concepts of process and structure (rather than 'facts' of positive science) shaped within historically and geographically specific sets of social relations.[15] The development of biomedicine cannot be separated from the organization of society, and the ways in which medicine is practiced will be related to social and economic change.

The practice of medicine over this century has become less holistic in approach, with increasing reliance on technological and pharmaceutical interventions following tremendous advances in biomedical knowledge.[16] As a belief system, medicine has marginalized social and economic factors in favor of a focus on universal disease categories, definitions of normal and abnormal states of being, and treatment of individual pathologies. Certain environments may be recognized as unhealthy and contributing to disease, but treatment of the individual remains the predominant concern and is carried out in isolation of the context of the 'sick' individual's life. Such a model has, as Rathwell and Phillips comment, "only limited relevance in the exploration and amelioration of socially and personally-based phenomena."[17] Despite this failing, biomedicine constitutes a dominant and powerful belief system in organizing knowledge about health and health care provision. The cultural and political hegemony of its ideas infiltrate everyday life as commonsense understandings of the world and how to live in it.[18] This is exemplified through the concept of 'medicalization' that denotes the increasing areas of everyday life that have been claimed as a province of medical interpretation and intervention,[19] the incorporation of biomedical ideas into lay understandings of health and illness[20] and, indeed, into the conceptual framing of academic disciplines such as medical anthropology and medical geography.[21] However, while biomedical knowledge plays a dominant role in current understandings of health and illness as a meaning system, it is not necessarily accepted uncritically. Biomedical interpretations may counter or accord with lay views, and may be resisted by disadvantaged and subordinate groups.[22] There is evidence that particular concepts of health, illness, disease and disability vary cross-culturally, according to class and in the context of particular places.[23]

Differences in conceptions of health, illness and health-care related knowledge are of concern in practice-based disciplines because health care professionals working with ethnic minorities may experience difficulties in communication and problems of 'non-compliance' with advocated treatment on the part of clients.[24] Such differences have also attracted considerable study in medical sociology and medical anthropology, evidenced by an extensive literature on the patient-physician relationship. As culture is deemed important in shaping health beliefs and practices, it has been a major focus in analysing health encounters involving visible minorities. However, the idea of a 'nesting' of contexts within which the patient-practitioner encounter is embedded, including the institution in which health care encounters take place, and the overall hierarchical organization of the health care system, suggests the complexity of health care outcomes and the part played by culture.[25]

CONTEXT, CULTURE AND GENDER

In this section the main thrusts of two predominant research themes pertinent to the understanding of the situation of immigrant women and their health and health care experiences are presented. First is a body of study focusing on cross-cultural interaction in the clinical setting; second is research from a critical perspective that emphasizes the significance of structural processes in forging the situation of immigrants in general - and, therefore, their health experiences. Both contribute to the explanation of health related phenomena, but the emphasis on the individual on one hand, and structure on the other, tends to underemphasize the interweaving of different levels of context in shaping human ideas and action. A third approach is offered that focuses on the context of everyday life, and is informed by theory that attempts to integrate the common dualism of agency and structure in analysis. In the course of this discussion, gender is highlighted as a fundamental social division to be considered in health research, together with class and race.

The Clinic

Kleinman's 'explanatory model' has been influential in medical anthropology in attempts to denote the cultural shaping of health beliefs and actions.[26] This model has been used to analyse the physician-patient encounter, to avoid misunderstandings between parties in attempts to attain effective negotiation of treatment procedures. The model is based on the presupposition that people's explanations of their illness and their expectations of medical help will be shaped by culturally-shaped health beliefs and meanings attached to illness. Thus, as patients meet health practitioners, differences in beliefs and practices may result in competing modes of health care practice, suggesting alternative pathways to health. Each may have different diagnoses, interpretations of medical history, treatment expectations and ways of defining a particular medical problem.

Support is lent to Kleinman's model by study findings which show that where there are incongruences between patients and practitioners in the evaluation of a disorder or expectations of treatment, there are likely to be lower levels of client satisfaction and adherence to treatment regimes.[27] Certainly it has been observed that when patients perceive that they have some control of the patient-practitioner interaction, satisfaction and compliance are greater.[28] Practice-based disciplines such as rehabilitation and nursing, which are premised on taking a holistic view of health, have adopted similar approaches, focusing on areas of 'mismatch' between beliefs, values and ways of communication in understanding treatment difficulties and ineffectiveness.[29]

The ability of the patient to actively participate in decision-making may be especially difficult in the face of cultural difference. As Weidman notes, "It is in the clinical setting of patient and health professional interactions that we see clearly the nature of the adaptive task facing a patient whose health cultural background is different from that of the legally sanctioned medical system."[30] Her concept of 'health cultures' incorporates the type of social organization of a health care system (which may range from care within social networks to the highly organized, hierarchical form of many western health care institutions) together with beliefs and values in an attempt to acknowledge contexts beyond that of the clinical health care encounter. Lazarus points out that, in addition to the patient-practitioner encounter, investigation of the timing and spacing of clinic organization is important to understanding patients' experiences of health service delivery, for it is at the level of institutional care that exploration of the significance of power relations in western health culture can begin.[31]

The power differential between lay and biomedical knowledge as cultural constructions are considered critical to the practice of medicine. Although Kleinman's explanatory model was developed in response to the reality of power differentials, Pappas claims, "his analysis of relationships between healers and patients, institutions, history, and politics are analyzed primarily as contexts, without integrating them into the analysis of action."[32] Kleinman's "excesses of agency"[33] in what is essentially a negotiation model is criticized for providing an insufficient account of the 'uneven playing-field' on which patient and practitioner meet; that is, its inadequate incorporation of the political economy in understanding not only the structuring of the nature of the physician-patient encounter, but also the support by medicine of a dominant ideology that supports class divisions.[34]

While a political-economy approach reveals the importance of class to clinical encounters, research is also beginning to show the significance of gender. Gender divisions within the power relations between practioner and patient have been identified in the process of labelling illness and prescribing treatment, particularly when involving women's mental health.[35] For example, women suffering from stress-related illness are more likely to be labelled neurotic than men, with the symptoms of social stress 'pathologized' and treated with tranquillizers. In addition to omissions of political economy and gender in conventional analysis of patient-practitioner relationships, racial power differentials also tend to be ignored in such explanations.[36]

Despite an interest in cross-cultural encounters, 'conventional' medical anthropological work on the patient-practitioner interaction has been inadequate in linking local level behavior with broader sets of social and economic relations. It is criticized for becoming medicalized itself, because framing the issue in terms of bridging the culture gap between the two parties essentially accepts the dominant ideology of biomedicine.[37] Social, political and economic relations of society remain essentially a backdrop to action rather than processes intimately integrated into the structuring of health and health-care.[38] A close focus on the clinical setting has not, in general, been able to integrate the spaces beyond the clinic that are crucial to health and illness experiences, and the responses to them. Nor is the significance of the process of becoming culturally different as an immigrant addressed. In order to bring greater understanding to both the health and health-care experiences of immigrant women, it is necessary to move beyond the clinic to other settings and circumstances of people's lives.

Beyond the Clinic: Structural Relationships

A move away from the patient-practitioner relationship and the institutional setting to other areas of a person's life introduces another 'level' of context, both empirically and theoretically, in understanding health and illness experiences. Causes and effects of ill health are of more concern than the dynamics of health care encounters, although all are part of the complex process of 'achieving health'. The change in empirical focus also signals a further arena in which culture takes on meaning. A body of literature on the experiences of immigrant women from different cultural groups, particularly in relation to their work and health, forcibly presents a picture of social and economic disadvantage.[39] It is pointed out that the commonality of circumstances of immigrants' lives in general have more to do with their life experiences, including their health, than do specific cultural characteristics and customary lifestyles. Thus, while the majority of research on health problems of specific ethnic groups has focused on 'exotic' diseases, or lifestyle oriented causality (for example, mental and physical illnesses intepreted in terms of pathological family structure, poor cultural adjustment, deficient diets, or other 'unhealthy' facets of lifestyles that are seen to be culturally based), recent work suggests that a neglect of social and material circumstances fosters a 'victim-blaming' view and the labelling of particular groups as problem or deviant.[40] Instead of focusing attention on characteristics specific to particular cultural groups, further exploration of the challenges that many immigrant women face and the significant relations that shape these is advocated.[41] Class, gender and race are found to interwine in complex ways in the social construction of the health experiences of immigrant women.

Class position in society is well recognized as a predictor of health, while gender divisions in illness are also well documented.[42] Indeed, the centrality of women's reproductive role, together with contemporary social and economic conditions, are seen as important influences in the observed increase in, and specificity of women's physical and mental health problems.[43] Although there is a movement of women into managerial and other senior white collar positions, occupational segregations based on gender cast women in general as a 'vulnerable' workforce engaged in low-skilled and low-paid labor, with consequences for their access to opportunities and resources for maintaining health.[44] Working class women, for instance, especially those who are lone parents, are more vulnerable to illhealth created by low income levels that restrict their access to the material resources.[45] Stress may also accompany the often conflicting demands of child care, home responsibilities and paid work.[46] Class and gender may also influence the effectiveness of health service provision for women. For example, employment status and gender role affect access to and use of health facilities,[47] while the structuring of health services in both developed and developing countries discriminate against women.[48] Working class women may also experience difficulty in following complex or regular treatment regimens due to the time and energy demands of their everyday lives.[49]

While class and gender are recognized as important to health experiences, less attention has been paid to the ways in which racial divisions combine with these.[50] Both the formal policies and informal practices of white society are seen as fundamental determinants of visible minorities' life experiences, and consequently those of health and access to services, with women facing special difficulties.[51] In Canada, the consequences of immigration policy, for instance, are crucial to women's access to opportunities and resources. As many women enter the country in the family class and 'assisted relative' categories, they are denied government sponsored language and job training programs and supported day care, making translation of previous skills to the Canadian labor market difficult.[52] Nevertheless, many enter the paid labor force in response to economic circumstances, and usually take low-paid jobs, working as, for example, kitchen helpers, cleaners, hospital aides and workers in factories and the garment trade, in a labor force that is stratified by ethnicity as well as by gender.[53] Despite programs intended to improve the job circumstances of immigrant women, upward mobility is far from assured. Ng argues that women may remain channelled as an available immigrant workforce through the unintended consequences of a program's everyday practice, if funding remains targeted on job placement rather than upgrading skills or job training.[54] Regional comparisons also suggest that the position of immigrant women in the labor force has more to do with informal practices within the workplace, including recruiting practices and barriers to advancement discriminating according to race, sex and age, than with individual or 'ethnic' characteristics.[55] Immigrant women engaged in the lower ranks of the labor force may experience particularly poor living and working conditions that are known to place people in high risk categories for poor health. Immigrant women may also lack awareness of the organization of health services and have difficulty in finding out where appropriate care may be obtained when they do face health problems.[56] When access to health services is gained, subsequent evaluation and the following of health management procedures can be difficult, particularly in the face of language and educational barriers.[57] Such women, in addition to vulnerability to the double workload of home and paid job, may have restricted access to information about, and controversies over, different aspects of health care.[58]

Investigation of structural relationships and processes that shape the experiences of immigrant women suggests that interpreting culture only in terms of beliefs, values and customary practices is inadequate. Lacking is an appreciation that the everyday economic and social circumstances of immigrants' lives, which affect their health and health care needs, are forged within, and consequential to, the social construction of the 'culturally different' individual. But just as an overemphasis on agency in studies of the practitioner-patient interaction is seen to neglect structure, focus on the 'macro-context' is in danger of neglecting human agency.

This neglect fosters an interpretation of immigrant women as 'victims'. Furthermore, while differences in health care knowledge have been acknowledged in the setting of the clinical encounter, connections between this and the broader sets of economic, political and social relations shaping the context of everyday life are not usually made. Qualitative studies, however, show close connections between local contexts, health-related ideas, and the use of different types of health care knowledge. It is through investigation of 'everyday worlds' that the recursive shaping of ideas and action in particular places is observed and can usefully be added to discussion regarding interlinking of culture and health-related phenomena.

Everyday Worlds

Extensive changes in both economic and social practices may occur following immigration. For the female immigrant, maintaining a household in Canada can be dramatically different from her previous experience, whether or not she is working outside the home.[59] Women may be more dependent on their husbands and more isolated once in Canada. What is described as 'traditional' behavior may have more to do with current conditions, including, for example, changed access to friends and other social contacts, lack of mobility, unfamiliarity with English, and changed methods of work, than with previous lifestyle.[60] Downward occupational mobility when skills are not transferable to the Canadian context, changed household composition, and movement from a rural area to a highly urbanized Canadian city are further aspects of immigrant women's experience in their 'new country'. These may produce stress, reduced sociability and uncertainty of ways of handling health problems, childrearing and job finding.[61] As yet, relatively little is known of the links between coping strategies immigrant women employ and the specific localities in which they settle. There is, however, a growing body of evidence that women are not merely passive receivers of their circumstances, but are actively involved in attempting to meet their health needs as they cope with negative economic and social changes. For example, women from many cultural backgrounds search for new health care services, build female-based social support networks, and redefine cultural expectations of appropriate female behavior.[62]

The utility of 'new' knowledge and new recipes for health action has also been found to constantly undergo reappraisal in the face of new experiences.[63] Evaluation of this knowledge takes place within the social networks of family, community and work. Place-specific experiences open up arrays of ideas that may enter stocks of accumulated knowledge to be drawn upon in future action. Commenting on the experiences of black people in Britain, Donovan states that:

> The form and pattern of [their] recipes for action will depend....on the individual's stock of knowledge, which will include family and cultural remedies, traditional theories, and a varying degree of westernisation. The degree of westernisation and the presence of traditional recipes will allow [them] to have quite different recipes from, for example, an individual with several generations in Britain and a member of the middle class.[64]

A similar mingling of traditonal and new knowledge has been found in the case of Chinese immigrant women with chronic illness in Vancouver, who may draw on both lay remedies and professional procedures in managing their health, fitting these in with the demands of home and workplace.[65] Hunt et al., however, suggest that lay knowledge is strongly adhered to. Professional knowledge gained in the clinical encounter may be incorporated into lay explanations only to the extent that it 'fits' with previous understandings and relevance to everyday experience.[66] Work by Conrad on people with epilepsy also shows the importance of everyday social experience to how professional medical knowledge and advice is responded to as patients attempt to reduce stigma, keep up 'normal' social lives and maintain independence from medical regimes as much as possible.[67] Although little research has been conducted specifically about the immigrant's acquisition and use of professional medical knowledge, there is evidence that a redefinition of health problems may occur in the context of the specific organization of health services and social policies of the receiving society.[68] Thus, there is growing evidence that people do define their health, illness and appropriate responses situationally, and that 'recipes for action' concerning health care will be interpreted in the context of particular places.

Without cognisance of the embeddedness of health experience and help-seeking behavior within relationships at different but interconnected levels of context, health professionals are likely to continue to define problems in conveying information and with non-compliance as being ethnic in origin.[69] There may be awareness of and sensitivity to a patient's life circumstances and lifestyle on the practitioner's part, but problems in the therapeutic process tend to be ascribed to the patient's non-western ways, values and stereotypical views of family composition. For instance, common characterizations of immigrants' lifestyles and family structure as "traditional and family oriented" or "backward" as compared to Canadian women tend to be used to explain the practical and organizational problems immigrant women may have in recreating household organization in response to the demands of the workplace, school and social agencies.[70] This is not to deny the existence of culturally distinct ideas and practices, or the importance of cultural sensitivity on the part of health professionals and policy makers, but it is to argue that how cultural difference is interpreted and managed is crucial. Cultural negotiation, 'adding on' cultural experts and interpreters, although important, do not significantly change the power relations embedded in class, gender and race divisions that are fundamental in constituting people's everyday worlds, concepts of health, illness, disability and the ways in which medicine is practiced.

DISCUSSION AND IMPLICATIONS FOR RESEARCH

A continuing critique of the social relations and underlying political and economic structures producing health and constraining health-seeking behavior must be part of a research agenda concerned with analyzing the health situations of minority, immigrant women. In addition, it would appear to be imperative that we look more closely at the everyday experience of people's lives as carried out in the community where they live, for it is there that health will be defined and the consequences of illness and disability will acquire meaning. In doing so, there is an evident tension between individual circumstances and general relations that seem significant to people's lives, which poses a methodological challenge in research. Associations are recognized between social and economic circumstances and ill-health and inequalities in access to health care. But the actions that individuals or groups take in local settings, and the influence of the political economy in shaping environmental conditions that might lead to ill-health, tend to be viewed as different types of phenomena not readily integrated in a unitary approach to health care problems. The individual and society are commonly understood as interacting, but essentially separate, entities.

Recognition of the embeddedness of health, illness and health care systems in societal structures and processes suggests that space, social relations and action need to be theoretically integrated in research. The issue of focus on either the micro- or macro-level is one dimension of the difficulty in addressing levels of context in research. However, as Giddens posits in his exposition of structuration theory, division of the micro- and the macro- is artificial. Attention to the constitution of ideas and action within the timing and spacing of everyday life is an avenue to approaching the individual and society within one framework.[71] Contextuality is a central notion of structuration. This includes a concern with the spatial dimension of social processes, and with structures underwriting the constitution of human action and ways of viewing the world. People are viewed as active agents shaping their lives within the events and relationships of contemporary society, but the 'stretching' of social institutions over time and space delimits the particular sets of resources that are present, or alternatively absent, as people go about their everyday lives. In essence, resources comprise the social and material conditions of everyday life, and accumulated stocks of knowledge that guide action. It is in the locales or settings of interaction, making up the contexts of an individual's daily routines, that such resources will be drawn upon. The distribution in time and space of such locales will shape the specificity of both the constraints and opportunities facing the individual.

A focus on the inherent spatiality of social life provides a valuable avenue for developing knowledge about how 'broader' power relations and local contextual factors interweave in constituting the resources and knowledge available to the immigrant woman. The availability of job opportunities, local occupational segregations, the state of the housing market and nature of housing policy, residential clustering according to class or ethnicity, and the provision of social and health services, are all aspects of the local context that give meaning to people's everyday lives. At the same time, they are also manifestations of a complex of national and international social, political and economic processes which are understood to operate unevenly, and vary both temporally and spatially.[72] As Jones and Moon assert, "Space and locality have a crucial role to play in explanation [of health-related phenomena] because the various processes concerned can be manifested in different ways at different places."[73] The fact that immigrant communities tend to be marginalized spatially, socially and economically means that area of residence can be crucially important to the availability and possible creation of resources, including the structure of local networks and the formal provision of health care services.[74]

Thus the issue is not whether the focus of research is on large-or small-scale phenomena, but the ability of a chosen method to explicate an interweaving of different levels of context in the constitution of health-related phenomena. Figure 1,10, adapted from Smith's notion of institutional ethnography,[75] diagramatically represents the interweaving of different

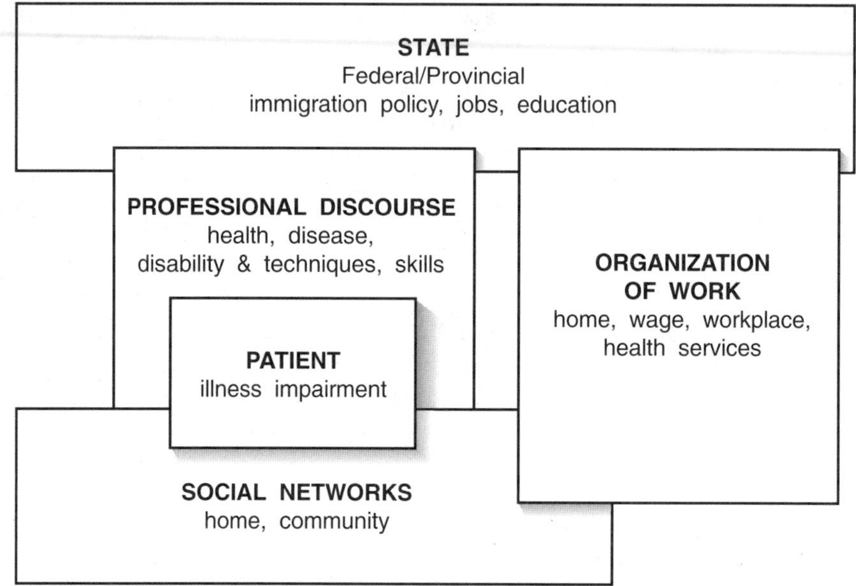

Figure 1,10 Contexts of Health Care Knowledge (modified from Smith, 1986)

levels of context that shape health experiences, health care encounters and responses to illness episodes. It indicates the interacting sets of relationships relevant to analysis of immigrant women's health experiences. This model permits empirical research of local level phenomena, such as practitioner-patient interactions, health and illness experiences and understandings and the lifeworlds of immigrant women, without losing sight of the complexities of relationships which are constitutive of people's life experiences. Rather than operating independently, these sets of relationships shape each other so that, when investigating 'the everyday', we are also able to investigate how state policies and practices, the organization of the workplace (including that of the institution), family and community networks and both lay and professional discourses all are relevant to the phenomenon in question. The options of women and the range of alternatives open to them, including concepts and experiences of health, will be influenced through the whole range of meanings that gender, class and ethnicity as dimensions of social identity may come to entail within these sets of relationships as they are manifest in a specific time and place.

A range of research strategies is advocated to incorporate both structural factors and human experiences, perceptions and abilities.[76] And while research concerned with immigrants' health within the above approach can usefully be conducted at any physical scale of context, attention to the settings of interaction of women's daily routines would seem to be particularly appropriate for discovering how their health ideas and practices, as cultural knowledge, are both created and shaped. Qualitative studies at the 'local level' have further value in that differences among immigrant women can be discovered. This is important, for such women are not necessarily an interacting group within a particular place; they may share similar attributes, but may not have actual connections. A geographically proximate group may be heterogenous and there may be considerable variation in behavior. Understanding the social construction of women's lifeworlds and means of agency requires knowledge of their functioning relationships in everyday life.

An additional advantage of an ethnographic approach is its ability to incorporate the standpoint of the subject into research. This is important on two related counts; one being concern over the production of knowledge, and the other relating to the power differential between researcher and researched. Currently, there is considerable sensitivity to the part played by a white researcher in the concerns of minority groups. This is an important matter as the consequences of research are not wholly predictable, and the unintended consequences may result in perpetuating inequities. It is especially crucial when socially constructed categories of people are associated with disadvantage, such as poor women and poor, visible minority women. Feminist and disability researchers have been particularly cognizant of the differential power of knowledge in defining and circumscribing life chances and options, and the need for developing a subject-centred body of knowledge. Constructing knowledge from subjects' perspectives is not just a matter of sensitizing health professionals and policy makers to the problems and circumstances of different groups of people's lives; it is fundamental to generating theory within a paradigm that counters the dominance of the biomedical knowledge framing western health care organization. This is important to how health knowledge develops and is used, for theoretical constructions not only influence how health problems are perceived and interventions chosen, but also frame priorities in research. Thus, adding the voices of those suffering health problems is considered essential in developing theoretical knowledge about all aspects of health, illness and disability.[77] That there will be a multiplicity of voices to be represented in subject-centred research is not to question the quality of such research, but rather indicates the range of social and regional divisions significant to people's health experiences and different ways of constructing knowledge. These divisions need to be addressed explicity, with the interconnections between class, gender and ethnicity being explored rather than being treated as separate axes of disadvantage or subordination.[78]

A plurality of methods may be used according to the specific concern of a particular research project, but qualitative methods and an interpretive approach are particularly suited to reaching subject-centred definitions and concerns. Qualitative research also has the advantage of allowing deconstruction of those everyday, taken-for-granted understandings such as culture, race, immigrant women and socio-economic disadvantage which in everyday usage may contribute to the reproduction of ideological constructions that reflect and feed the power relations in society. For instance, 'race,' 'immigrant woman' or 'culturally different' are social constructs with real meaning for people, in that ethnic minorities may experience pervasive disadvantage and discrimination, which is not necessarily intentional but is caught up in everyday processes.[79] As the author selects the way in which subjects' narratives are represented, how this is done and the account of the researcher of his or her part in the research process are aspects of qualitative research that take on particular significance, and readers have special responsibility for assessing claims made and quality of data in the context of specific researcher and researched relationships.[80] Research findings also need to be related to theory, such as theories of power and resistance, of ethnicity and feminist theory, if they are not to remain at the level of description, and thus not necessarily challenge the dominant framework within which health-related phenomena are commonly interpreted. Attention to terminology and guiding concepts that direct research must be of major concern if the operation of hegemonic ideas in shaping commonsense understandings are to be discovered.

Some form of participatory research, in which local communities have input into the development of a study and its questions, is an approach also amenable to addressing concerns of power differentials in knowledge construction. Such research needs to take account of internal differences of interest in what may be seen as an interacting cultural community. Cultural groupings and ethnic communities may include, for example, divisions of class, gender, region and religion that affect health and health care experiences. So that, while there may be certain universalities of immigrant experiences, the nature of local needs cannot be assumed.

Which health problems are identified and how are these interpreted among different social groupings within a particular locally based 'community'? What are the everyday consequences of illness or disability? What sort of experiences do people encounter in using the formal health care system? These are essential questions, for the ability to define one's own needs and act upon that understanding is central to achieving health.[81] More information is also needed on how people work out their current situation, including ways of recreating practices, redefining roles, identifying resources, and sharing and circulating information. Certainly, any research concerning minorities must avoid studies that may contribute to the labelling of specific groups as 'problem' groups, or that foster ethnic minority labels and categories that detract from or minimize the political dimensions of their situation.[82] To this end, studies concerned with either the manner in which individuals achieve health or cope with chronic illness and disability needs to take explicit account of the meaning of health and illness behaviors within specific community contexts, together with an analysis that can take account of broader sets of relationships.

CONCLUSION

The plurality of approaches and concerns in current health-related research signals a tension between the more traditional models of positivist enquiry and a critical or social perspective that seeks to illuminate health, disease and medicine as social phenomena, rather than as objective or value-free entities and practice. In the social sciences disciplinary boundaries are blurred as the embeddedness of health, illness and health care systems in societal structures and processes is emphasized in analysis. Together, recent work within different disciplines brings a deeper understanding to the significance and meaning of context to health and health-related phenomena. Views of the world, health and health care experiences and coping strategies, although framed and circumscribed within broader societal relations, are given meaning within the local conditions and home, community and work settings of particular places. Thus, while immigrant women may share similar attributes and some circumstances, there are also likely to be considerable variations in experience. The concept of 'immigrant woman' will be situationally defined, with its meaning and content mutable, just as the immigrant woman's health experiences will not be independent of her class, age, gender and ethnicity. Future research needs to focus on the interweaving of agency and structure as this constitutes the immigrant woman and her possibilities for achieving health. Moreover, while cultural change may be seen as a major shaper of immigrant women's experiences, caution must be used in how the term 'immigrant woman' is interpreted, for it constitutes a taxonomic category that may foster distorted generalizations about women who are immigrants. Within the health promotion mandate, questions concerning the effectiveness and transmission of ideas of western medicine and largely unexamined concepts of health promotion, whether phrased in terms of individual lifestyle or community health, are likely to be inadequately framed without greater understanding of the complex structuring of people's everyday worlds.

REFERENCES

[1] Epp, J. 1986. *Achieving Health for All: A Framework for Health Promotion.* Ottawa: Health and Welfare Canada.

[2] *Ibid.*; Epp, J. 1988. *Mental Health for Canadians:Striking a Balance.* Ottawa: Health and Welfare Canada.

[3] Pipe, A.L. 1990. *Politics, Pathology and the Public's Health.* The John F. McCreary Lecture, Faculty of Medicine, University of British Columbia, Vancouver, October 15, 1990.

[4] Crawford, R. 1980. Healthism and the Medicalization of Everyday Life, *International Journal of Health Services* 10(3), pp. 365-388.

[5] Canada. 1986. *Review of the Literature on Migrant Mental Health.* Ottawa: Health and Welfare Canada.

[6] Canada. 1988. *After the Door Has Been Opened: Mental Health Issues Affecting Immigrants and Refugees in Canada.* Ottawa: Health and Welfare Canada.

[7] Bandaranayake, R. 1986. Ethnic Difference in Disease - An Epidemiological Perspective, in Rathwell, T. and Philips, D. (eds.), *Health, Race and Ethnicity.* London: Croom Helm, pp. 80-99; Berkanovic, E. and Telensky, C. 1985. Mexican-American, Black-American and White-American Differences in Reporting Illnesses, Disability and Physician Visits for Illnesses, *Social Science and Medicine* 20(6), pp. 567-577; Keefe, S.E. 1982. Help-seeking Behavior Among Foreign-born and Native-born Mexican Americans, *Social Science and Medicine* 16, pp. 1467-1472; Lin, T.Y., Tardiff, K., Donetz, G. and Goresky, W. 1978. Ethnicity and Patterns of Help-seeking, *Culture, Medicine and Psychiatry* 2, pp. 4-13; Lipton, J.A. and Marbach, J.J. 1984. Ethnicity and Pain Experience, *Social Science and*

Medicine 19(12), pp. 1279-1298; Mirowsky, J. and Ross, C.E. 1980. Minority Status, Ethnic Culture and Distress: A Comparison of Blacks, Whites, Mexicans and Mexican Americans, *American Journal of Sociology* 86, pp. 487-495.

[8] Employment and Immigration Canada. 1987. *Annual Report 1986-87*. Ottawa: Employment and Immigration Canada Cat. No. MP1.

[9] Meininger, J. 1986. Sex Differences in Factors Associated With Use of Medical Care and Alternative Illness Behaviours, *Social Science and Medicine* 22, pp. 285-292; Verbrugge, L.M. 1985. Gender and Health: An Update on Hypotheses and Evidence, *Journal of Health and Social Behaviour* 26 , pp. 156-182.

[10] Zambrana, R.E. 1988. A Research Agenda on Issues Affecting Poor and Minority Women: A Model for Understanding Their Health Needs, *Women and Health* 12(3/4), pp. 137-160.

[11] Spradley, J. 1980. *Participant Observation*. New York: Holt, Rinehart and Winston.

[12] Kapferer, B. 1988. Gramsci's Body and a Critical Medical Anthropology, *Medical Anthropology Quarterly* 2, pp. 426-432.

[13] Geertz, C. 1983. *Local Knowledge*. New York: Basic Books.

[14] Comaroff, J. 1982. Medicine: Symbol and Ideology, in Wright, P. and Treacher, A. (eds.), The Problem of Medical Knowledge: Examining the Social Construction of Medicine, Edinburgh: Edinburgh University Press, pp.49-68; Eyles, J. and Woods, K.J. 1983. *The Social Geography of Medicine and Health*. London: Croom Helm; Mishler, E.G. (ed.) 1981. *Social Contexts of Health, Illness and Patient Care*. New York: Cambridge University Press,1981.

[15] Cunningham-Burley, S. 1990. Mothers' Beliefs About and Perceptions of Their Children's Illnesses, in Cunningham-Burley, S. and McKeganey, N.P. (eds.), *Readings in Medical Sociology*. London: Routledge, pp. 85-109; Navarro, V. 1980. Work, Ideology and Science: The Case of Medicine, *Social Science and Medicine* 14c, pp. 191-205; Waitzin, H. 1983. *The Second Sickness: Contradictions of Capitalist Health Care*. New York: Free Press.

[16] Shorter, E. 1985. *Bedside Manners: The Troubled History of Doctors and Patients*. New York: Simon and Schuster.

[17] Rathwell, T. and Philips, D. (eds.) 1986. *Health, Race and Ethnicity*. London: Croom Helm.

[18] Frankenburg, R. 1988. Gramsci, Culture, and Medical Anthropology: Kundry and Parsifal? or Rat's Tail to Sea Serpent?, *Medical Anthropology Quarterly* 2, pp.324-337, 1988; Gesler, W.M. 1991. *The Cultural Geography of Health Care Delivery*. University of Pittsburg Press; Kapferer, *op. cit*.

[19] Crawford, *op. cit*.

[20] Cornwell, J. 1984. *Hard-earned Lives: Accounts of Health and Illness from East London*. London: Tavistock; Eyles, J. and Donovan, J. 1986. Making Sense of Sickness and Care: An Ethnography of Health in a West Midlands Town, *Transactions. Institute of British Geographers* 11, pp. 415-427.

[21] Jones, K. and Moon, G. 1987. *Health, Disease and Society: An Introduction to Medical Geography*, London: Routledge and Kegan Paul; Scheper-Hughes, N. 1990. Three Propositions for a Critically Applied Medical Anthropology, *Social Science and Medicine* 30, pp.189-197.

[22] Eyles and Woods, *op. cit.*

[23] Blaxter, M. 1986. *The Meaning of Disability*. London: Heinemann; Calnan, M. 1990. Food and Health: A Comparison of Beliefs and Practices in Middle-class and Working-class Households, in Cunningham-Burley, S. and McKeganey, N.P. (eds.), *Readings in Medical Sociology*. London: Routledge, pp.9-36; Cornwell, *op. cit.*; Eyles and Donovan, *op. cit.*; Kleinman, A. 1978. Concepts and a Model for the Comparison of Medical Systems as Cultural Systems, *Social Science and Medicine* 12, pp. 85-93; Mullen, K. 1990. Drink is All Right in Moderation: Accounts of Alcohol Use and Abuse from Male Glaswegians, in Cunningham-Burley, S. and McKeganey, N.P. (eds.), *Readings in Medical Sociology*. London: Routledge, pp. 138-156.

[24] See, for example, as a response to such concerns, Waxler-Morrison, N., Anderson, J. and Richardson, E. (eds.) 1990. *Cross-cultural Caring: A Handbook for Health Professionals in Western Canada*. Vancouver: University of British Columbia Press.

[25] Mishler, *op. cit.*

[26] Kleinman, *op. cit.*; Kleinman, A. 1980. *Social Origins of Distress and Disease*. New Haven, Ct.: Yale University Press.

[27] Canada, 1986, *op. cit.*; Jenkins, J.H. 1988. Conceptions of Schizophrenia as a Problem of Nerves: A Cross-cultural Comparison of Mexican-Americans and Anglo-Americans, *Social Science and Medicine* 26, pp. 1233-1243; Kleinman, 1978, *op. cit.*

[28] Eisenthal, S., Emery, R., Lazare, A. and Udin, H. 1979. Adherence and the Negotiated Approach to Patienthood, *Archives of General Psychiatry* 36, pp. 393-398.

[29] Dyck, I. 1989. The Immigrant Client: Issues in Developing Culturally Sensitive Practice, *The Canadian Journal of Occupational Therapy* 56, pp. 248-255.

[30] Weidman, H.H. 1979. The Transcultural View: Prerequisite to Interethnic (Intercultural) Communication in Medicine, *Social Science and Medicine* 13B, pp. 85-87.

[31] Lazarus, E.S. 1988. Theoretical Considerations for the Study of the Doctor-Patient Relationship: Implications of a Perinatal Study, *Medical Anthropology Quarterly* 2(1), pp.34-58.

[32] Pappas, G. 1990. Some Implications for the Study of the Doctor-patient Interaction: Power, Structure and Agency in the Works of Howard Waitzin and Arthur Kleinman, *Social Science and Medicine* 30, pp. 199-204.

[33] *Ibid.*

[34] Waitzin, H. 1984. The Micropolitics of Medicine: A Contextual Analysis, *International Journal of Health Services* 14, pp. 339-377.

[35] Davis, D.L. and Low, S.M. (eds.) 1989. *Gender, Health and Illness: The Case of Nerves*. New York: Hemisphere Publishing; Doyal, L. and Elston, M.A. 1986. Women, Health and Medicine, in Beechey, V. and E. Whitelegg, E.(eds.) *Women in Britain Today*. Milton Keynes: Open University Press, pp. 173-209; Miles, A., *The Neurotic Woman: The Role of Gender in Psychiatric Illness*. New York: New York University Press.

[36] Pappas, *op. cit.*

[37] Singer, M. 1990. Reinventing Medical Anthropology: Toward a Critical Alignment, *Social Science and Medicine* 30, pp. 179-187.

[38] See also for further discussion on the contributions of critical medical anthropology to the analysis of the context of medicine: Baer, H., Singer, M. and Johnsen, J. (eds) 1986. Toward a Critical Medical Anthropology, *Social Science and Medicine* 23(2); Baer, H.A. 1990. The Possibilities and Dilemmas of Building Bridges Between Critical Anthropology and Clinical Anthropology: A Discussion. *Social Science and Medicine* 30, pp. 1011-1013; Padgett, D. and Johnson, T.M. 1990. Somatizing Distress: Hospital Treatment of Psychiatric Co-morbidity and the Limitations of Biomedicine, *Social Science and Medicine* 30, pp. 205-209; Scheper-Hughes, *op. cit.;* Singer, M., Baer, H.A. and Lazarus, E. 1990. Critical Medical Anthropology in Question, *Social Science and Medicine* 30, pp. V-V111.

[39] See, for example: Alberro, A. and Montero, G. 1976. The Immigrant Woman, in Matheson, G. (ed.), *Women in the Canadian Mosaic*. Toronto: Peter Martin Associates,1976; Arnopoulos, S. 1979. *Problems of Immigrant Women in the Canadian Labour Force*. Ottawa: Canadian Advisory Council on the Status of Women; Ng, R. and Ramirez, J. 1981. *Immigrant Housewives in Canada*. A report available from the Immigrant Women's Centre, Toronto; Zambrana, *op. cit.*

[40] Donovan, J. 1986. *We Don't Buy Sickness, It Just Comes: Health, Illness and Health Care in the Lives of Black People in London*. London: Tavistock, 1986; Pearson, M. 1986. The Politics of Ethnic Minority Health Studies, in Rathwell, T. and Philips, D. (eds.), *Health, Race and Ethnicity*. London: Croom Helm, pp. 100-116.

[41] Nyakabwa, K. and Harvey, C.D.H. 1990. Adaptation to Canada: The Case of Black Immigrant Women, in Dhruvarajan, V. (ed.), *Women and Well-Being*. Montreal: McGill-Queen's University Press, pp. 138-149; Szekely, E.A. 1990. Immigrant Women and the Problem of Difference,

in Dhruvarajan, V. (ed.), *Women and Well-Being*. Montreal: McGill-Queen's University Press, pp. 125-137.

[42] Clarke, J. 1983. Sexism, Feminism and Medicalism: A Decade Review of Literature on Gender and Illness, *Sociology of Health and Illness* 5, pp. 62-82.

[43] Begin, M. 1990. Redesigning Health Care for Women, in Dhruvarajan, V. (ed.), *Women and Well-Being*. Montreal: McGill-Queen's University Press, pp. 3-13; Dally, A. 1982. *Inventing Motherhood*. London: Burnett Books Ltd.

[44] Cairncross, F. 1986. The Vulnerable Workforce, *Times Literary Supplement*, 4 April ,1986.

[45] Graham, H. 1984. *Women, Health and the Family*. Brighton, Sussex: Wheatsheaf.

[46] Dyck, I. 1989. Integrating Home and Wage Workplace: Women's Daily Lives in a Canadian Suburb, *The Canadian Geographer* 33(4), pp. 329-341.

[47] Coupland, V. 1982. Gender, Class and Space as Accessibility Constraints for Women with Young Children, in *Contemporary Perspectives on Health and Health Care*, Occasional Paper no. 20, London: Department of Geography, Queen Mary College, pp.51-69; Pearson, M. 1989. *Awareness and Use of Well Woman Services in Liverpool*. Occasional Paper No. 1., Department of General Practice, University of Liverpool.

[48] Kobayashi, A.L. and Rosenberg, M.W. 1990. The Universal Right to Health: An Ideological Discourse, Paper presented to the Conference on Health and Development, Kingston, Jamaica, December, 1990.

[49] Blaxter, M. 1983. The Causes of Disease; Women Talking, *Social Science and Medicine* 17, pp. 59-69.

[50] Graham, *op. cit.*; Szekely, *op cit.*

[51] Bodnar, A. and Reimer, M. 1979. *The Organization of Social Services and its Implications for the Mental Health of Immigrant Women*. Report produced for the Secretary of State, Canada; Dhruvarajan, V. (ed.) 1990. *Women and Well-Being*. Montreal: McGill-Queen's University Press.

[52] Anderson, J.M. and Lynam, M.J. 1987. The Meaning of Work for Immigrant Women in the Lower Echelons of the Canadian Labour Force. *Canadian Ethnic Studies* 19, pp. 67-90.

[53] Ng, R. 1988. *The Politics of Community Services*. Toronto: Garamond.

[54] *Ibid.*

[55] Hom, S. 1979. Working in a Cannery, *Multiculturalism* 11(4), pp. 20.

[56] Nyakabwa and Harvey, *op. cit.*

[57] Anderson, J.M. 1986. Ethnicity and Illness Experience: Ideological Structures and the Health Care Delivery System, *Social Science and Medicine* 22, pp. 1277-1283.

[58] Zambrana, *op cit.*

[59] Ng and Ramirez, *op cit*; Westwood, S. and Bachu, P. (eds.) 1988. *Enterprising Women.* London: Routledge.

[60] Cassin, M. and Newton, J.L. 1979. Family and work, *Multiculturalism* 11(4), pp.21-22.

[61] Dyck, I., *Living in a New Country: Childrearing and Coping Strategies of Indo-Canadian Women in Vancouver.* Research in progress.

[62] Whelehan, P. and contributors. 1988. *Women and Health: Cross-cultural Perspectives.* Massachusetts: Bergin & Garvey.

[63] Donovan, *op cit.*

[64] *Ibid.*

[65] Anderson, J. M., Dyck, I. and Lynam, M.J., *Acquiring and Using Professional Health Care Knowledge: The Experiences of Chinese-Canadian Women with Chronic Health Conditions.* Research in progress.

[66] Hunt, L.M., Jordan, B. and Irwin, S. 1989. Views of What's Wrong: Diagnosis and Patients' Concepts of Illness, *Social Science and Medicine* 28, pp.945-956.

[67] Conrad, P. 1985. The Meaning of Medications: Another Look at Compliance, *Social Science and Medicine* 20(1), pp.29-37.

[68] Sachs, L. 1987. Evil Eye or Bacteria: Turkish Migrant Women and Swedish Health Care, *Resarch in the Sociology of Health Care* 5, pp.249-301.

[69] Szekely, *op cit.*

[70] Cassin and Newton, *op cit.*

[71] Giddens, A. 1984. *The Constitution of Society.* Cambridge:Polity Press; Giddens, A. 1987. *Social Theory and Modern Sociology.* California: Stanford University Press.

[72] Gregory, D. 1981. Human Agency and Human Geography, *Transactions, Institute of British Geographers* N.S. 6(1), pp.1-18; Sayer, A. 1984. *Method in Social Science: A Realist Approach.* London:Hutchinson; Soja, E.W. 1987. The Postmodernization of Geography: A Review, *Annals of the Association of American Geographers* 77, pp.289-323.

[73] Jones and Moon, *op cit.*

[74] Curtis, S.E. and Ogden, P.E. Bangladeshis in London: A Challenge to Welfare, *Revue Europeenee des Migrations Internationales* 2(3), pp.135-149; Graham, *op cit.*

[75] Smith, D.E. 1986. Institutional Ethnography: A Feminist Method, *Resources for Feminist Resarch* 15, pp. 6-13.

[76] Lowe, M.S. and Short, J.R. 1990. Progressive Human Geography, *Progress in Human Geography* 14(2), pp1-11.

[77] Gerber, D.A. 1990. Listening to Disabled People: The Voice of Authority in Robert B. Edgerton's "The Cloak of Competence", *Disability, Handicap and Society* 5(1), pp.3-24; Wendell, S. 1989. Toward a Feminist Theory of Disability, *Hypatia* 4, pp. 104-124.

[78] Smith, S.J. 1990. Social Geography: Patriarchy, Racism, Nationalism, *Progress in Human Geography* 14(2), pp. 260- 271.

[79] Ng, *op. cit.*; and Rathwell and Phillips, *op. cit.*

[80] Further discussion on the role of the researcher in qualitative research, the presentation of data, and special issues in research concerned with people with disabilities can be found in: Atkinson, P. 1990. *The Ethnographic Imagination*, London: Routledge; Biklen, S.K. and Moseley, C.R. 1988. "Are you retarded?" "No, I'm Catholic": Qualitative Methods in the Study of People with Severe Handicaps, *Journal of The Association for Persons with Severe Handicaps* 13(3), pp. 155-162; Hammersley, M. and Atkinson, P. 1983. *Ethnography: Principles in Practice*. London: Tavistock.

[81] Lord, J. and Farlow, D.M. 1990. A Study of Personal Empowerment, *Health Promotion* 29, pp. 2-8.

[82] Curtis and Ogden, *op. cit.*; Pearson, *op. cit.*

11 HOUSING FOR PERSONS WITH HIV INFECTION IN CANADA: Health, Culture and Context

Sharon Manson Willms
School of Social Work
University of British Columbia

INTRODUCTION

In March 1990, an interdisciplinary research team consisting of a social economist/social worker, housing expert and health geographer associated with the Centre for Human Settlements at the University of British Columbia joined forces to examine the housing system for persons with HIV across Canada. The study was funded by the National Welfare Grants division of Health and Welfare Canada following the recommendation of a National Task Force of the Canadian Association of Schools of Social Work. Four constituencies were identified for study purposes: gay men; injection drug users; sex trade workers; and hemophiliacs. Pediatric AIDS cases were excluded. During the course of the study a separate category emerged, which consisted of women with AIDS. Over 150 interviews, in both French and English, were conducted in every province and territory in Canada, identifying the formal and informal organizational responses in urban and rural contexts as shown in Table 1,11. As well, individuals with HIV were interviewed and asked to describe their own experiences with respect to their housing situations.

This chapter reports on the national investigation of the housing needs of persons with HIV infection; the community where persons with HIV are located; and the regional and national themes that emerged. The relationship between the health status of individuals, the community that produces the local culture, and the political and economic context of the various regional impacts on the individual experiences are examined, as well as the actions that may be taken to ameliorate the pain of the person living with HIV infection.

Table 1,11 Housing for Persons with HIV: Interviews (March 1991)

	Community-Based Services	*Health & Government*	*Housing*	*Individuals*	*Totals*
B.C.	7	4	8	8	**27**
Alberta	4	1	1	0	**6**
Saskatchewan	5	2	0	2	**9**
Manitoba	3	0	1	6	**10**
Ontario	25	9	16	0	**50**
Quebec	13	6	7	1	**27**
Nova Scotia	3	0	1	0	**4**
New Brunswick	4	2	0	1	**7**
Newfoundland	2	1	0	0	**3**
P.E.I.	2	0	0	0	**2**
N.W.T.	1	2	0	0	**3**
Yukon	1	3	0	0	**4**
TOTALS	68	31	35	18	**152**

There are a variety of issues that impact on housing needs that are complicated by the presence of illness, especially a terminal illness. The adequate, accessible, affordable and available housing which together are the constituents of appropriate housing varies with the health status of the individual, the culture of the community and the interaction between the two. Acquired Immune Deficiency Syndrome (AIDS) and its precursor, Human Immunodeficiency Virus (HIV), have social, political, medical, economic and cultural dimensions which impinge on the ability of the individual to keep or acquire appropriate housing and the ability of the community to provide the necessary supports to make an appropriate range of housing options available. Political and social response to HIV/AIDS is complicated by homophobia, heterosexism and fear of disease transmission based on ignorance. As persons with HIV are living longer through medical intervention and changes in lifestyle, housing needs have changed. Early literature focussed on hospice and palliative care, rather than self-care in the community.[1]

In order to understand housing issues, it is important to briefly describe HIV disease and its progression; the social problems associated with public response to AIDS; and the implications for housing. The themes that have emerged from this national investigation are reported and discussed.

HEALTH STATUS INDICATORS

Physical Characteristics

HIV is a communicable disease that is transmitted through bodily fluids - semen, blood and breast milk. Sexual intercourse and sharing needles with an infected person are the primary forms of transmission. Blood transfusions that took place before 1985 were responsible for infecting 943 people in Canada.[2] The majority are in the hemophiliac group. Deep needle sticks are rare, generally experienced only by health professionals, and the probability of contracting HIV from a needle stick is much less probable than contracting hepatitis for example.[3]

Without laboratory testing at the asymptomatic stage, knowledge of one's own HIV infection or identifying the presence of the virus in others is very uncommon. Individuals who carry HIV may remain asymptomatic for several years before displaying any of the signs of AIDS, and may only become aware of their HIV status when symptoms typical of AIDS appear, prompting a clinical examination or blood test. Women are especially vulnerable to late diagnosis because typical lists of symptoms are based upon the male experience of HIV and do not include gynecological symptoms experienced by women.[4] High risk behaviors for contracting HIV include intravenous drug use, blood transfusion prior to 1985, practicing homosexual or bisexual behavior, multiple sexual partners and/or having sexual intercourse with a person who has HIV.[5] Diagnosis of HIV or AIDS is dependent upon the individual's ability to discern their status, or upon a medical diagnosis based upon typical symptoms. Typical physical symptoms of HIV include lymph node swelling of unknown origin, involuntary weight loss, unexplained fever over 38 celsius, chronic diarrhea, weakness, fatigue, deteriorated ability to work and night sweats.[6] These symptoms can continue for several years before progressing to AIDS, which is defined by the presence of AIDS specific diseases such as pneumocystis carinii pneumonia (PCP), cryptococcal meningitis, toxoplasmosis or various abnormal immunological laboratory test results.[7]

Figure 1,11 records the list of diseases associated with HIV and AIDS and the estimated time of progression of the disease.[8] In patients who have developed AIDS and died, the progression of the disease from infection to death has been about 10 years. There are various opportunistic infections which characterize the early stages, including thrush, shingles, severe athletes foot, and oral leukoplakia. As the helper cell count in the blood diminishes, there is a concurrent rise in the risk of life threatening opportunistic infections.

Psycho-Social Issues

The diagnosis of a terminal illness can trigger psycho-social problems in any person. There are, however, several complicating factors specific to HIV. The three main points when persons with HIV are at risk for developing psycho-social problems are at the time of:

1. diagnosis of seropositivity;
2. transition from asymptomatic to symptomatic; and
3. diagnosis of AIDS.

As shown in Table 2,11, the diagnosis of HIV in any of its forms may precipitate depression, suicide ideation, suicide attempts and suicide, and/or exacerbate any pre-existing mental illnesses.[9] The long incubation period and inability to predict when the infection will progress can leave the individual in a constant state of stress and uncertainty; physical decline and loss of attractiveness negatively affects self-esteem; the possibility of infecting others may result in self-imposed isolation and feelings of guilt; fear of reaction from lovers, friends, family, community and the workplace may result in an inability to seek appropriate forms of support. The experience of discrimination for many persons with HIV is not unusual if they have already identified with the gay community or are an injection drug user. Discrimination escalates with a diagnosis of AIDS and previously supportive networks may disappear. In cases where homosexual behavior has been closeted, the fear of being 'found out' may create grave stress and anguish for the individual who fears being 'out of the closet'. Internalized homophobia, found to be common in gay men

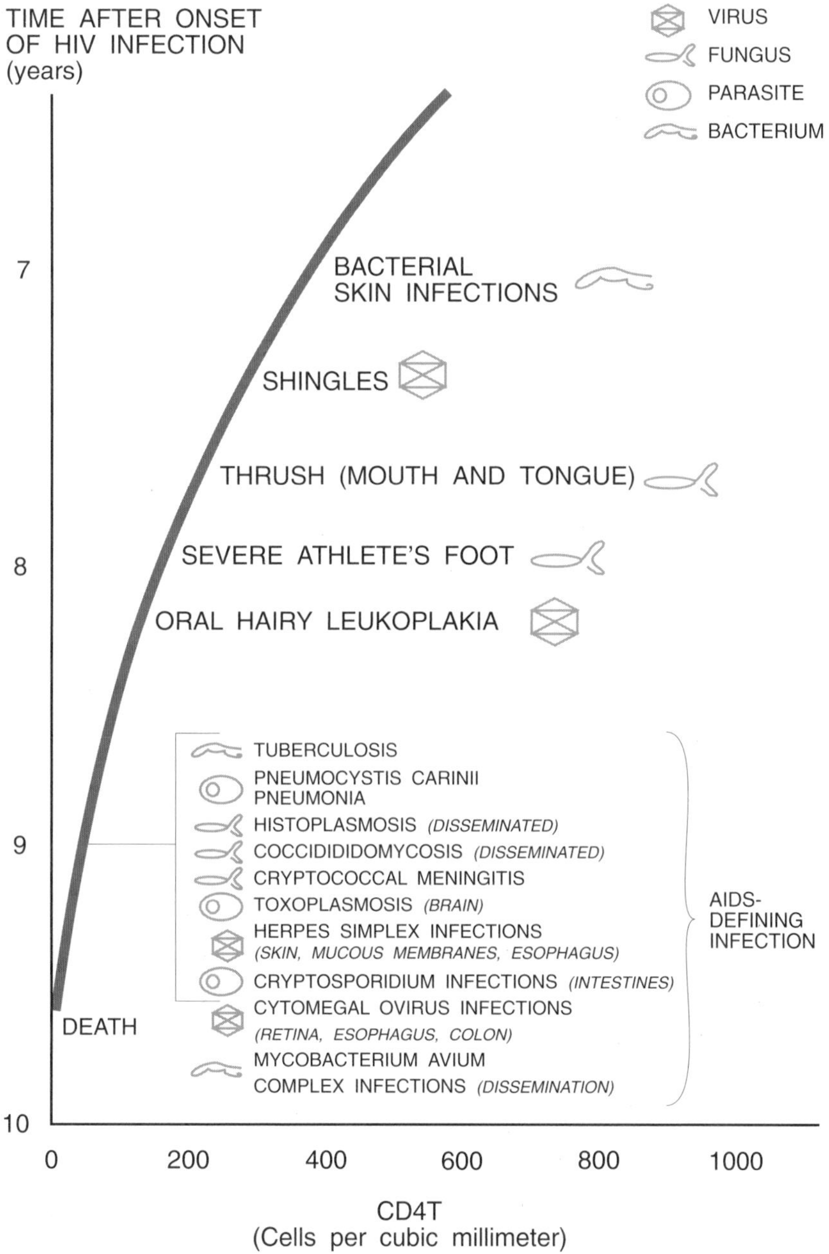

Figure 1,11 Diseases Associated with HIV and AIDS and Time to Progression (adapted from Scientific American, August 1990, p. 53).

who are diagnosed as HIV positive and subsequently seek support, is seen as a response to long term negative social conditioning about homosexuality.[10] The combination of homosexuality and a diagnosis of a terminal illness in the prime of life (the modal age group of HIV diagnosed persons is 30 to 39 years of age, as shown in Figure 2,11[11]) makes it difficult for some individuals to resist the notion that it is a punishment rather than a disease that affects some people who have engaged in high risk behavior. The new taboo is coming out of the closet as HIV positive, especially with other gay men.[12] Further, given that the vast majority of cases in Canada and the United States have affected gay men (85 percent in Canada[13]), most gay men have friends who have died of AIDS. This personal experience arms the newly diagnosed individual with the knowledge of the intractability of the end, and an understanding of the losses that are yet to come.

Table 2,11 Psycho-Social Issues Associated With HIV Infection

- Uncertainty of disease progression
- Loss of some physical capacity
- Loss of some sexual capacity
- Loss of self-esteem
- Feelings of guilt
- Fear of reaction from community
- Objectification and discrimination
- Self-imposed isolation
- Overshelming sense of loss
- Fear
- Depression
- Despair
- Suicide

For women the psycho-social issues related to HIV are often associated with their status as care givers in the family. Many women learn of their seropositive status at the time of pregnancy and are faced with the dual prospect of diminished capacity to care for themselves at the same time as the increased demands related to the birth of a child. Women

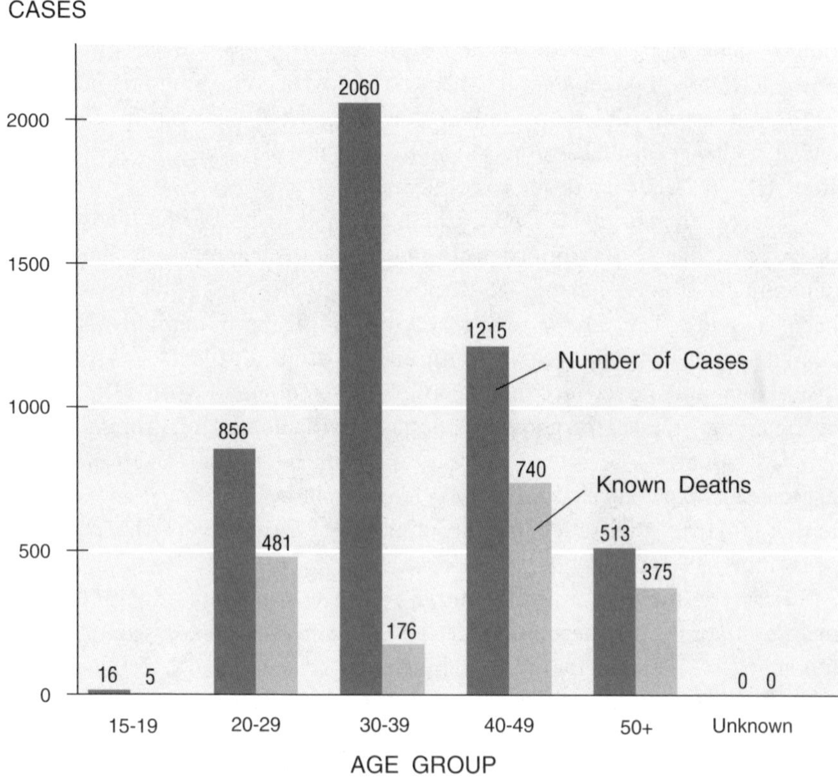

Figure 2,11 Adult Cases of HIV Diagnoses (February 4, 1991).

interviewed in this study expressed a severe sense of isolation, had fewer support networks than men, and admitted to disclosing their HIV status to fewer people than men typically do.

Hemophiliacs confront similar problems as other seropositive men. Given that hemophilia is genetic, many families experience HIV first through the extended family or hemophiliac community. Homophobia is an explicit issue for many men infected through blood transfusion. They are concerned that they will be identified as homosexuals because of their seropositive status. There is a strong association between HIV and homosexuality in the minds of most Canadians and, as discussed earlier, there is evidence to support this association. In most hemophiliac communities outside of the three major cities in Canada, there is an avoidance of the use of the word AIDS as a descriptor for an individual's condition.

For injection drug users the diagnosis of HIV can serve as a catalytic turning point and a decision to stop drug use. Diagnosis of HIV may offer an opportunity to intervene and begin a detoxification program. According to subjects in this study, this is not always the outcome of diagnosis, and if a change in illegal drug use does not take place then the progression from HIV to AIDS to death is accelerated.

The experience of HIV mixes both sex and death on the landscape of the psyche. These two primal forces in human existence are frightening to contemplate, thus a number of taboos to regulate public discussion have been created. The taboos, while serving the needs of individuals in the community to deal with fear of birth and of death, operate to restrict open discussion and hence acceptance of persons infected with HIV. Fear becomes the operant response, of both individuals and the community to the individual. The attendant consequences of actions/reactions based on fear are experienced by the individual in terms of psycho-social problems, and then reinforced by the community response to the problems presented.

Fear for the future is a natural response, and stress coupled with uncertainty may further diminish the immune system's capacity of the individuals. These factors may impinge on the individual's ability to cope with the disease and limit the support available within the community for similar reasons. And they impinge on the ability of the individual to seek and utilize appropriate resources.

HEALTH AND HOUSING

In general, the presence of a terminal illness demands some change in housing conditions. The diagnosis of AIDS is complicated by uncertainty in the progression of the disease, as not all persons will experience the symptoms discussed previously, while some persons experience many of them. Individuals can be very ill and then recover enough to resume previous patterns of living. The fact that the disease does not progress in a linear fashion means that the need for medical, social and support services is variable. Treatment regimes are dependent upon close monitoring of CD4 T cells and primary prophylactic action. Once a major opportunistic infection has occurred, then secondary prevention techniques are required. As none of the treatments thus far are cures, there is a need for continuous treatment, many of which are administered in hospitals or specialized clinics. These factors affect the demand for housing around hospitals, specialized clinics or centres which serve the person with HIV.

Issues in Housing

Housing represents more than a physical building in our culture. It is a basic element in ensuring safety and security. It locates one in the community and gives a sense of place and belonging. Housing facilitates routinized behavior -- going to work, preparing meals, having a familiar place to sleep, etc. It is an anchor from which services are accessed and received. The requirements imposed by HIV disease and the societal response to persons who are infected interact to affect the housing market for persons who are HIV positive. Four aspects used to determine the appropriateness of housing (namely adequacy, accessibility, affordability and availability) are considered from the perspectives of the individual, the community and the state response (government at all levels: municipal, provincial and national). The picture that results is the reality of the limited choices offered to persons with HIV infection in the housing market.

A report by the Resource Information Service of the London Special Needs Housing Group describes the standards for housing for persons with HIV infection as:

> Their housing should be warm, easy to move around in and keep clean. It should offer privacy and control over their own lives. The effect of stress on their health means that their accommodation should be as stress free as possible. It should offer protection from harassment and physical attack, security of tenure and be manageable financially. Good housing is also housing that is responsive to what people want, which offers them some choice about the way they wish to live and is flexible to changing circumstances. Housing for people with HIV infection needs to take into consideration the support and care they are likely to require if their condition worsens.[14]

These criteria are offered as instruction to British social housing planners who are attempting to respond to the social housing needs of persons with HIV infection. The situation for most Canadians is much different as, for the vast majority, housing is obtained on the open market, not through a non-market state subsidized sector. When searching for accommodation, the person with HIV infection must consider their needs against what is available in the private market because 95 percent of Canada's housing stock is in the private sector.[15]

Adequacy

Housing must be warm, dry, easy to keep clean and offer a safe and secure environment. While the list may seem obvious to every Canadian, the importance of warm, dry, mold free and simple to keep hygienically clean housing is of paramount importance to persons who are immuno-compromised. The risk of a cold or flu developing into PCP is significant. The presence of innocuous bacteria normally associated with food and sanitation facilities may prove deadly to a person with HIV infection.

The community has an interest in the provision of adequate housing for persons who are HIV infected, especially for individuals who continue to engage in high risk behavior. Of particular concern are homeless injection drug users. Without an address, they are more difficult to serve, monitor and provide ongoing prevention education. Similarly, without a place to sleep at night, it is more likely that the injection drug user will use drugs to stay awake and keep moving, thus further compromising their health.[16] Or, as studies on safe sex practice have indicated, relapse to non-safe sex practices are 50 percent higher when drugs or alcohol are involved.[17] The epidemic rise of STD's in inner city populations where sex is traded for drugs also gives rise to concern for the increased risk of HIV infection.[18]

In the case of the state's role in providing adequate housing, the issue may be framed in terms of social justice where there are social rights to housing. Canada has no such commitment to the right to housing, and the Constitution Act, while providing for equal treatment before the law, does not enshrine economic or housing rights, and excludes sexual orientation.[19] The state must balance the competing interests between the social rights of the individual and the collective good of the community.

Accessibility

The entrance to the home must be accessible for a person who is weak and in a state of chronic fatigue. A level entry or few stairs to climb are important to a person's ability to stay in their own home. A third floor walk-up cannot be considered accessible to a person carrying the day's groceries when they are already fatigued.

Inside the home, there must be room enough to allow a person who is physically handicapped to move about, and to move easily from bath, to bedroom, to eating area without significant impediments. The bathroom should have supports for access to the toilet and into the bathtub. Space for a care giver to spend the night and/or move around the living area is important to being able to maintain the person in their own home during bouts of fatigue or opportunistic infections that do not require hospitalization, but require increased amounts of care. Many interviewees noted quiet location as an important factor in maintaining their health. Privacy is also important for both psychological and physical health reasons.

Housing must be reasonably close to the services required, such as shops and transportation, and the medical and social services needed. Proximity to services has been a factor in demand for housing in Seattle and San Francisco around the major hospital centres. In Vancouver, the location of St. Paul's Hospital in Vancouver's West End coincides with a greater density of gay men than other areas of the city. This clustering of the gay community preceded the discovery of HIV infection. In Toronto and Montreal, there is a similar finding. Toronto's Casey House and 127 Isabella are located in the College and Church area, which has been the centre of the gay community and is also a centre for hospital services. In Montreal, Le Village is recognized as a gay area and Le Plateau as the anglophone gay area. Both are located in downtown Montreal. In Winnipeg there seems to be no one area where people who are HIV positive congregate, despite the presence of the Village Clinic which offers integrated care and service for gay men and women, and particularly persons with HIV.

The geographic distribution of persons with HIV infection is primarily found in Canada's three largest urban areas as shown in Figure 3,11. HIV is mostly an urban phenomenon. This finding is most likely skewed by the fact that many persons who suspect that they are HIV positive go to the major centres for testing. Epidemiological data related to seroprevalence are collected on the basis of the diagnosis site, not the person's residence.

There appear to be advantages for individuals with HIV if they live near one of the three big medical and social service centres. The community which has formed to respond to the needs of persons with HIV is more accessible, and additional support is often available on the basis of the person's close proximity to this community of interest. In many cases service delivery is more efficient as there are shorter distances to be covered for both the care giver and the person seeking care. A critical mass of people concerned about the impact of HIV on people is formed and the necessary energy to develop services can be channelled productively.

At first glance, an integrated services centre is more efficient, as it concentrates resources and expertise in a single location. The competing scenario is that of ghettoizing the population of interest into a particular

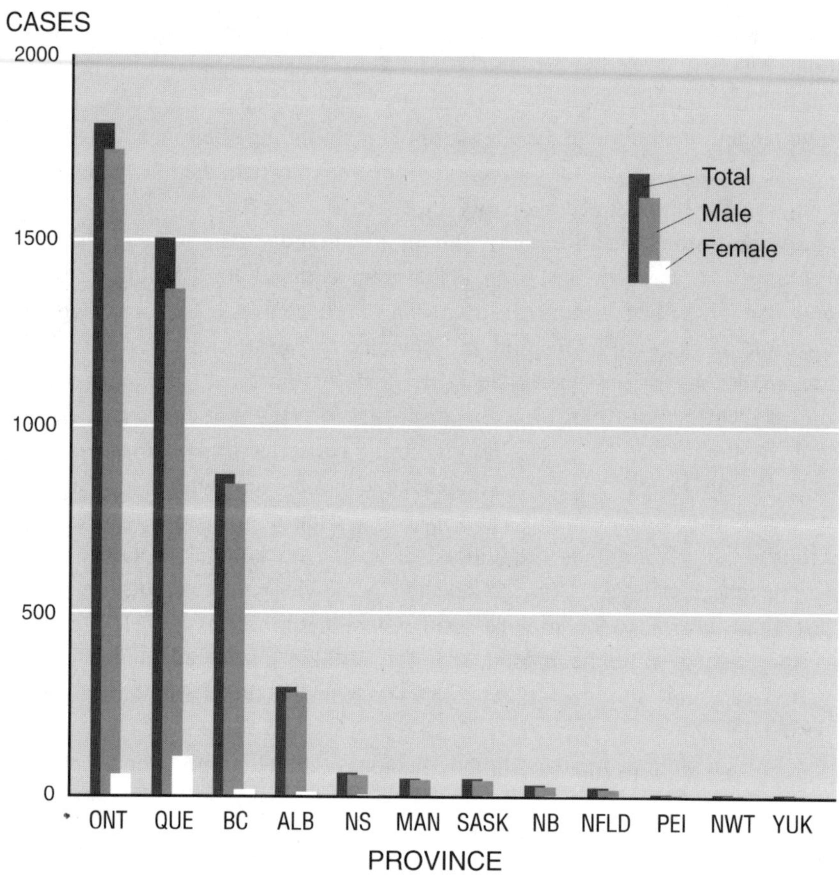

Figure 3,11 Geographic Distribution of Persons with HIV Infection (February 4, 1991).

geographic location. Further, by situating services in a single location, persons who do not live in the immediate vicinity are underserved. For example, persons with HIV in rural areas have little or no support systems and even those persons who reside outside the three big urban areas find that they do not receive the same calibre of service. In Victoria, for example, persons with HIV outside of hospital must come to Vancouver to have a blood test. In Regina, there is one health care professional who is designated as responsible for the patient population with HIV . . . and all other patients with infectious diseases.

Affordability

"Most Canadians who are disabled are poor." [20] In studies on homelessness, a strong correlation between onset of a disabling condition and subsequent transition to homeless has been observed.[21] The facts related to income distribution support this observation in the context of persons with HIV. Single adult earners are in a poor position compared to dual earner families, and are at a higher risk for sinking below the poverty line than almost all other groups with the exception of single parent women.[22] And for women with HIV infection, many have children and thus are at a very high risk for poverty.

The exception is the group who have benefitted from the federal government's Catastrophic Relief Program. This program pays $120,000 cash over four years to persons who contracted HIV as a consequence of blood transfusion prior to 1985. Approximately 943 persons have applied for benefits under the Catastrophic Relief Program,[23] the majority of whom are identified as hemophiliacs.

Others are faced with an accelerated decline into poverty as they are forced to leave their jobs due to illness. For those that are able to maintain their attachment to the labor force, the attachment becomes more like a shackle than a voluntary association because many are afraid to move jobs for fear of being uncovered as seropositive as a consequence of medical tests required for life insurance; or they realistically fear that the added stress of taking on a new job may imperil health.

Income needs for persons who are disabled are generally more than those experienced by persons who have good health as a resource. The extra expenses associated with illness or the services required for daily living can be substantial, and these needs vary over the course of the disease. Many people experience severe illness, recuperate and are again able to participate in daily living activities, including paid work.

Given that the average age for the largest group with HIV infection is in the 30 to 39 year bracket, most have not built up savings that are sufficient to carry them through to the end of their disease. As the most frequently reported risk for acquiring HIV in Canada is through homosexual contact,[24] the majority are not in a relationship that resembles marriage as an economic union. Single men in their mid-30s who are identified as homosexuals are prone to stigmatization as they do not fit within the cultural norm of 'deserving poor', despite the fact that they are indeed sick and in need of support. Therefore, the experience of many interviewees is one of having to persistently fight for any or all entitlements provided by the state, even when they are entitled by law. Income assistance is stigmatizing to most recipients, however, being single, male, mid-30s and gay is a combination that is destined to ascribe marginality in our culture without significant activism by supportive members of the community.

The situation for women with HIV, especially those with children, is very difficult financially. While they may receive income assistance, it is usually insufficient to meet both shelter and support needs. Mothers on income assistance who are also HIV positive may be unable physically to access free services such as food banks to supplement their inadequate incomes.

Injection drug users, when faced with spending money on housing or drugs to support their habit, will satisfy short term needs. Affordability as an issue for injection drug users is experienced more at the level of the community of interest who are trying to provide appropriate housing at a price that is within reach of the agencies concerned. Whoever the constituency, there is an obvious gap between income substitutes and the poverty line, whether the money comes from disability pensions or income assistance, as shown in Table 3,11.[25] The lack of adequate core income programs condemns most disabled persons to live in abject poverty.

Both the community of interest and the geographic community have a concern in affordable housing for persons with HIV. The community of interest may be those persons who are active advocates on behalf of persons with HIV and include seropositive members. The geographic community, who are the majority, are the neighbors in the community where persons with HIV live, and in particular the city dwellers of Toronto, Montreal and Vancouver. There is pressure not to raise taxes to pay for services for persons with HIV by rate payers. Similarly, the local population does not want to have its local hospital beds filled with persons who are there because there are inadequate housing supports in the community. In short, there are competing interests within the community, and the management of these interests is dependent upon the ability of the advocacy group to stage manage the concerns to support the direction for spending.

The government position with respect to affordability issues and HIV infection is related to tax transfers between the three levels of government. Who pays for what, and how much, is a complicated issue because housing in Canada is a private market phenomena. There is very little control over housing markets, especially rental housing, other than regulating the market. The attempts thus far have been inadequate to solve the housing affordability problems for urban poor, and even less benefit to persons with HIV, other than the recipients of Catastrophic Relief payments who are a very small proportion of the persons with HIV.

Table 3,11 Available Income Transfer Programs Compared to Poverty Lines by Province

Province	*Average Monthly Welfare Income 1988*	*Yearly Welfare & Benefit Payment*	*Poverty Line 1988*	*Average Yearly Disability Pension Age 35-39*
Newfoundland	710.50	8526	10984	6139
Prince Edward Island	641.50	7698	9526	6139
Nova Scotia	736.35	8836	10984	6139
New Brunswick	613.00	7356	10984	6139
Quebec	592.00	7104	11564	6139
Ontario	901.25	10815	11564	6139
Manitoba	595.50	7146	11564	6139
Saskatchewan	764.17	9170	10984	6139
Alberta	889.17	10670	10984	6139
British Columbia	637.17	7646	11564	6139
Yukon	569.17	6830	11564	6139
Northwest Territories	-	-	-	6139

Low Income Cutoffs 1989 Statistics Canada by Size of Household and Population Size of Area

Size of Household	*500,000*	*100,000*	*30,000*	*<30,000*	*Rural*
1	13414	11788	11511	10493	9135
2	18192	15968	15602	14228	12382

Availability

Housing availability is closely related to the issues of affordability and accessibility. For the individual, many live close to health and social service centres, which puts them into higher priced neighborhoods than they might otherwise have chosen. The urban core, which is the site of the largest medical service centres in the three largest cities in Canada, means that the person with HIV is competing in a market place that is characterized by low vacancy rates, generally reported as less than three percent, and in some cases less than one percent.[26] In Toronto and Vancouver the vacancy rates hover around one percent or less.[27] In Montreal the vacancy rates are higher,[28] however the income assistance rates are among the lowest in Canada.[29] Table 4,11[30] presents selected average rents for one bedroom apartments and vacancy rates.

Table 4,11 Average Rents for One Bedroom Apartments and Vacancy Rates for Selected Cities (April 1990)

City	*Rent*	*Vacancy Rate*
Charlottetown	\$386	4.7%
Quebec	\$410	5.0%
Montreal	\$419	4.5%
Toronto	\$559 all units \$693 vacant units	1.0%
Calgary	\$456	2.0%
Edmonton	\$413	1.8%
Vancouver	\$548	0.9%

In most cases that were examined, with the exception of persons with hemophilia, the majority of persons with HIV interviewed were not homeowners. Homeowners must struggle to maintain themselves in their own home, while renters must seek to attain accommodation in the private rental sector.

Two out of five households in Canada are renters, and they are concentrated in the three largest cities in Canada, where fully 60 percent are renters.[31] The National Council of Welfare estimates that approximately 37 percent of renters live below the poverty line, as compared to less than 10 percent of homeowners.[32] Availability of rental accommodation for persons with HIV is constrained by competition on the open market where vacancy rates do not rise above 3 percent; the need for accessible housing as previously defined; and the most desirable locations limited to certain neighborhoods in each of the three largest cities. The aggregate vacancy rates do not take into account the actual vacancy rates in the rental units in the bottom one-third of the price field, which may be under considerably greater pressure than the overall rental market. This observation is underscored by the recent report on distribution of incomes by Statistics Canada which states that "among renters, rent ate up at least 30 percent of income for more than one-quarter of families while for the lowest-income families almost 55 percent of income went on rent".[33] This is the situation for persons who wish to remain, and are able to maintain themselves, in their own household. The availability of resources for other than independent living is further constrained by lack of resources appropriate to the needs of persons with HIV. Across Canada there are limited resources for home-care support, group care residence, congregate care such as the planned Helmcken House in Vancouver, and for existing facilities such as Casey House in Toronto, Les Habitations Jean-Pierre-Valiquette in Montreal, and Karos House in Edmonton. Other cities, such as Winnipeg, are at the needs assessment phase in planning for residences. Hospital beds across the country are under pressure, and the lack of available resources for persons being discharged from hospital means that persons with AIDS may be in hospital longer than is desirable. Palliative care beds in the hospital are used for persons with AIDS in the final stage. It has been noted by several health care professionals, however, that due to the lack of available housing in the community, it appears that persons with AIDS have longer stays in palliative care units than other patients. Hospices are another choice, and appear to be appropriate for some persons. The issue for the majority of hospices is the lack of adequate core funding to accomplish the task of supporting people to the end of their lives.

The community has an interest in ensuring that there are various housing services available to meet the needs of persons with HIV infection over the course of their disease. If there are inadequate services to support individuals who are in the chronic stages of the disease, then the alternatives are to utilize acute care services, which are generally more costly. For the community of interest, the family or supportive network involved with individuals with HIV, there is a burden of care associated with a lack of service. If the only available choices are living at home or hospital care, then the supportive family or friends may be unable to provide the level of support required to meet, on an ongoing basis, the physical and emotional demands of a person who has HIV, without respite care and/or other support services. The geographic community may counter any proposals with NIMBY attitudes (not in my backyard) which have created a dilemma for many residential services, whether they are AIDS related or not. Communication with the neighborhood and involvement of key persons of influence in the community are essential steps in order to make the community neutral, at least, towards the idea of housing.

The state role is complicated by the nature of the bureaucratic organization of the services required for the person with HIV infection. Persons with HIV require health care, social services and housing support, which are not provided under one integrated Ministry in any province. The issue is further complicated by the nature of the distribution of responsibility for each of these issues to the federal, provincial or municipal governments. There is no uniform pattern of service delivery or locus of responsibility for services to persons who are HIV infected in Canada. By virtue of the Constitution Act,[34] responsibility for housing, social services and health care services are a provincial responsibility. The logic of distribution within each province is, for the most part, idiosyncratic and related to the historical development of services in each individual province. Availability of services and the nature of the system of service delivery is provincial, and decided upon individually. The role of the federal government has been to stimulate and initiate the development of new services and service forms, not to provide ongoing support for operational costs in what is seen as a provincial responsibility. The three major urban cities, Toronto, Montreal and Vancouver, have all pledged municipal support for addressing some aspect of the service needs of persons with HIV. The ability of municipalities to respond is also limited by the provincial government. Provincial government response is in some cases constrained by lack of available funds for transfer from the federal government, and in other instances the political will to act on this issue clearly seems to be lacking.

APPROPRIATE HOUSING FOR PERSONS WITH HIV: SOME CONCLUSIONS

Appropriate housing is housing that is suitable to the individual. It maximizes choice, independence, and a sense of belonging to the community. True choice means that services must be available that are appropriate to the person in their particular situation. Thus, if the desired state is to maintain the individual in their own home, what are the services that are required to facilitate that choice? And how do needs change as the disease progresses? Just as there is a continuum between independence and dependence, there must be a continuum of care for the person with HIV. As the disease does not have a clear linear progression, the continuum of care must be flexible and responsive to the changing needs of the individual. Both health protection and health promotion strategies are needed in order to prevent further transmission of the disease. Prevention is associated with being able to provide security, and security is related to housing where one can be sure that the person with HIV has a permanent address where they can be located. Shelter is more than a house in our culture; it is the anchor from which one locates oneself in the community, receives and accesses services, and exercises basic citizenship rights. Health promotion approaches view health as a resource for control over one's life, and meeting basic needs is essential before one can introduce effective health promotion interventions. Preventing further spread of the disease is dependent upon lifestyle changes that everyone has to make. Preventing rapid transition from seropositive to AIDS is dependent upon adopting healthy lifestyle choices that are designed to promote wellness, reduce stress, and increase supportive contacts within the community. Housing has an integral role in both health promotion and health protection.

Housing for persons with HIV in Canada poses serious questions about how our culture and institutions respond to the most marginalized persons in our society. Housing for persons with HIV is an excellent surrogate indicator of the health of our communities, if healthy communities are described as communities that enhance the quality of life for all Canadians by involving citizens in ensuring that health is a primary factor in political, economic and social decision making.[35]

As displayed in Table 5,11, there are competing forces that must be considered in the provision of appropriate housing for persons with HIV infection in Canada. The individual's health status, the culture in which

Table 5,11 Housing Issues for Persons with HIV Infection in Canada

Housing Issues	*Individual*	*Community*	*State*
Adequate	- Warm - Easy to clean - Dry - Safe - Secure	- Protection from further disease transmission	- Social justice - Social rights - Health protection - Health promotion
Accessible	- Level entry - Supported bath - Space for caregiver - Proximity to services	- Proximity to support systems - Critical mass of concerned individuals	- Specialized services *vs* ghettoized services - Rural *vs* urban
Affordable	- Poverty level incomes, except hemophiliacs - Income needs vary over course of disease	- Local tax rates - Hospital bed pressure - Less costly for home care	- Liability in blood products - Tax transfers - Minimal market control over housing
Available	- Rental housing - Low vacancy rates - High demand in inner city	- Range of housing and support services needed - NIMBY	- Multiple centres of services: health, social and housing

we live and the political, economic and social contexts are important factors in planning for the immediate and future needs of persons with HIV. Action to solve the problem must be based in a clearly defined set of values and principles that address the rights and needs of the individual with HIV, and support the community in meeting the needs and honoring the rights of the individual. Finally, government must enact programs and policies which respect the rights and responsibilities of individuals in the community to achieve health for all.

REFERENCES

[1] Manson Willms, S., Hayes, M. and Hulchanski, J.D. 1991. *Housing Options for Persons with Aids: An Annotated Bibliography*. Housing and Community Planning Papers, Vancouver, B.C.: UBC Centre for Human Settlements.

[2] Canadian Hemophilia Society. Personal communication, February 8, 1991. Figures quoted are accurate to January 14, 1991.

[3] See discussion in: Brandt, E. 1987. AIDS Social, Scientific and Health Policy Perspectives, in Gottlieb, M. et al. (eds.), *Current Topics in AIDS Volume I*. Chichester, England: John Wiley & Sons, pp. 20-21; and, Pfaffl, W. 1988. Medical Bases, in Jager, H. (ed.), *AIDS and AIDS Risk Patient Care*. Chichester, England: Ellis Horwood Ltd., p. 31.

[4] See, for example, Froschl, M. and Braun-Falco, O. 1988. Women and AIDS, in Jager, H. (ed.), *op. cit.*

[5] Wagman, L., Community Health Educator, Vancouver Health Department. Sensitivity Training Seminar held at the Centre for Human Settlements, University of British Columbia, Vancouver, B.C., June, 1990.

[6] Mills, J. and Masur, H. 1990. AIDS-Related Infections, *Scientific American*, August, pp. 50-57.

[7] *Ibid.* and Pfaffl, *op. cit.*, pp. 28-47.

[8] Mills and Masur, *op. cit.*, pp. 50-57

[9] See, Paquin, C. 1990. The Importance of Psychological Support at the Time of HIV Testing, *Intervention* 86, pp. 32-38; and, Templeton, K. 1990. Issues Facing HIV Positive Gay Men as Revealed Through the Counselling Process. Paper presented at the HIV Research Seminar Series held October 10, 1990, at the Centre for Human Settlements, University of British Columbia, Vancouver, B.C.

[10] Henderson Baumgartner, G. 1985. *AIDS: Psychosocial Factors in the Acquired Immune Deficiency Syndrome*. Springfield: Charles C. Thomas Publisher.

[11] Federal Centre for AIDS, Ottawa, Canada. Personal communication. Figures are accurate to February 4, 1991.

[12] Hayes, W. 1991. To Be Young and Gay and Living in the '90s, *Utne Reader*, March/April, pp. 94-100.

[13] Federal Centre for AIDS, *op. cit.*

[14] London Special Needs Housing Group. 1987. *AIDS: The Issues for Housing*. London: Resource Information Group, pp. 13-14.

[15] Hulchanski, J.D. 1990. Canada, in Van Vliet, W. (ed.), *International Handbook of Housing Policies and Practices*. New York: Greenwood Press, pp. 289-325.

[16] Turvey, J., Street Youth Worker, Downtown Eastside Youth Activity Society, Vancouver, B.C., personal interview, September 5, 1990.

[17] Marchand, R. 1990. Needs Assessment: Gay Community Prevention Programs. Vancouver: AIDS Vancouver.

[18] Aral, S.O. and Holmes, K.K. 1991. Sexually Transmitted Diseases in the AIDS Era, *Scientific American* 264(2), pp. 62-69.

[19] The Constitution Act of Canada, Section 15, April 17, 1985.

[20] Torjman, S. 1988. *Income Insecurity: The Disability Income System in Canada.* Toronto: G. Allan Roeher Institute.

[21] Rossi, P. and Wright, J. 1987. The Determinants of Homelessness, *Health Affairs*, Spring, pp. 19-32.

[22] Ross, D.P. and Shillington, R. 1989. *The Canadian Fact Book on Poverty 1989.* Ottawa: Canadian Council on Social Development.

[23] Canadian Hemophilia Society, *op. cit.*

[24] Federal Centre for AIDS, *op. cit.*

[25] Table 3,11 is adapted from Torjman, S. *op. cit.*, pp. 32-33; and Table 5,11, p. 43.

[26] Hulchanski, *op. cit.*

[27] Canada Mortgage and Housing Corporation, *Toronto CMA Rental Market Report*, October 1990.

[28] Canada Mortgage and Housing Corporation, *Metropolitan Montreal Area Rental Market Report*, Spring 1990.

[29] Torjman, *op. cit.*

[30] Table 4,11 is adapted from Canada Mortgage and Housing Corporation, *Rental Market Survey Prince Edward Island*, April 1990; *Region Metropolitaine de Quebec*, Avril 1990; *Metropolitan Montreal Area Rental Market Report*, Spring 1990; *Toronto CMA Rental Market Report*, October 1990; *Calgary CMA and Southern Alberta Centres*, October 1990; *Edmonton CMA and Northern Alberta Centres Rental Market Survey Report*, April 1990; *Rental Market Survey Report B.C.*, April 1990.

[31] Torjman, *op. cit.*

[32] National Council on Welfare. 1988. *Poverty Profiles*. Ottawa: National Council on Welfare.

[33] Vancouver Sun, Middle-Income Canadians Losing to Rich, February 5, 1991, p.B3.

[34] The Constitution Act of Canada, *op. cit.*

[35] Canadian Healthy Communities Project. 1989. *Mission Statement*. Ottawa: Canadian Healthy Communities Project.

12 THE STATE OF HEALTH OF THE CREE AND THE INUIT OF NORTHERN QUEBEC (NUNAVIT)

Jean-Pierre Thouez

Geography Department
University of Montreal

INTRODUCTION

There are several major obstacles involved in measuring the state of health of the Cree and Inuit populations of Northern Quebec. First, it is necessary to clarify the object of the analysis; that is, to define what is meant by health. As is the case with many other native peoples, the native populations of Northern Quebec have experienced serious political, economic, cultural and structural changes since the 1950s. According to Dubos,[1] it is the rapid pace of such changes, and the constant struggle to adapt to new conditions, that have led to physical, psychological and social disturbances among these peoples. Any meaningful definition of health must take into consideration these elements.

Reference to the concepts of (a) geographic "reduction" (the placement of native people on reservations, or keeping them in specific enclaves), b) economic "reduction" (the maintenance of small unstable local reservations), c) juridical-administrative "reduction" (reflected in the administrative and legal dependency upon the provincial and federal governments) and, d) ideological "reduction" (consisting of ethnic ideology)[2] offers more specific explanations for social change. Without entering into the details of the theoretical approach in formulating these "reductions" (or elements for analysis), it is important to note that the consequence of change, the increased dependency of native peoples, has brought about a situation in which the traditional symbolism that heretofore helped to explain their worldly presence no longer provides a meaningful spiritual structure for human activity. Once deprived of the framework for interpreting their universe, native people find themselves in a world devoid of its original meaning. They are caught up in new models of activity which are so different that they bear no resemblance to the social conventions internalized through the centuries.

The James Bay hydro-electric project and the Treaty bearing the same name illustrate the ambivalence of policies aimed at these populations. On the one hand, the natural effects of the project and the legal terms of the Treaty have rendered moot many aspects of aboriginal land claims in Northern Quebec. This treaty effectively extinguishes aboriginal rights to the territory of Northern Quebec. Under the terms of the treaty, the territory is subdivided into three categories of land. Category I lands (about 1 percent of the territory) remain the property of the Inuit and the Cree. These peoples also retain exclusive hunting and fishing rights on Category II lands (about 14 percent of the territory). Category III lands (about 85 percent of the territory) become public lands. In compensation for the loss of their property rights, the native peoples receive indemnity in money, administered by new regional organizations. The inhabitants of Povungnituk, Ivujivik, and Salluit have refused to accept the agreement; they have formed an organization called Inuit Tungavingat Nunamini.[3] On the other hand, interesting new programs have been created which guarantee income for hunters, fishermen and trappers, and which create regional organizations in the areas of education, health and management, among others, which are supervised by the natives.[4,5]

In this author's opinion, the notion of health is an integral part of human development. The concept of health in these populations, like that in the Third World, should be defined with reference to a holistic paradigm which considers man in his wholeness and not just in the organic and disparate aspects represented by his pathology. "State of health" should be considered in the context of the interrelationship of ecological, economic, social, cultural, sanitary and political levels. The political dimension can interact with the other levels, obliging us to identify the strategic areas of intervention.

A second obstacle in measuring health involves the choice of method and the difficulties of obtaining accurate and appropriate data. For example, for the Inuit there are three official sources of information: the Canadian census; statistics regarding births and deaths (published until 1952 by the former Federal Statistics Office, and today by Statistics Canada); and the register of the beneficiaries of the James Bay Treaty. All of these sources pose special problems. The first because of under-reporting and the impossibility of analysing by category of age with any accuracy. The second no longer exists. The third has not been subjected to quality testing by those responsible. For the Cree, as for the other Indians of Canada, there are several sources of statistical information. Certain sources only cover the registered Indian population. Others relate to the whole Indian population. Therefore, these various types of sources are not comparable. The criticisms made with respect to the Inuit statistics are applicable to those statistics relating to the Indian populations as well. It is also significant that the discrimination against Indian women, which leads to loss of their Indian status when they marry a non-Indian, skews any statistics regarding the Indian population. Moreover, the small size of the population leads to significant annual variability in demographic phenomena.

Bearing in mind these two areas of concern throughout, this chapter will first set forth some geographical information which has an impact upon the subject. Thereafter, it will describe the mortality and morbidity data available. Finally, a few risk factors or determinants of health will be discussed.

GEOGRAPHICAL ASPECTS OF THE STUDY AREA

Located north of the 49th parallel, the region of Northern Quebec (Nouveau-Québec) is bounded to the west by James and Hudson bays, to the north by Hudson Strait, and to the east by the border with Labrador (Figure 1,12). Half of its nearly one million square kilometres is located above the 55th parallel. Thus, Northern Quebec amounts to slightly over two thirds of the total area of the Province of Quebec. Distances between inhabited sites are considerable and vary between 80 and 360 kilometres. Ivujivik lies 1,300 kilometres from Quebec City and 1,200 kilometres from Montreal, while the southernmost town of Northern Quebec, Waswanipi, is about 700 kilometres from Montreal.

The terrain is characterized by medium and low altitude plateaus covered by glacial and post-glacial material with occasional higher landforms and marine plains. From the south to the north, the main vegetation zones are boreal forest, taiga and tundra. The principal landscape characteristic is the absence of trees. The soil is permanently frozen north of Kuujjuaq. About 20 major rivers flow though the territory, some in the

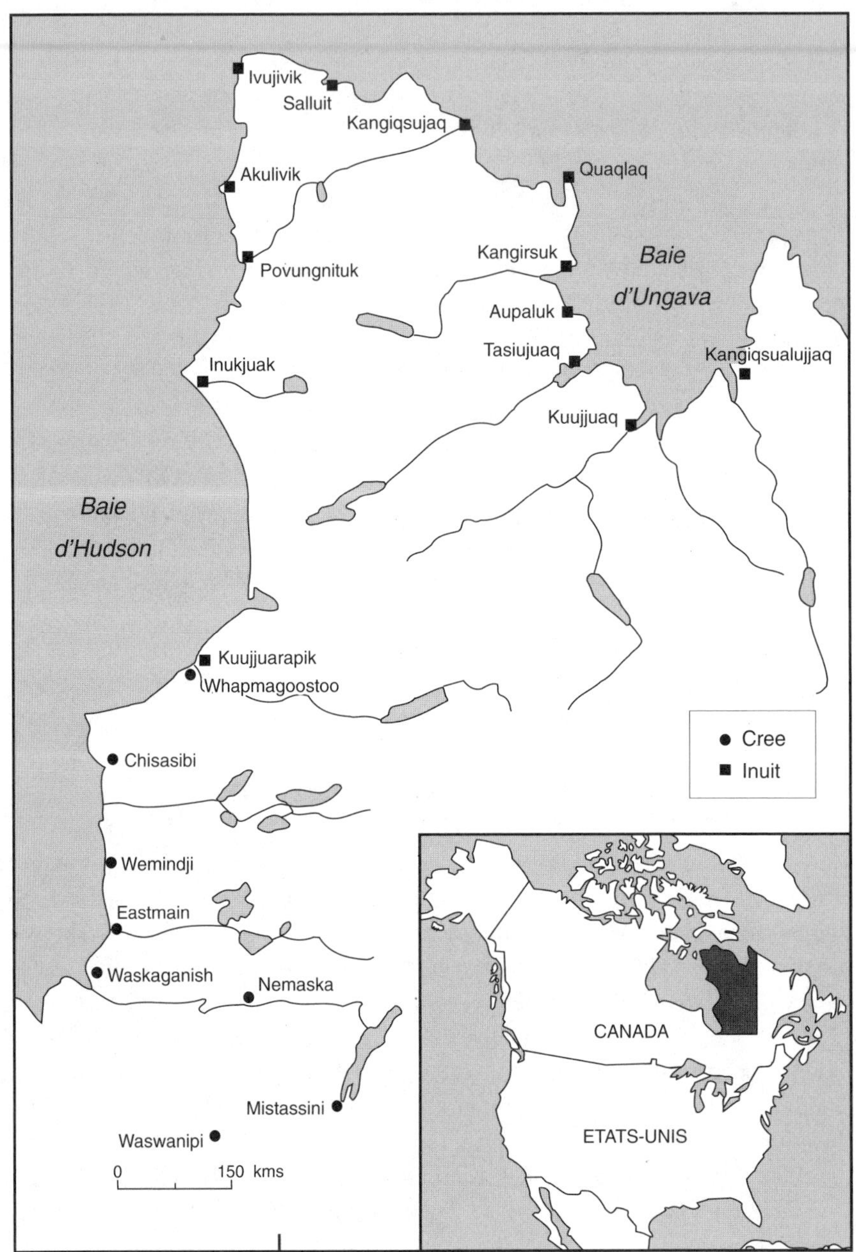

Figure 1,12 Northern Quebec

direction of Ungava Bay, others in the direction of Hudson Bay or James Bay. Hunting, fishing and trapping activities, as well as maritime and even air traffic closely follow the freeze-flow cycle of rivers and lakes. The average temperature varies from 1.1°C in the southern part of the region to -9.4°C in the northern area of the Ungava peninsula. Because of maritime influence, the temperature variation is less along the coastline; however, latitude also plays upon the climate.[6]

When compared to Southern Quebec, the population of Northern Quebec is considerably smaller and younger. These statistics are drawn from the list of official beneficiaries of the James Bay and Northern Quebec Agreement because there are significant mistakes, as well as omissions, in the 1981 Canadian Census. In 1983, the population was estimated as comprising 13,000 natives and 1,000 non-natives. The Cree population was estimated at 7,158, but many hundreds of individuals affiliated with one of the eight bands but residing outside of the villages should be added to this number. At the same time, the number of Inuit were estimated at 5,618, including a number of related non-natives who shared the dwellings of one or more natives. Many of these cases represent informal or formal marriages where there are children.

As early as the end of the 18th century, a White population had settled in Northern Quebec on lands which could generally be called Indian lands. In 1663, French merchants from Montreal arrived in Nemaska. Thereafter, between 1663 and 1778, the Hudson's Bay Company also ran a trading post at that location. The Company established permanent posts on the sites of most villages that exist today: Chisasibi (Fort George, 1837), Waskagamish (Rupert House, 1776), Whapmagoostoo (Poste-de-la-Baleine, 1901).

Later, Revillon Frères, merchants from France, tried to compete with the Hudson's Bay Company operating in Fort George. In 1905, Revillon opened a trading establishment on what may generally be considered to be Inuit territory at Tasiuujaq (Leaf Bay) and another at Salluit (Saglouc between 1900-1961). Thus, the opening of trading posts on Inuit land occurred later than on Indian land, dating only from the beginning of the 20th century. The establishment of commercial outposts demonstrates the importance of hunting, fishing and trapping activities.

Missionaries, particularly protestant missionaries, also played an important role in the development of Northern Quebec from the mid-19th century, with regard to the evangelization and schooling of the population. Missionaries introduced syllabic writing and translated the Bible to Inuktitut.

In the first half of the 20th century, famine and contagious diseases led to sedentarization of the native peoples. The settling process was particularly encouraged by the federal government during the 1950s, and then by the provincial government during the 1960s.[7] The sedentarization of the Cree, which began in the 1940s, may now be considered complete (even though they often move through the "bush" for long periods). In this connection, the relocation of Waswanipi to its present site at the confluence of the Opiwika and Waswanipi Rivers occurred in 1974. The new location is about 30 kilometres from the old location. Nemaska was established in 1977 on a new site near Lake Champion. The Cree population varies from village to village: about 2,000 inhabitants in Mistassini and Chisasibi, 300 in Nemaska. Except for Whapmagoostoo, located near the Inuit village of Kuujjuarapik, all of the Cree communities are located south of the 55th parallel. All are at the mouth of one of the rivers which flow into James Bay, except for Waswanipi, Mistassini and Nemaska, which are located inland near lakes.

The Inuit communities are located north of the 55th parallel. Five are on the coast of Ungava Bay (41.8 percent of the total Inuit population), and eight may be found along the Hudson Strait and Hudson Bay (58.2 percent of the population). These settlements most often lie deep inside a bay or a fjord. Kuujjuak (Fort Chimo) is the largest community, with a population of 950 natives and 315 non-natives. The smallest settlements are Quaqtaq, Aupaluk and Tasiujaq, all of which have less than 200 inhabitants. In the case of the Inuit, the settling process began during the 1950s and is almost complete today. In 1986, part of Kuujjuarapik's population was relocated to a new site, Umiujaq, near Lake Delisle. Most non-native individuals live in either Kuujjuaq or Kuujjuarapik.

Overall, the population of Northern Quebec has been growing at a steady pace since 1970 with an annual growth rate of about 2.5 percent. However, a progressive drop in natality has been observed in the last few decades, following the introduction of modern birth control methods. For the period 1979-1983, the total fertility rate (TFR) of the Inuit was 4.3 children per woman, while it was 1.62 in 1981 for the entire province.[8] The birth rate was 36.4 per 1,000 live births among the Inuit and 30 per 1,000 for the Cree for the period 1975-1981. The percentage of the population below 20 years of age was 52 percent for the Cree and 60 percent for the Inuit, while it was 31.5 percent for the Quebec population as a whole.

The general state of health of the Cree is still significantly inferior to that of the rest of the population of Quebec. This is even more so for the Inuit. Whereas it seems logical that variations in health status should be observed between populations, communities and individuals, some traditional values with beneficial health consequences still influence the behavior of the Inuit and Cree, but others, which have their origins in modern North American society, are more or less well integrated into Inuit and Cree lifestyles.

INFANT MORTALITY

In Canada and Quebec, the Infant Mortality Rate (IMR) was about 8.0 per 1,000 live births at the beginning of the 1980s. The tendency of this rate to fall since the 1930s accelerated during the 1960s. This improvement is due to the reduction of neonatal mortality (85 percent of neonatal deaths in 1931), the reduction of risks of mortality due to a congenital abnormality (a drop of 24 percent from 1966 to 1983), and the decrease in cases of below average weight at birth.[9] Other factors in this evolution are better overall living conditions of the population, the considerable progress in treating infectious disease (the development of antibiotics) during the 1940s and 1950s, and, more recently, the creation of perinatal programs.

Among the Cree, the average IMR for 1975-1982 was 37.0 per 1,000[10] and 17.2 between 1982 and 1986.[11] Perinatal mortality was reduced from 9.6 to 8.4 per 1,000 between these two periods, neonatal mortality from 8.0 to 5.2 per 1,000, and post-neonatal mortality dropped to 14.6 per 1,000 from 29.0. The level of perinatal and neonatal mortality is comparable to that of Quebec and Canada, but the rate of post-neonatal mortality is clearly higher.

Among the Inuit, for the period 1974-1978, the IMR was 47.0 per 1,000 (58.2 according to Labbé[12]) and 42 per 1,000 for 1979-1983 (69.0 between 1976 and 1982 if the under reporting of deaths and births on civilian rolls is taken into account).[13] According to Choinière,[14] the most significant drop in the infant mortality rate appears to have occurred before 1971. Since then, the IMR has dropped 20 percent among the Inuit, while there was a 54 percent reduction in the general population of Quebec. In other words, the IMR among Inuit, according to the latest data, is more than twice as high as that of Quebec. As was the case for the Cree, the post-neonatal period is when mortality reaches high levels among Inuit children.[15]

Post-neonatal mortality is a sensitive indicator of lifestyle and environmental factors. However, two thirds of deaths may be explained by causes originating in the perinatal period, namely congenital malformations (11.76 percent), sudden infant death syndrome (SIDS, 23.53 percent) and genetic and hereditary diseases (familial leukodistrophy and encephalopathy, 29.41 percent). The mortality rate among male children is higher than for females, but Courteau[16] notes a persistent risk throughout childhood for female children. We do not possess detailed data for the Inuit; however, Labbé[17] noticed that the greatest number of deaths was due to gastroenteritis, meningitis, pneumonia, and epidemics of childhood diseases, including measles in 1975, thus explaining the higher mortality rate observed among these populations.

Moreover, clear geographic variations in the IMR and its components were observed: the IMR of the Inuit was twice that of the Cree. Infant mortality was five times higher in the Ungava Bay communities than in the Hudson area between 1976 and 1978.[18] These tendencies seem to persist, even though the gap between regions has narrowed.[19,20]

In addition, life expectancy among the Cree was 71.8 years for the period 1975-1981 (69.8 years for men, 73.2 years for women) and 71.9 years for the period 1982-1986 (70.9 years for men and 73.0 years for women).[21] In the case of the Inuit, the average life expectancy appears to have fallen to 60 years from 64 years. This occurred for both sexes during the periods of 1974-78 and 1979-83.[22] Following a peak life expectancy of 66 years in 1969 to 1973, this decrease might reflect the deterioration of the state of health of the population in general, but it could also be a random event. Furthermore the decrease could be explained in part by the improvement in the recording of deaths among the Inuit. Thus, between 1971 and 1981, the decrease in infant mortality among the Inuit had only a negligible impact on life expectancy.

CAUSE OF MORTALITY

Among the Cree, the main causes of death for the period 1982-1986 were diseases of the circulatory system (20.44 percent), injuries and poisoning (19.89 percent), neoplasms (17.13 percent), respiratory infections (11.05 percent), and digestive diseases (3.31 percent). Death from injuries, violence and poisoning represented almost one fifth of total Cree deaths, compared to 8 percent for Canada.[23] These patterns tend to persist after statistical procedures such as age standardization are implemented.

Between 1979 and 1983, the Inuit showed the following mortality profile: circulatory diseases (22.3 percent), trauma and poisonings (22.8 percent), respiratory diseases (16.0 percent), neoplasms (14.1 percent). According to Choinière,[23] violent deaths have the strongest impact on Inuit life expectancy.

The SMRs for cancer in native populations were consistently below the national average, but the risk of cancer was assumed to be rising following the gradual modification of nutritional habits, increase in cigarette smoking and environmental changes. Cree females had a higher incidence of mortality from cancer than males; they died mainly of digestive tract cancers (pancreas etc.) rather than from cancers specific to their gender.[25] Lung cancer mortality is the leading cause of death among all neoplasms, but the rate among native peoples appears to be lower than that for Canada in general.[26] Dufour[27] provides data on the prevalence of cancer among the Inuit. Lung and colorectal cancers are most frequent among both sexes (36.8 percent); uterine cancer appears to be more prevalent among women and nasopharynx cancer predominates among men.

HOSPITAL MORBIDITY

The health care resources are similar from one village to another, with the exception of three communities that are the sites of regional hospitals: Kuujjuaq (Regional Hospital of Ungava Bay); Povungnituk (Regional Hospital of Hudson Bay); and Chisabibi (Cree Regional Hospital). The Cree of the mainland also have access to the hospital of Chibougameau. When there is a serious health problem, the patient is evacuated to the regional hospital or to the hospitals in Montreal or Quebec. Thus, there is no major difference in the physical accessibility to health care between the Cree and the Inuit, or between their respective villages.

The study of hospitalization diagnosis gives us information regarding the main causes of change in the average state of health of a population. In fact, if childbirth (the main cause of hospitalization for the Cree and Inuit) is put aside, the main causes of hospitalization for native peoples are the same as those for the general population, namely, those which disable and those which result in mortality. The rate of visits to a hospital was higher for children less than one year old and for the elderly (over 65 years) in both populations studied.

According to Pelchat and Wilkins,[28] the largest number of hospitalizations in the Cree community were related to complications at childbirth and during pregnancy (22 percent of total), respiratory diseases (15 percent), diseases of the digestive tract (10 percent), and accidents and poisoning (8.7 percent), based on data collected from 1983 to 1984. Inuit women show a similar profile: childbirth and pregnancy complications (18 percent), respiratory diseases (11.2 percent), digestive tract diseases (10.5 percent), trauma and poisonings (7 percent).[29]

Hospitalization due to respiratory problems is twice as high among the Cree and three times as high among the Inuit as in the general Quebec population. Pneumonia and other acute respiratory problems account for at least 50 percent of the cases.[30,31] The incidence of tuberculosis is still high even though the tendency has been toward a decrease in the last few decades.[32,33] Digestive tract diseases (teeth problems, appendicitis and hernias, biliary lithicis etc.) are also significant causes of hospitalization.

The incidence of infectious disease, while remaining a problem, appears to be decreasing, while chronic diseases are becoming more prevalent. However, the latency period of most cancers is very long, and excess in cancer incidence may not show before the next decade. When compared to the rest of Quebec, hospitalization for respiratory diseases remains low, and longitudinal data on the prevalence of cardiovascular risk factors among the Cree and Inuit populations are still scarce. Hospitalization for cardiovascular diseases is more frequent than for ischemic diseases of the heart. Changes in nutritional habits, lifestyle and smoking habits are probably too recent to have a significant impact on the incidence of heart disease.

Statistics derived from the clinics of each community are very useful in evaluating morbidity that does not lead to hospitalization: respiratory problems (colds, angina, etc.), skin problems, nervous and sensory diseases (including otitis), digestive tract problems (including oral conditions), accidents and routine visits account for 80 percent of consultations. The reasons vary depending on the age and sex of the patients. In both populations, there are frequent visits by children, particularly those less than one year of age. Within that age group, nearly half of the consultations were due to ear infections, colic and skin problems.[34,35] In the category of accidents, there is a predominance of males: for the Inuit between 15-34 years old, one out of two men (50 percent) will consult for that reason.[36] Among the Cree, however, it seems that the accident rate is lower than that observed in other native groups.[37]

DETERMINANTS OF HEALTH

Climate

The adaptation to cold weather by the Inuit and the Cree is considered to be a phenomena that can be explained by their repeated exposure to the northern climate. Moreover, until now the survival of these peoples in a harsh environment was favored by cultural adaptation (type of dwelling, clothing, food). However, the cold remains a problem: it can be related to the etiology of lung disorders such as emphysema and chronic bronchitis, and to infections of the upper respiratory tract frequent in these populations. Physical factors may also be held responsible for chilblains and hypothermia. Growing penetration of ultraviolet rays into the northern atmosphere and their reflection on the snow (especially in the spring) cause diverse opthalmological problems such as corneal burns.

Pollution of the Environment

There are man-made (mines, paper mills) and natural (erosion of sedimentary and volcanic rocks) sources of inorganic mercury in the region. Inorganic mercury in the environment is transformed by bacteria into organic mercury (methyl) which reaches high levels in predatory fish and sea mammals. Acidification of lakes increases the mercury content in fish. Studies carried out on the Cree in 1978 showed a correlation between mercury levels and mild neurological findings; cases of severe methyl poisoning were also found.[38] Mercury intoxication was feared for the population of Salluit, but a later study showed that there was no major problem in this regard. However, the high rate of mercury found in species of fish frequently eaten, particularly those found in the James Bay dam waters, means that caution should be exercised in eating these fish, particularly in the case of pregnant women. The Cree hunters have been alarmed by notices sent recently by government authorities concerning cadmium and cesium contamination of bush food (moose and caribou).

At the village level, waste causes a pollution problem. The frozen soil cover slows down biodegradation and encourages accumulation of waste in dumps. In certain locations, these dumps lead to the contamination of the drinking water. Microbial pollution of water at its source or during storage may cause diarrhea among the very young and the elderly. The fact that water supplies are often limited does not encourage adequate hygiene and skin infections are common.

Housing

Even with the housing construction of James Bay, and renovation programs, the overall housing conditions are poor. The dwellings are often too small and overcrowded. The ambient air is generally hot and dry in winter, and often polluted by cigarette smoke. Crowding and other general conditions of the home certainly favor the development of respiratory conditions and the fast propagation of infections (colds, streprocies of the pharynx and of the skin, and meningitis).

Living Habits

The proportion of bush food and fish in the diet is more limited because of sedentarization, the increase in hired labor as opposed to hunters and fishermen, and demographic pressure. According to Duhaime,[39] the proportion of hired workers is often based on the size of the village, but the share of part-time jobs is greater in Inuit villages which number less than 500 inhabitants, and there are more full time jobs in the larger communities. Public administration is the main employer. The share of public transfer payments (family allowance, old-age pension or welfare) in the Inuit's monetary income has changed little, but if direct and indirect effects are taken into account, public funds have increased in relative importance. However, the share of income from the craft industry (sale of furs, sculpture, etc.) has largely decreased. Also, evaluation of the importance of self-consumed hunting products remains difficult, even though it is known that spinoffs from programs encouraging hunting, fishing and trapping (started among the Cree and Inuit following the James Bay Agreement) constitute the main source of income for many families which do not have regular salaries. The switch to a money economy has brought about deep changes in the behavior of native populations. More specifically, it has led to social segmentation based upon the type of work and mode of consumption.

A significant part of the daily diet of native people currently comes from goods purchased. The considerable increase in carbohydrates in the present native diet is considered to be largely responsible for dental cavities, anemia and probably a reduced resistance to infections, as well as the emergence of "new" diseases such as diabetes.[40] The proportion of Cree who are obese or have high cholesterol levels is considerable, particularly among women. The incidence of high blood pressure has increased dramatically among the Cree.[41] Even though the frequency of cardiovascular disease remains lower than for the Quebec population as a whole, it should be on the rise in coming years, this growth being fueled by high alcohol and tobacco consumption.[42] Alcohol is also a prime factor in problems of family violence, accidents and criminality.

Bottle feeding has become a frequent habit for adopted children. This type of feeding has been associated with a higher occurrence of respiratory diseases, otitis, gastroenteritis, anemia and dental cavities. Passive smoking also leads to respiratory disorders.

The distance of the villages from fishing or hunting regions eventually gives rise to health problems. Accidents related to these traditional activities, especially drowning, are frequent. In addition, noise pollution (snowmobiles, high performance guns) may cause irreversible lesions of the auditive nerve.

Contact with the values of North American culture, its consumer products and technology, combined with the change from traditional lifestyles, has brought about a serious crisis among natives. This may partially explain the present psychological disturbances and political problems.

DISCUSSION

In order to describe the state of development of a population, two models have been used: demographic transition and epidemiological development. Of the two native groups, the Cree have reached a more advanced stage, rather close to that of Caucasian populations according to the classic indicators of the demographic transition model. Expectancy of life at birth was clearly higher among the Cree than among the Inuit. As a result, passage from infectious to chronic diseases is more acute, which does not mean that the first (gastroenteritis, epidemics, tuberculosis, venereal diseases, etc.) are underestimated, for they still threaten public health. In order to cope with health problems, a health care system has been established which resembles in its organization the health care system in operation in the rest of the province. This includes three levels as follows: the first level consists of a community clinic in each village; the second level consists of hospitals in Kuujjuaq, Povungnituk, Chisasibi and Chibougameau; and the third level is evacuation to Montreal and Quebec hospitals. Also, the Kativik and Cree CRCSS (Regional Council of Health and Social Services) have the mission to plan and coordinate the services and resources in their territories, and district health centres (DSC) based in Quebec City (Projet Nord) and Montreal (Cree), perform the analysis of needs, and the programming and program evaluation. Furthermore, they coordinate the evacuation programs.

If the entire demand-supply relationships are briefly considered, many problems must be mentioned. The functioning cost of the health care system is high while gains in general health status are low; thus prevention should be stressed (immunization of children, follow-up of pregnant women, etc.) rather than curative activities. Encouraging native populations to recognize and accept some responsibility for their own health condition implies that all aspects of disease have been considered, and that people are kept informed of factors which influence their health and about those which can be controlled, either individually or collectively. It implies, also, that something is done so that when natives leave the village on a hunting or fishing trip, the people have some basic knowledge of first aid, thus reducing dependency upon the system. This can be accomplished only through the improvement of communication (verbal and written) with medical personnel, a reduction of the rotation of professionals, and greater native control in the health care system. The responsible groups should establish close ties with the communities through the use of health committees.

Finally, the cultural dimension must be taken into account. Contact with a different culture raises several challenges, such as presentation of illness and health care in a way that is relevant to native experiences.[43,44] Even today, the native explanation of certain disorders is completely different from an explanation that would be seen in southern Quebec in the use of language and expression. This is confirmed in the area of mental health. More than in other domains, the influence of cultural factors is such that the "eldest" are frequently called upon as leaders and/or decision-makers.[45,46]

The health status of native peoples, in particular that of the Inuit, is comparable to that of a third world nation. However, the resources mobilized to face the health problems in northern populations do not allow this comparison to be pushed very far. Models advanced to account for economic development and social change[47] are more relevant. Indeed, it can be said that the health care system is not aimed at the real causes of disorders, but only at their consequences. The causes are beyond the competence of the means of action of the system; they mainly lie at the level of maintaining sanitary conditions where the populations live, and of keeping abreast of the new habits they have acquired. As long as health is considered separately from other aspects of native life, and as long as the development of health care is in the hands of non-natives, "ethnicization" of native identity can be expected to constitute an ideological construct that will reinforce integration and dependency upon the south.[48]

REFERENCES

[1] Dubos, R. 1981. L'homme face à son milieu, in *Médecine et Société: Les années 1960*. Quebec: Éditions coopératives Albert St-Martin, pp. 53-81.

[2] Simard, J.J. 1982. *La révolution congelée: coopérative et développement au Nouveau-Québec Inuit*. Doctoral thesis. Montreal: Faculty of Social Science, Laval University, pp. 9-57.

[3] Peters, E. *Aboriginal Self-government Arrangements in Canada.* Institute of Intergovernmental Relations. Queen's University Background Papers, 15, pp. 67.

[4] Salisbury, R.F. 1986. *A Homeland for the Cree: Regional Development in James Bay 1971-1981.* Montreal: McGill-Queen's University Press, p. 172.

[5] Duhaime, G. 1983. *Le pays des Inuit: situation économique en 1983*. Montreal: Sociological Research Laboratory, Sociology Department, Laval University, p. 518.

[6] O.P.D.Q. 1983. *Le Nord du Québec: profil régional.* Quebec: Quebec Planning Office, p. 600.

[7] La Rusic, I. 1985. Profil des communautés nordiques du Québec, *Recherches amérindiennes du Québec*, November, 1985, p. 109.

[8] Choiniere, R., Levasseur, M. and Robitaille, N. 1988. La mortalité des Inuit du Nouveau-Québec de 1944 à 1983: évolution selon l'âge et la cause de décès, *Recherches amérindiennes au Québec* 26(1), pp. 29-37.

[9] Shulman, E. 1987. *Progress in the Health Status of Canadians*. Ottawa: Health Services Directorate, Health Services and Promotion Branch, draft copy.

[10] Robinson, E. 1988. The Health of the James Bay Cree, *Canadian Family Physician* 34, pp. 1606-1613.

[11] Courteau, R. and Robitaille, N. 1988. La fécondité des Inuit du Nouveau-Québec Depuis 1931: Passage d'une Fécondité Naturelle à une Fécondité Contrôlée, *Population* 43(2), pp. 427-450.

[12] Labbe, J. 1987. *Les inuit du Nord Québecois et leur santé*. Ministére de la Santé et des Services Sociaux, p.87.

[13] Choiniere, Levasseur and Robitaille, *op. cit.*

[14] *Ibid.*

[15] Labbe, *op. cit.*

[16] Courteau and Robitaille, *op. cit.*

[17] Labbe, *op. cit.*

[18] *Ibid.*

[19] Courteau and Robitaille, *op. cit.*

[20] Labbe, *op. cit.*

[21] Courteau and Robitaille, *op. cit.*

[22] Choiniere, Levasseur and Robitaille, *op. cit.*

[23] Courteau and Robitaille, *op. cit.*

[24] Choiniere, Levasseur and Robitaille, *op. cit.*

[25] Courteau and Robitaille, *op. cit.*

[26] *Ibid.*

[27] Dufour, R. 1987. Le cancer au Nouveau-Québec Inuit: résultats d'une enquête préalable à la création d'un registre des cancers, *Rev. Can. Santé Publique* 78(4), pp. 267-270.

[28] Pelchat, Y. and Wilkins, R. 1986. *Fréquentation hospitalière de la population autochtone de la Baie James 1981-82 à 1984-85, Rapport.* Montreal: Community Health Department, Montreal General Hospital, p. 75.

[29] Morissette, M. and Tourigny, A. 1986. *Les services médicaux du Québec arctique (région 10-A). Analyze de la problématique et recommandations, rapport d'étape.* Quebec: Projet Nord, DSC-CHUL, p. 155.

[30] Labbe, *op. cit.*

[31] Pelchat and Wilkins, *op. cit.*

[32] Robinson, *op. cit.*

[33] Labbe, *op. cit.*

[34] *Ibid.*

[35] Robinson, E. 1985. *La santé de Cris de la Baie James.* Montreal: Department of Community Health, Montreal General Hospital, draft report, p. 126.

[36] Therien, F. 1984. *Health Problems in Northern Quebec.* Quebec: Projet Nord, DSC-CHUL, draft report, p. 22.

[37] Robinson, (1988) *op. cit.*

[38] McKeown *et al.*, 1983a and b, *op. cit.*

[39] Duhaime, *op. cit.*

[40] Thouez, J.-P., Foggin, P. and Rannou, A. 1989. Hypertension et "modernité" chez les Cris et Inuit du Nord du Québec, *Le Géographe Canadien* 33(1), pp. 19-31.

[41] Thouez, J.-P., Foggin, P. and Rannou, A. 1989. The Other Face of Development: Native Populations, Health Status and Indicators of Malnutrition. The Case of the Cree and Inuit of Northern Quebec, *Soc. Sc. Med.* 29(8), pp. 965-974.

[42] Foggin, P. and Lauzon, H. 1987. *État de santé et facteurs de risque: Les Cris du nord québécois.* Montreal: Geography Department, University of Montreal, preliminary report, p. 108.

[43] Morissette and Tourigny, *op. cit.*

[44] Thouez, J.-P., Foggin, P. and Rannou, A. 1990. Correlates of Health Care: The Inuit and Cree of Northern Quebec, *Soc. Sc. Med.* 23D, pp. 1-10.

[45] Wenzel, G. 1981. Inuit Health and the Health Care System: Change and Status Quo, *Etudes/Inuit/Studies* 5(1), pp. 7-14.

[46] O'Neil, J.D. 1981. Beyond Healers and Patients: The Emergence of Local Responsibility in Inuit Health Care, *Etudes/Inuit/Studies* 5(1), pp. 17-26.

[47] Allahar, A.L. 1989. *Sociology and the Periphery: Theories and Issues*. Toronto: Garamond Press, p. 168.

[48] Dorais, L.J. 1986. Les autochtones Canadiens et leur identité, in Guillaume, P. et al. (under the direction of), *Minorités et états*. Montreal: Bordeaux-Laval University Press, pp. 89-100.

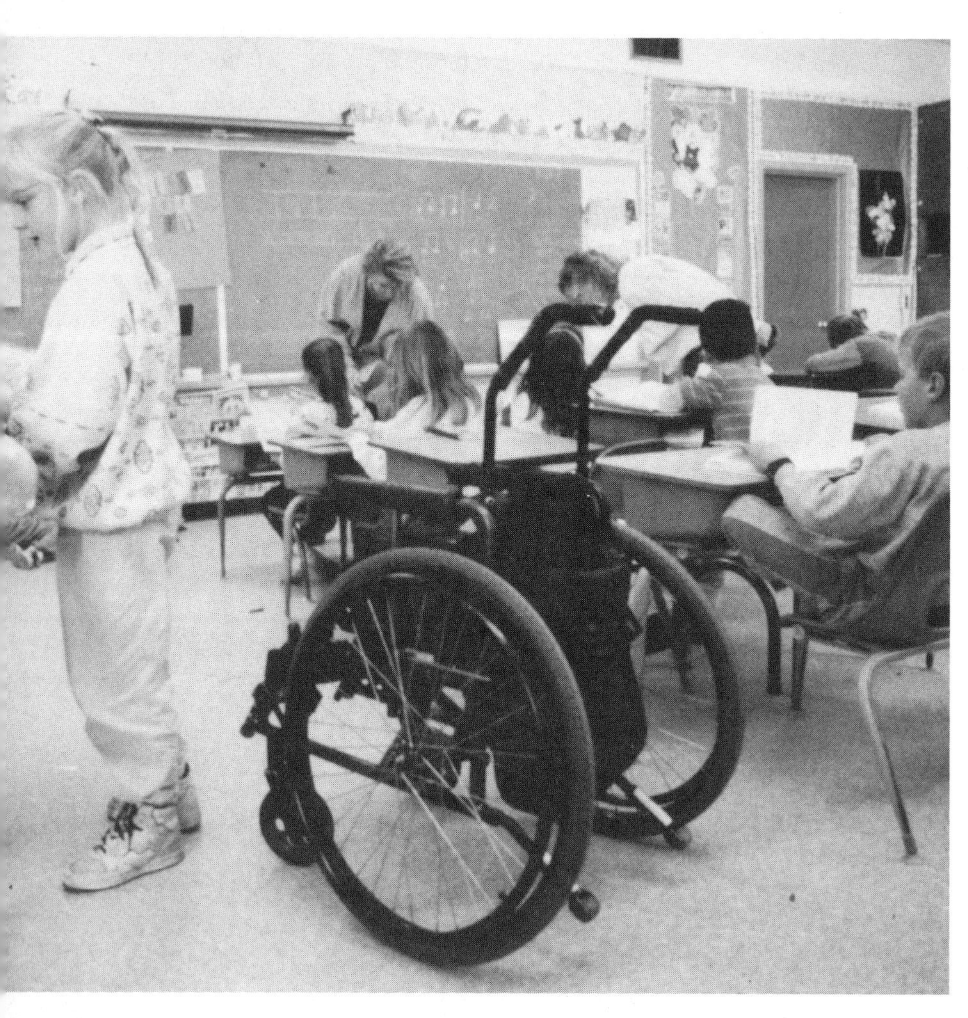

13 PUTTING UP OR SHUTTING UP: Interpreting Health Status Indicators from an Inequities Perspective

Clyde Hertzman

Department of Health Care and Epidemiology
University of British Columbia

and

Michael Hayes

Department of Geography
Simon Fraser University

INTRODUCTION

Health promotion has emerged as the major health policy thrust of the 1980s. Health promotion is defined as "the process of enabling people to increase control over, and improve, their health".[1] Health is viewed as a resource for daily living, which gives people the ability to manage and even change their surroundings. The Epp Report[2] specifies a (now familiar) Framework for Health Promotion, in which three health challenges are identified: reducing inequities in health status; increasing prevention; and enhancing coping. It is the first of these challenges that is addressed in this chapter. If a concerted effort to reduce inequities is to be made, then the analysis of population based measures of health status must be interpreted in light of the underlying causes of inequities.

The idea of healthy communities has gained rapid appeal as an organizing concept for health promotion in Canada and other parts of the industrialized world. Just as rapidly, concerns about how to measure the healthfulness of communities have been raised by participants in the healthy communities process and by intellectuals providing critical support for the field. These concerns have been succinctly summarized elsewhere.[3] Concerns about community health status indicators are extremely broad in scope, ranging from the inadequacy of training of local community organizers to be able to identify and collect community health status information, to the lack of a theoretical base for health promotion. It would be pretentious to suggest that any one contribution could adequately address all of the issues associated with the identification and measurement of community health status indicators. The purpose of this chapter is to present an organizing framework for community-based measures of health status explicitly linked to inequities in health. The intention in linking measures directly to inequities is to maintain a focus on what the authors believe to be the most important aspect of health promotion. The framework attempts to identify three dimensions of inequities: the stage in the lifecycle at which inequities merge; associated population characteristics; and the source(s) of inequity.

In this discussion, health status will first be approached from a "societal perspective". Then the field of vision will be reoriented so that the implications of what has emerged can be seen from the local perspective. The choice to adopt a broad societal view is based primarily on the best current knowledge of how health status is determined in our society. The principal determinants of health have their origin in processes that occur on a much wider scale than that of the local community: at the regional, provincial, national and international levels; and across decades and generations. Processes involve both the socially constructed realms of economics, socio-cultural practises and beliefs, and politics; and the influences of the natural environment (weather patterns, distribution of natural resources, etc.); and combinations of the two (transport of pollutants through air, water and soil). Of course, these processes operate simultaneously at the local level, both as proximate threats to health and as health outcomes. Impacts occur at the local level in the experience of individuals in specific places at specific times. But from a societal perspective, the inquiry should begin with the greatest geopolitical breadth and historic depth, thus putting local factors in this context.

The intention is not to downplay the role of the local community in addressing local concerns. It is taken as a given that local organizations should take action, participate, establish and nurture community development initiatives around such issues. These initiatives will be, implicitly or explicitly, related to concepts of health and well-being which might be quite different from the one(s) motivated by those with a vested interest in a broader societal perspective. We must allow for the simultaneous operation of these different levels, not confuse them, and evolve a language which will help to keep concepts about them clear. In general terms, the societal perspective is conducive to a centralized planning function within higher levels of the state apparatus, the purpose of which is to address social policy issues at the macrogeographic scale of province or nation, whereas the local perspective is more conducive to addressing microgeographic scale issues borne out of the experience of individuals within specific contexts.

In this chapter, health status is represented by the traditional indicators of mortality and morbidity. The purpose is not to engage in debate about the use of specific indicators, but rather to situate measures that are commonly used within a framework that draws attention to inequities. Therefore, discussion is confined to narrow, frequently measured and easily understood outcomes, so that the range of their potential determinants can be very broad. For instance, when longevity is used as an outcome, the full range of determinants of longevity can be brought to bear in answering the question "why do some people live longer than others?". If, however, health is defined in terms of well-being, a conceptual confusion arises in which well-being can no longer be spoken of as a determinant of health, since it has become part of the outcome. Ironically, this actually reduces the opportunity to highlight the role of well-being in health.

The strategic choice taken here is to narrow the concept of health to disability free life years, thus allowing well-being to function as a determinant. This choice should not be taken to mean that well-being could not be called a legitimate dimension of health. It could well be. But the crucial issue here is the possibility that well-being and longevity have biological pathways which link them. In other words, the quality of life (i.e. well-being) undoubtedly influences the quantity of life through various pathways, and it is just such pathways that the authors wish to uncover with the framework. Thus, separating well-being and longevity will assist in understanding the relationship between them while combining them in a spirit of "holism" will lead to confusion. It could also be argued that longevity is the ultimate holistic indicator, since the body's survival is ultimately the only unequivocal indicator of well-being.

This discussion proceeds from a sample of startling and intriguing observations from population-based sources which contain important insights into the determinants of inequities in health. For instance, large social class gradients in life expectancy by income, education, and occupational class have persisted everywhere they have been looked for in the industrial world during the 20th century. These differences cut across all major disease processes.[4,5] Similar differences have persisted across census tracts in urban Canada during the 1970s and 1980s.[6] Among the major industrialized countries, five managed to maintain unemployment rates below five percent during the economic crises of the early 1970s and 1980s. Among these five countries are four of the top five in the world in average life expectancy for both males and females. The gradient in life expectancy across social classes is greater in Britain (a high unemployment industrialized country), than in Sweden (a low unemployment industrialized country).[7,8]

In addition, within the lowest income countries in the world, there are places such as Costa Rica and Kerala province of India which manage to achieve life expectancies at or above 70 years, while across the developing world, average life expectancies remain below 60.[9] These places are also characterized by relatively small disparities in the distribution of material wealth and a strong commitment to investing available resources in education and welfare.

The gradual collapse of the political economies of the East European countries over the past 20 years has been accompanied by declining life expectancies and increasing gaps in expectation of life between Eastern and Western Europe.[10] In Canada, the life expectancy of married men is 8 to 10 years longer than it is for those never married, widowed or divorced. Among women, the analogous difference is three to four years.[11] In general, socioeconomic differences in mortality cut across all major disease processes. Canada's native population is a low socioeconomic status group by all measures of education, income and occupation, but its lower than average life expectancy is due to high rates of accidents and respiratory diseases only. The native population does not have increased rates of cancer or circulatory disease, even after its age distribution has been taken into account.[12]

These observations are complemented by other, equally provocative ones which come from studies of samples rather than whole populations. Among male British civil servants, there is a three-fold mortality gradient for heart disease, from the highest to the lowest ranks of the service, after controlling for age, smoking, systolic blood pressure and plasma cholesterol.[13] In a nine-year mortality follow-up of residents in a federally designated poverty area in Oakland, California, it was found that they experienced higher age, race, and sex adjusted mortality rates than residents of non-poverty areas, even after multivariate adjustment was made for baseline health status, race, income, employment status, access to medical care, health insurance coverage, smoking, alcohol consumption, physical activity, body mass index, sleep patterns, social isolation, marital status, depression and personal uncertainty.[14]

In a longitudinal study of children born on the Hawaiian island of Kauai, it was found that the early childhood developmental problems associated with severe perinatal stress were overcome by early adulthood in those from high socioeconomic status families, but not in those from low socioeconomic status families.[15]

In another study, a 19 year followup of a randomized controlled trial of an early childhood enrichment program in an inner-city ghetto in the United States demonstrated that the intervention group did better than the controls in the following areas: a higher proportion graduated from high school and went on to college; less than half as many were ever classified as mentally retarded; 40 percent fewer were ever arrested or detained; 50 percent more were employed at age 19; 45 percent fewer were on welfare; and teenage pregnancies among the female study subjects occurred at half the rate.[16] In studies of screening tests for dementia in the elderly using the mini-mental state examination, investigators cannot seem to avoid the problem that those who "fail" it are four to six times more likely to have had eight or less years of schooling in early life than those who do not fail.[17]

Each of the above observations is based on a difference in health status between two or more whole populations, or population samples, but cannot be attributed to individual-based, traditionally defined epidemiologic risk factors. How do we organize the universe of observations of this type in a way which is ultimately relevant to the needs of communities, defined either on the basis of societal or local perspectives? To begin with, there is a need for a broad framework which addresses the social, environmental and biological determinants of health status. This framework must not only include a very broad range of possible determinants of health, but also resist the usual research doctrine of viewing causation as an entirely disease-specific phenomenon. Failure to do so may obscure the underlying determinants of health status which cut across a wide variety of disease processes, thereby preventing us from seeing the best opportunities for intervention.

A FRAMEWORK FOR THE INVESTIGATION OF HETEROGENEITIES IN HEALTH

The conceptual and accounting framework presented here is based on explaining "heterogeneities" in health status. Heterogeneity, here, means that the total population can be divided into subgroups, according to a number of different characteristics, such that the average health status of the members of different subgroups will show significant differences from one another. The model is presented as a cube with three axes representing the three key dimensions for analysis (Figure 1,13). These are labelled "stages of the life cycle", "partitions by population characteristic", and "sources of heterogeneity". Each of these is divided into discrete levels and any combination of individual levels forms a box within the overall cube. Each box then represents a unit of investigation and accountancy.

The dimensions identified in the figure were chosen specifically to address the issue of heterogeneity, and do not purport to constitute a general model of health and disease. In particular, they presuppose a measure or measures of health status which can be attached to each member of the population, as a precondition for analysis of the heterogeneity of such a measure. Our discussion is limited to health status in industrialized countries, although the cube could be adapted for application to lesser developed countries.

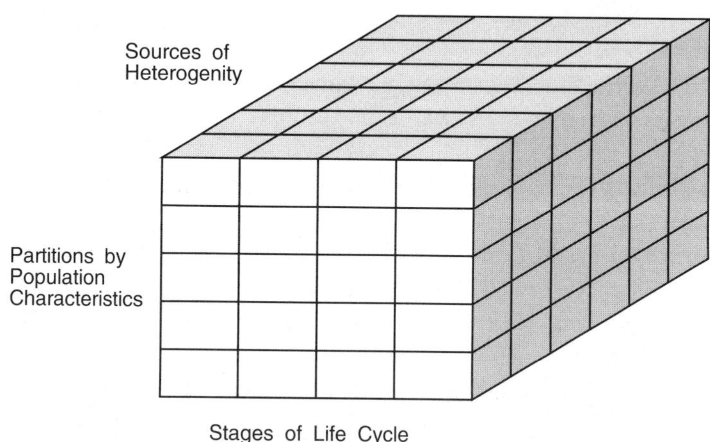

Stages of Life Cycle

1. Pre-term to 1 year
2. 1-44 years
3. 45-74 years
4. 75+ years

Characteristics

1. Socioeconomic Status
2. Ethnicity/migration
3. Geographic
4. Special Populations

Sources of Heterogenity

1. Reverse causality
2. Differential susceptibility
3. Individual lifestyle
4. Physical Environment
5. Social Environment
6. Differential access to/response to health care services

Figure 1,13 Model for Investigation of Heterogeneities in Population Health Status

Stages of the Life Cycle

The life cycle is fundamental to the study of heterogeneities in health status because it is the basis of biological change in all organisms, and because a complete accounting of heterogeneities must take into account events throughout the cycle. This model divides the life cycle into four time periods during which different diseases or conditions are predominant. The perinatal period is self-evident and has been the source of population health indicators (e.g. infant mortality rate and birth weight distribution) for many years. During the period up to 44 years of age, the principal threats to health in industrialized countries are not disease-based, but include accidents, violence and suicide. The period from 45 to 74 years is when heart disease, cancer and other chronic diseases are the principal threats to health, function and life. The final period begins at age 75, when the late effects of chronic degenerative disease, multiple organ system dysfunction and behavioral problems unique to the elderly become predominant. In reality the life cycle is continuous, and dividing it into discrete intervals is arbitrary. But what is lost in conceptual purity by doing so is regained in usefulness as an heuristic device.

A distinction must be made between stages in the life cycle at which the array of diseases and conditions is different, and the time at which various determinants of health status have their effect. Since the framework accounts for heterogeneities in health status it must, by definition, begin with a recording of the expression of health status and work backwards in time to explore its determinants. This does not imply, however, that specific determinants always operate in the same stage of the life cycle as the heterogeneities they create. In fact, there may be time lags of the order of 50 years or more.

For instance, Barker et al.[18] have recently shown that male birth weight and weight gain before one year of age, measured between 1911 and 1930, are powerful and independent predictors of coronary death in mid-to-late adult life, as measured between 1951 and 1987. Some might argue that these findings do not necessarily imply a unique causal pathway. Rather, they may merely reflect the extraordinary diversity of material deprivation's health effects over time (i.e. both weight in early life and subsequent heart disease risk are themselves caused by poverty, operating through a variety of mechanisms, of varying latency, over a lifetime). Nor is it certain that specific determinants will always produce the same outcome. Material deprivation, for example, may manifest in a series of outcomes, just as the influence of smoking may result in heart disease, cancer or asphyxiation through fire. According to some, such non-specificity of outcome is, itself, a threat to an inference of causation.

The "diagnostic tests of causation", familiar to epidemiologists, make specificity a criterion to be considered in the evaluation of cause and effect.[19] Factors such as socioeconomic status, which seem to correlate with many health outcomes, are relegated to the category of nuisance variables and are removed from consideration by a variety of statistical techniques. But from the perspective of the authors, this non-specificity is what is most attractive. Any factor that can seemingly influence so many outcomes must, common-sensically, be important and worthy of study. While this point may seem obvious, trivial, or obscure to the lay person, it illustrates a fundamental philosophical dichotomy among investigators working at the interface of science and health. Those prepared to mix pragmatism and common sense with empiricism will tend to focus on broad determinants of health, and view the relationship between health and social development as crucial. Others of a more positivist bent will stick to those factors which are best amenable to vigorous empirical study: individual-based, monocausal, and short-term in their biological effect.

Population Partitions

A population can be partitioned according to numerous different characteristics, but the interesting partitions are those which demonstrate heterogeneity of health status across their sub-groups. For instance, socioeconomic status might be divided into quintiles based on income, occupation or education (or combinations thereof). A number of studies have found significant differences in life expectancy and health expectancy between such quintiles.[20] Other partitions which demonstrate interesting heterogeneities include ethnicity/migration, geographic location, gender and "special populations". For example, Verbrugge[21] has pointed out that gender differences in health status might be explained by the fact that, while both sexes suffer from and die of similar conditions, men develop the fatal conditions with greater frequency than females, while females develop the non-fatal disabling conditions at a faster rate than males.

"Special populations" were included in Figure 1,13 to take account of populations which might not be defined primarily by geography, ethnicity or class, but might have special characteristics expressed through health status differentials. Such populations would include various "faith groups" (e.g. Mormons, Seventh Day Adventists, Buddhists) and persons with specific behavioral traits (e.g. intravenous drug users, vegetarians, alcoholics, tea-totallers). The health status of special populations may be relatively better than average (Mormons, Seventh Day Adventists), or remarkably worse (intravenous drug users).

Sources of Heterogeneity

These are meant to include the "universe" of causal pathways which might explain the heterogeneities in health status which are found within each population partition, across the life cycle. Each type of causal pathway has radically different implications for how we think about the determinants of heath, and for policies to address inequities. They are:

1. *Reverse causality* - That is, changes in the subgroup of a given partition to which a person belongs may follow changes in health status, rather than the reverse. The more true this is, the less important it would be to study the given partition on health grounds. The classic example of this is the reaction that "mortality differentials by income level can be explained by the fact that the sick become poor, not that the poor become sick".

2. *Differential susceptibility* - That is, given subgroups within a partition represent differences in inherited biologic potential or that changes in the level of the population characteristic are dependent, in part, on factors which also determine health status. For instance, the argument may be made that upward mobility is based on characteristics of the individual which, at the level of the organism, also determine better prospects for health status. This argument has been suggested to explain observations of upward mobility among tall people born in lower socioeconomic groups.

3. *Individual lifestyle* - That is, the health habits and behaviors of those in different sub-groups lead them to have different levels of risk of particular life threatening and/or disabling conditions. The socioeconomic status gradient in smoking behavior is a good example.

4. *Physical environment* - That is, the effects of exposure to physical, chemical and biological agents from the natural and the human-made environment. For instance, heterogeneities in the rates of occupational disease, motor vehicle injury and flood-related outcomes across a partition would be accounted for here.

5. *Social environment* - This includes the effects of relative social isolation, deprivation, stress/coping situations, etc., and encompasses, broadly

speaking, all aspects of social organization which might affect health status. It is complementary to individual lifestyle because of the interactive relationship between social conditions and the behavioral response of individuals that represent what we have come to see as lifestyle.

6. *Differential access/response to health care services* - That is, heterogeneities in health status related to care seeking behavior, health services in remote areas, drug costs and compliance with treatment regimes. Also included here would be differential survival rates for a given disease between populations when effective therapy has been offered equally to all.

Sources 1 and 2, above, differ fundamentally from 3 to 6. The first two sources are really points of skepticism which must be overcome before consideration of the other four is warranted. Within these latter four, there are significant boundary-line issues and issues of hierarchy. If, for instance, smoking habits differ between subgroups within a partition, should they be thought of as lifestyle-related or as a function of the social environment? At its present state of evolution, the cube highlights these problems but does not provide a neat solution to them.

When the partitions and sources of heterogeneity are intersected, relatively few unallowable combinations are found. The only one identified so far is reverse causality in relation to the gender partition.

CONSTRUCTING THE FRAMEWORK

Each combination of stage of life cycle and partition represents an opportunity for differential health status to emerge. Figure 2,13 begins with a two-dimensional cell from the cube representing a specific stage of the life cycle and partition. Within the cell is found a graph, showing a gradient of health status associated with changes in the attribute of a characteristic, Y. The area under the line represents the universe of health status differences which might be explained by differences in characteristic Y at life stage X. While the relationship shown here is a monotonic gradient, other informative possibilities might certainly present themselves. For instance, a U-shaped curve has been discovered for the relationship between cholesterol level and total mortality in men, such that a theoretical minimum risk is found at approximately 180 to 220 mg percent. Above this level, increasing risks of coronary heart disease mortality predominate, below it, hemorrhagic stroke, cancer and a variety of miscellaneous causes of death are more common.[22,23] It has been suggested that for physiologic correlates of health, homeostatic processes within the body may well lead to "U" or "J" shaped relationships, whereas this would be much less likely with determinants whose origin is external to the body. In any case, the accounting framework can accommodate any conceivable relationship.

Health status may be expressed in numbers of excess deaths, potential years of life lost, quality years lost or other measures. In general, the most useful summary measures of health status would combine a prevalence-based measure of disability with a life-table, which is incidence-based. It is necessary that the health status differences be expressed in a quantitative manner which will allow for an accounting of how much of the total can be explained in different ways. The bottom part of Figure 2,13 shows the same health status gradient divided into five subsections. These are meant to graphically represent the contributions of the various sources of heterogeneity in a simple fashion. The overall study of heterogeneities in population health status is represented by the activities which quantify the health status differences related to each characteristic at each stage.

There are, of course, outstanding issues in applying the framework which cannot be addressed at this stage. At the operational level there is a lack both of appropriate data to fill the various cells in the cube, and a consensus on how to interpret, explain and parcel-out the relative impact of various sources of heterogeneity. This latter aspect reflects a much deeper problem: lack of a common language and logic with which to link the diverse, multifactorial, dynamic relations which seemingly underly inequities.

USING THE FRAMEWORK

The underlying principle of the cube framework is that determinants of health are best understood by exploring differences in health status between sub-groups of a population. These subgroups represent, conceptually, a form of community imposed from above, as individual members of a given sub-group are seen to share one or more characteristics which are significant in determining their health status. In many cases, these will be communities "without propinquity"; in other words, they will not share location or kinship. However, in community health promotion,

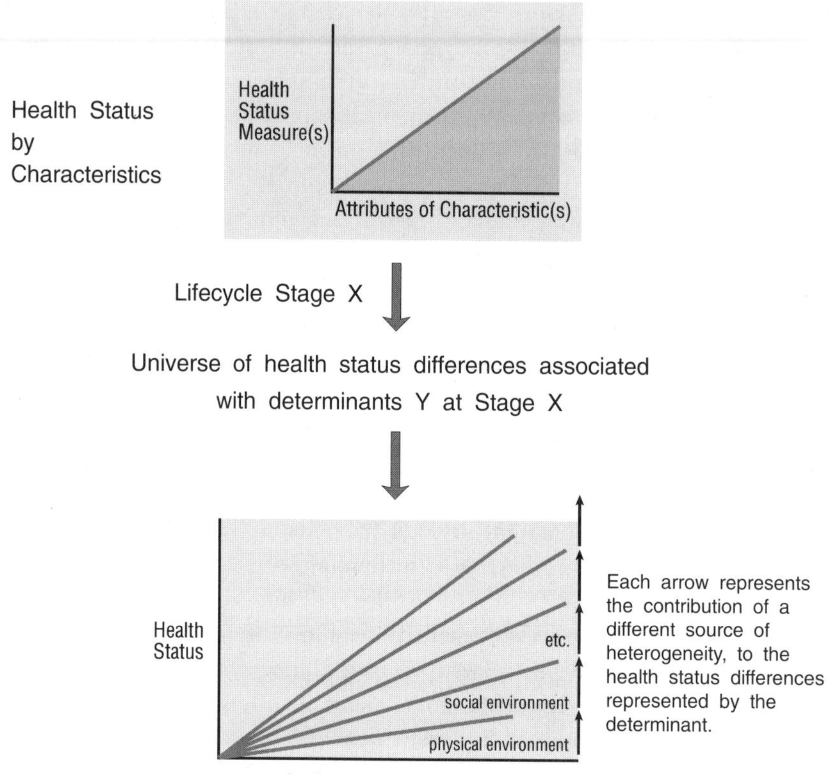

Figure 2,13 Quantifying the Effects of Determinants of Health

propinquity is assumed to exist, either because of physical closeness (community as local area) or voluntarism (community as network of like-minded individual organizations). This is not a trivial matter because the programmatic agenda of community health promotion depends upon concrete, not abstract, links between individuals. Thus, community as defined by community health promotion is a subset of community as conceived by the cube framework; however, the concepts are not in conflict.

To employ the cube, populations would first be divided into communities (with or without propinquity) whose health status, narrowly defined, showed marked differences. Second, the sources of the differences would be identified using the best available research evidence. Third, a basket of community based "indicators" would be put in place which would allow charting of progress in modifying those determinants which are deemed to be modifiable. Fourth, programs would be put in place which would help bring about these modifications. Each of these steps is treacherous, laden with confusion, uncertainty and conflict. What follows is a step-by-step examination of the issues associated with implementing this organizing framework, using British Columbia as the backdrop for illustration.

From a societal perspective, every individual in British Columbia could be ascribed to many different "communities". From the cube perspective these are simply the nominal categories known as population partitions. The two communities with the greatest appeal at the outset are the geographic and socioeconomic ones. Everyone could be assigned an identity within each of these communities, with greater or lesser difficulty; each community contains differences in health status by sub-group which are durable and significant. Census tracts have served as an excellent source of geographic definition and information. In several Canadian cities, including Montreal and Hamilton, efforts have been made to evaluate mortality differentials by census tract, grouping the geographic units by sociodemographic similarity.[24,25] Investigations of mortality by income level in urban Canada in 1971 and 1986 also follow this model.[26] Their findings of large and durable health status differences between census tracts in urban communities are a strong empirical basis for motivating national health policy in Canada.[27] Moreover, the census is a rich source of information on housing and other sociodemographic variables which are important determinants of health expectancy. It is not yet as clear whether or not census tract-based definitions of community are equally useful in the rural environment, but this could be explored in British Columbia without great difficulty.

The socioeconomic community poses greater difficulty for the concept of community health promotion. The socioeconomic status of the population is determined primarily by four factors: on an individual level there is the triad of educational background, occupation and income, while on a collective basis there is the level of function of the local economy. In each local area, there is great diversity in these factors, and, moreover, the educational system, wage structure and availability of work are primarily determined by factors operating at the provincial, national and international level, and are only under local control to a small degree. Often, school and work take place outside one's geographic area of residence. If socioeconomic communities were ignored or down-played in service of a purely local community perspective on health promotion, the opportunity to address many of the most fundamental determinants of health status would be sacrificed.

Socioeconomic communities would be created by ignoring location and grouping solely by specific personal characteristics -- income, education, occupation, etc. The major problem here is obtaining non-spatially based sources of this information. The census is reported by area, not by specific household or family. Either census data would have to be made available by individual (through an anonymous process of "record linkage"), or special population-based surveys would be needed to obtain the required information. However, we lack an institutional framework, analogous to local government, to represent socioeconomic communities and health. This was painfully evident when the Royal Commission on the Economy and Unemployment Insurance failed to highlight the important health aspects of these policy areas, and similarly, when the discussions about a national daycare policy failed to consider the long-term health consequences of early childhood experiences.

The next stage would involve defining differences between subgroups of the communities using a narrow definition of health status. Health expectancy, which integrates life expectancy with disability, is an excellent summary measure that has been used successfully in other studies in Canada and the United States. Simply put, health expectancy would have to be calculated for both geographic and socioeconomic communities on an ongoing basis in order to monitor improvement and deterioration over time, and the progress made in redressing differences within these communities across the province. Thus, health expectancy serves as a societal outcome indicator for the process of health promotion across all communities, however defined. Because it would likely react slowly to intervention, health expectancy could usefully be supplemented by a measure of the proportion of infants born of ow birthweight (less than 2500 grams). These measures, taken together, would provide a feasible and cost effective way to monitor progress in redressing inequalities in health status.

Using this method to track the process of redressing inequalities will have more validity in relation to socioeconomic communities than local areas. After all, when individuals move they take most attributes of their socioeconomic status with them, but when local areas are redeveloped their character can change completely, and in ways which will influence community health status. A process like this may very well unfold in the Downtown East Side of the City of Vancouver, with traditional residents being displaced in favor of those who can afford accommodation in trendy new housing developments. On paper, this will appear as a remarkable improvement in birth weights, life expectancy, etc., whereas all it will really represent will be the displacement of the under-privileged. If they are displaced to a common location, the net difference in health status across local areas may be preserved; with the identities of the "healthy" and "sick" communities merely being swapped around. If, instead, the displaced are distributed evenly throughout the region, this may present itself as an equalizing trend in health status. Thus trend data comparing local areas will be uninterpretable without considering trends towards increasing or decreasing socioeconomic heterogeneity within each of the communities being compared.

The next step in the process would be to define the *modifiable* determinants of health status *differences* between subgroups. This process is fundamentally different from trying to define health status in a community from first principles. The philosophical perspective can best be stated as, "I don't know exactly what health is, but I know who has more of it than others". Yet the process could still become incredibly complex if a simplifying approach were not imposed at the outset. This could be done by asking the following questions:

- Are the health status differences (between local areas or socioeconomic groups) based on specific diseases or injuries, or do they cut across many different conditions?

- At what stages in the life cycle do they express themselves?

- Which of their known determinants impact on health status over the short-term (e.g. the wearing of seat belts) versus the long-term (downstream effects of child abuse)?

- What putative determinants can be ruled out on the basis of their being fundamentally non-modifiable or because there is convincing evidence that they are relatively unimportant?

Those that are left should be grouped according to the "sources of heterogeneity" in the cube framework, as this classification of determinants helps to understand subsequent implications for policy and action.

The principles of community organizing collide with the top-down approach at the following step, wherein community-based indicators are identified. After all, choosing indicators is virtually equivalent to setting priorities because programmatic responses will likely correspond very closely to them. At first, it would appear that the most logical approach would be to impose indicators which were closely linked to the apparent sources of differences in health status. For instance, if it were found that the potential years of life lost due to motor vehicle injuries were higher in one part of a large city than another, would it not be reasonable to simply use person-years of life lost due to motor vehicle injuries as an indicator? The problem with this approach is that it would force the community to make a priority of tackling motor vehicle injuries directly. However there might be substantial reason to believe that the problem of motor vehicle injuries was inseparable from a more complex web of community dysfunction, which was well understood at the local level. Here a community might strongly desire to use resources in a way which broadly strengthened community function and simply allow the more specifically defined outcomes of threat to life and limb to adjust themselves in response to healthy community development. Could this be legitimized within the framework of a program which was focussed primarily on health outcomes? The authors believe that the answer to this crucial question is "yes". At present there is a large and growing body of evidence which suggests that a strong social infrastructure, in and of itself, is a powerful determinant of health status that can cut across a broad range of diseases and misadventures. Therefore, there is an emerging scientific basis for accepting social development as a potentially effective "ecological strategy" for improving health status, even when health status itself is narrowly defined.

In operational terms, it would be legitimate for a top-down authority (such as the provincial government) to use a global measure of outcome, such as health expectancy, to monitor health status by community across the province, and do so quite independently of the ground-up process by which communities define their own indicators. From the top down, a great deal of information can be provided to local communities about the importance of housing quality, early childhood experiences, traditional epidemiological risk factors, and other determinants of health relevant to them. This can be integrated with understandings of the local context by community representatives to develop appropriate indicators for the individual community. It is true that such baskets of indicators will differ from group to group and likely will not be strictly comparable with one another. But from the top-down perspective this would not be that important, because the health expectancy measure (supplemented with a wide variety of other routinely-collected data, if necessary) would be the basis of global comparison. Most important, if health expectancy were used as the basis of allocating resources (giving more to the less-well-off communities), then a concrete link would be created between the survey measure of health status and community development activities.

Within the socioeconomic community the conflict between top-down and bottom-up does not exist in the same way. As mentioned previously, there is no group analogous to the local government which represents, in a comprehensive way, the interests of people defined in socioeconomic terms. Indirectly, this role is played by trade unions, industry organizations, the Workers' Compensation Board, school boards, the unemployment insurance system, etc. Mobilizing these agencies to focus on health differences across socioeconomic groups would likely not involve "elite" versus grass roots conflicts, but would simply bring traditional economic antagonisms into the health promotion process. However, these need not block the development of indicators across socioeconomic communities. It would be worth the risk of taking some initiatives from the top down in order to create a data infrastructure from which to proceed. Data produced in this way could be used to organize the constituency. For instance, the following project suggestions would be extremely helpful for the development of ongoing indicators.

1. Workers' Compensation Board records should be linked to the mortality file to construct a modified life table of those who have ever received Workers' Compensation.

2. There should be an ongoing link between the unemployment insurance file and the mortality data base so that the relationship between unemployment and mortality can be followed. This information should be compared to that emerging from Sweden on the relationship between labor market adjustment strategies and the decline in socioeconomic gradients in mortality.

The best work currently being done in the industrialized world in relation to the work process and health is that being carried out in Britain and Sweden. Marmot and Theorell showed that people in the lower ranks of the civil service tended to have sustained high blood pressure throughout the day, whereas those in the upper ranks of the civil service had blood pressure declines after work.[28]

Work like this, and other projects evaluating the relationship between work process and health, should be considered as part of an initiative to develop indicators for socioeconomic communities. Such a program is currently being considered for Ontario, under the auspices of the Ontario Worker's Compensation Institute.

REFLECTIONS

This chapter has advocated developing indicators which are based on population health status differences rather than beginning from first principles. It has emphasized social infrastructure over disease-based risk factors. It supports a three tiered approach to indicator development, in which there is an ongoing survey of healthy life expectancy which can be related to both local and socioeconomic communities; an improvement in record linkage and targeted studies to develop a basket of indicators for the socioeconomic community; and support for local communities in developing context-specific groups of indicators which highlight important relationships between social infrastructure and health status. This approach is quite different from one in which a large and complex provincial health status survey is made the starting point of the process. It does not support the notion of mass cholesterol or blood pressure screening as an approach to community health promotion. Nor does it directly address the idea of health promotion as a positive means of cost control for the formidable demands of the sickness care system. Instead, it emphasizes the deeper structures of society whose transformation, over the long term, will likely have the most significant impact on human health status. In this regard, two initiatives in other parts of Canada are worthy of note. First, in the province of Quebec, a plan is being developed to allocate approximately 10 percent of the provincial health care budget to different regions based on their health expectancy. In this scenario, areas with lower health expectancy will receive greater resources, while those with higher health status will receive fewer resources. Second, in the province of Ontario, the Ministry of Community and Social Services has begun an initiative regarding primary prevention of emotional and behavioral problems for children aged 0 to 8 years in economically disadvantaged communities.[29] This program is of particular interest because it is explicitly based on the results of longitudinal studies which demonstrate the powerful latent health effects of social disadvantage in early life.

REFERENCES

[1] World Health Organization. 1986. Ottawa Charter for Health Promotion. World Health Organization.

[2] Epp, J. 1986. Achieving Health for All: A Framework for Health Promotion. Ottawa: Minister of Supply and Services Canada.

[3] Hayes, M. and Willms, S. 1990. Healthy Community Indicators: The Perils of the Search and the Paucity of the Find. *Health Promotion International* 5, pp. 161-166; See also *Health Promotion International* 3(1), 1988.

[4] United Nations. 1982. Levels and Trends of Morality Since 1950. New York: United Nations.

[5] Marmot, M.G., Kogevinas, M. and Elston, M.A. 1987. Social/economic Status and Disease. *Annual Review of Public Health* 8, pp. 111-35.

[6] Wilkins, R., Adams, O. and Brancker, A. 1989. Changes in Mortality by Income in Urban Canada from 1971 to 1986. *Health Reports* 1, pp. 137-174.

[7] Lundberg, O. 1986. Class and Health: Comparing Britain and Sweden. *Soc. Sci. Med.* 23, pp. 511-17.

[8] Vagero, D. and Lundberg, O. 1989. Health Inequalities in Britain and Sweden. *Lancet* ii, pp. 35-6.

[9] Caldwell, J.C. Routes to low mortality in poor countries.

[10] Jozan, P. *Recent Mortality Trends in Eastern Europe*. Budapest: World Health Organization, undated.

[11] Nagnur, D. 1989. Unpublished data. Ottawa: Statistics Canada.

[12] Bobet, E. 1989. Indian Mortality. *Canadian Social Trends* 15, pp. 11-14.

[13] Marmot, M.G. 1986. Social Inequalities in Mortality: The Social Environment, in Wilkinson, R.G. (ed.), *Class and Health*. London: Tavistock Publications, pp. 21-33.

[14] Haan, M., Kaplan, G.A. and Camacho, T. 1987. Poverty and Health: Prospective Evidence from the Alemada County Study. *American Journal of Epidemiology* 125, pp. 989-98.

[15] Warner, E.E. and Smith, R.S. 1982. *Vulnerable But Invincible: A Longitudinal Study of Resilient Children and Youth*. New York: McGraw-Hill.

[16] Schweinhart, L.J. et al. 1985. Effects of the Perry Preschool Program on Youths Through Age 19: A Summary. *Topics in Early Childhood Special Education Quarterly* 5, pp. 26-35.

[17] Berkman, L.F. 1986. The Association Between Educational Attainment and Mental Status Examinations: Of Etiologic Significance for Senile Dementias or Not? *Journal of Chronic Disease* 39, pp. 171-174.

[18] Barker, D.J.P. et al. 1989. Weight in Infancy and Death from Ischaemic Heart Disease. *Lancet* ii, pp. 577-80.

[19] Schesselman, J.J. 1987. "Proof" of Cause and Effect in Epidemiologic Studies: Criteria for Judgement. *Preventive Medicine*16, pp. 195-210.

[20] Hertzman C. 1986. *The Health Context of Worklife Choice*. Canadian Mental Health Association.

[21] Verbrugge, L.M. 1989. Recent, Present, and Future Health of American Adults. *Annual Review of Public Health* 10, pp. 333-361.

[22] Isles, C.G. et al. 1989. Plasma Cholesterol, Coronary Heart Disease and Cancer in the Renfrew and Paisley Survey. *British Medical Journal* 298, pp. 920-4.

[23] Martin, M.J. et al. 1986. Serum Cholesterol, Blood Pressure and Mortality: Indications from a Cohort of 361,662 Men. *Lancet* ii, pp. 933-6.

[24] Wilkins, R. 1979. *L'esperance de Vie par Quartier a Montreal, 1976*. Montreal, Institute for Research on Public Policy.

[25] Liaw, K.L., Wort, S.A. and Hayes, M.V. 1989. Intraurban Mortality Variation and Income Disparity: A Case Study in the Hamilton-Wentworth Region. *Canadian Geographer* 3, pp. 131-145.

[26] Wilkins et al., *op. cit.*

[27] Epp, J., *op. cit.*

[26] Marmot, M.G. and Theorell, T. 1988. Class and Cardiovascular Disease: The Contribution of Work. *International Journal of Health Services* 18, pp. 659-674.

[27] Technical Advisory Group to the Coordinated Primary Prevention Initiative. Summary of Research and Recommendations: Primary Prevention of Emotional and Behavioural Problems for Children, Aged 0-8 Years in Economically Disadvantaged Communities. Ontario Ministry of Community and Social Services.

14 BETWEEN A ROCK AND A WET PLACE: Health Services Planning in the GVRHD

Lillian Bayne

Health Services Planner
Greater Vancouver Regional Hospital District

INTRODUCTION

The Greater Vancouver Regional Hospital District (GVRHD) is one of 29 Regional Hospital Districts in the province which share in the costs of building and equipping the province's hospitals. However, in terms of population size and density, economic activity, and proportion of the province's health care services which fall within its boundaries, it stands in a class of its own. The GVRHD accounts for considerably less than one half of one percent (0.36 percent) of the provincial land mass, but almost half the province's 3.1 million people reside within its bounds.[1] In the 1980s, close to 70 percent of the Region's net population gain came from international migration. Receiving about 80 percent of the province's net international migration, the GVRHD experiences a higher degree of cultural heterogeneity than elsewhere in the province.[2] Unconstrained by topography to the east, the GVRHD continues to expand as municipalities from up the Fraser River Valley grow in size and seek membership in the GVRHD. Eighteen of the province's 147 municipalities (12 percent) fall within its bounds.[3] The City and District of Langley only became members of the GVRHD in January 1989, and the Districts of Pitt Meadows and Maple Ridge are examining the potential of membership. There is every expectation that the GVRHD may someday extend to Mission. The major force constraining growth further up the Fraser River Valley to Hope, where a natural topographical boundary occurs (the "rock"), is the will of the province whose interests would not be served by creating such a sizeable political counter force.[4]

It is not surprising that almost 60 percent of the province's nurses and over 61 percent of its doctors live within GVRHD[5] as over half (5,482 beds or 50.6 percent) of the province's 10,828 hospital beds are located in the Region's 27 hospitals.[6,7] In planning for hospital capital, the GVRHD negotiates with 23 distinct hospital societies, including those of all of the province's fully affiliated university teaching hospitals.

The GVRHD is responsible for sharing hospital capital costs for the region with the Ministry of Health on a 40/60 basis so one has to see its role within the provincial context. Hospitals in the province will consume half of the $4.8 billion provincial health care budget for 1990/91 with a total of $158 million spent on construction and capital equipment acquisition.[8] Although the GVRHD contribution to hospital capital seems small in relation to these sums - a mere $34 million (less than one and a half cents (1.43) for every dollar the province spends on all aspects of hospitals) - it does represent a steadily growing burden on property tax payers in the Region. The most recent Ministry of Health Five Year [financing] Plan shows an increase from $90 million per year for the 1989-93 five year period to an estimated $122 million per year for the 1990-1994 period. Shared 40/60 between the GVRHD and the Ministry, this represents an increase of $65 million to the GVRHD over five years.[9]

Regional decision-makers have acknowledged the need to set limits on regional spending and to plan for expenditures within these limits. They are working now to move the GVRHD from being a "capital planning agency", overseeing the design and construction of hospital facilities, to being a "rational planning agency", setting forth realistic capital requirements for effective hospital-based care. After several years of negotiation, the GVRHD was recently given formal recognition and approval to act in this latter capacity by the Deputy Minister of Health.

As part of its efforts to develop its planning role, the GVRHD has had to come to a better understanding of the factors that inhibit or promote effective regional planning. Central to this planning is the conception of space. It is the spatial unit under consideration, after all, that distinguishes regional planning from central, provincial, local or hospital planning. A better understanding of regional planning may be sought by exploring the issues inherent in existing spatial relations. This paper considers three key issues, (1) jurisdictional - relating to authority and responsibility, (2) population dynamics - relating to the nature, size and growth potential of the region's population, and (3) distribution and utilization - relating to service location and access. The problems these issues create for planners seeking to rationalize hospital-based care in the region are illustrated with reference to their impact on planning obstetrical beds.

FACTORS INHIBITING EFFECTIVE REGIONAL PLANNING

Jurisdictional Issues

A 1989 review of the role of the GVRHD revealed the perception that it functioned mainly as a "speed bump", slowing down but not fundamentally affecting capital expenditures on hospital facilities.[10] This finding was not surprising when considered in the light of the origins of the Regional Hospital Districts (RHDs). The 1967 Hospital District Act which created the RHDs "empowers" them to:

> "establish, acquire, construct, reconstruct, enlarge, operate and maintain hospitals and hospital facilities".

Although the Act appeared to confer wide-ranging powers on the RHDs, it has been argued that the real intent of their formation was to introduce a bureaucratic buffer between the provincial government and municipalities wanting to take advantage of the National Health Grants program.[11] This program, which provided matched federal funding for hospital construction, fuelled the rapid growth of the hospital sector.[12] Seeing the economic benefits of local hospitals as sources of status and employment for them, municipalities sought to take advantage of the federal dollars, made available to match their own for hospital construction. The burden of operating costs continued to fall principally on the provincial government. At a time when back bench politicking was in full swing, the province needed a means of routinizing and slowing down the hospital construction project approval process. The RHDs served this purpose.[13]

As noted above, the financial contribution to hospital capital made by the GVRHD has increased dramatically in real dollar terms in recent years, as has public concern over hospital-based service and its costs. Reacting to this perception, regional decision-makers are seeking to be more publicly accountable for their expenditures, but the question of to whom they are to be accountable arises. Although a child of the province, RHDs are financed through a regional tax base. However, regional board members are not elected by a regional electorate but by their local municipal or area electorate and sit on the GVRHD Board by virtue of their status as municipal representatives. The GVRHD itself is made up of 21 member municipalities and electoral areas.[14] The regional identity, then, emerges not from some sense of responsibility to a regional constituency but from an aggregation of responsibilities to municipal constituencies.

Under these circumstances, it is not surprising that decisions regarding hospital construction and capital equipment acquisition are difficult. Members of the GVRHD board must put aside concern for their immediate electorate and consider the interests of the region as a whole for an effective regional plan to be forged. This may mean accepting that construction of the extended care beds that an individual member's own electorate has been demanding for years will have to wait until beds are provided to a relatively more needy area of the region. But, there is no immediate political reward in making regionally responsible decisions. The director who votes "against" local development may not be re-elected by those immediately affected by the decision to allocate resources, even if more fairly, to another municipality in the region. It was not surprising, then, that when the GVRHD's Advisory Committee last year sought reversal of approval for a project which would see the renovation/reconstruction of beds in a part of the region where they were not perceived to be needed, the directors from the affected municipality voted it down.

Population Dynamics

A second factor complicating planning is the rapidly changing population profile of the region. Not only are more people than ever before moving into the province - and particularly into the Greater Vancouver Region (more than 300,000 are expected in the next ten years) - but their settlement patterns are creating distinct sub-regions with different health care priorities within the GVRHD.

Figure 1,14, reproduced from the Planning and Statistics Division's (PSD) PEOPLE version 16, illustrates the changes expected for the region as a whole over the next 15 years. As illustrated in the first cell (top left), the region's population is expected to continue to grow steadily, but, as indicated in the second cell (top right), at a decreasing rate. As shown in the fifth and sixth cells (bottom left and right respectively), the proportion of over 65 year olds in the population is expected to increase, with the elderly dependency ratio overtaking the child dependency ratio in about 2010. As illustrated in the fourth cell (middle right), net migration is the largest single contributor to growth in the region, currently at almost three times the natural increase effect.

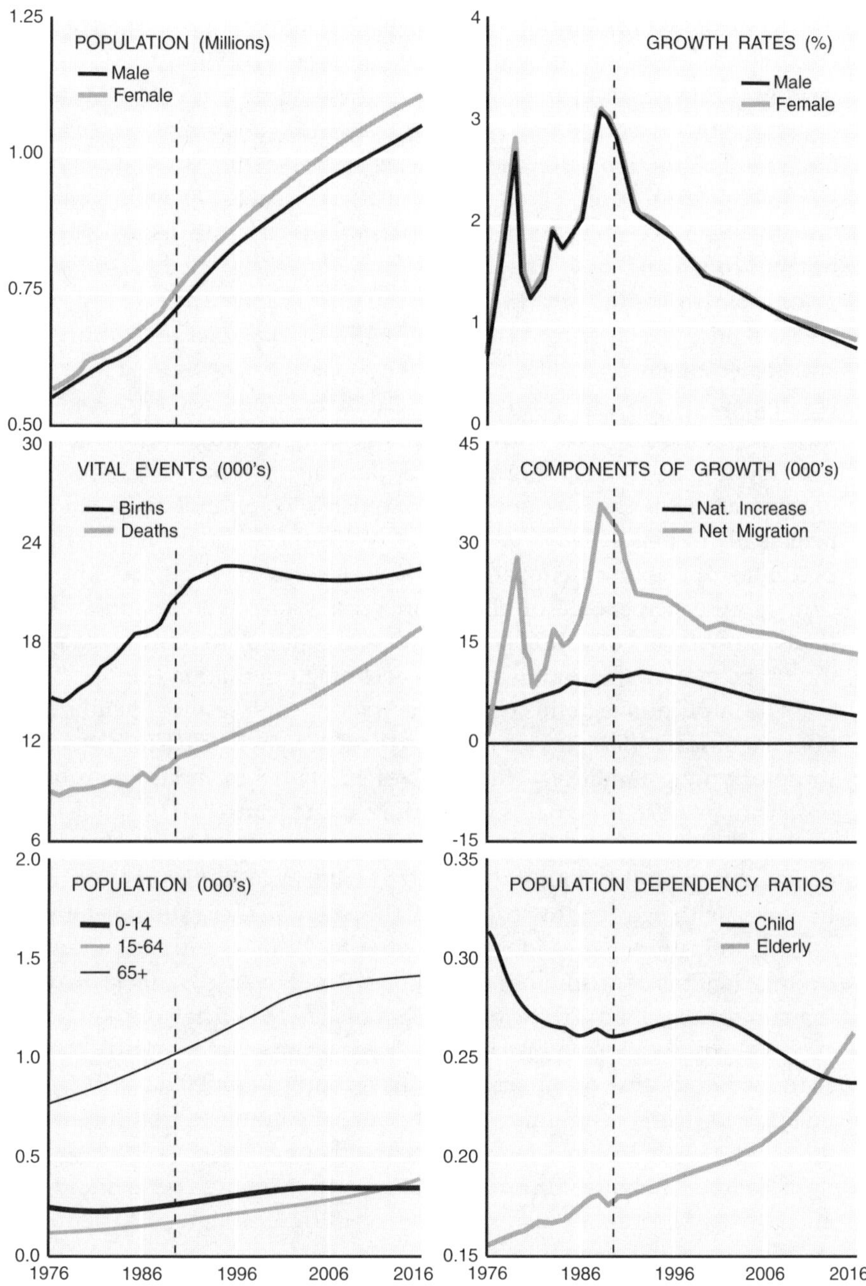

Figure 1,14 Population Profile: Greater Vancouver 1976 to 2016

While these population characteristics provide a good summary overview of the region, they do not hold for sub-regions of the GVRHD. Comparison of Figures 2,14 through 4,14, presenting data for three different Local Health Areas (LHAs) within the GVRHD, shows intra-regional differences.[15] Although differences in scale exist, creating differences in absolute impact of future changes, the relative impact of growing population, for example, can be seen by comparing curves in the first cells of each figure. While Langley's population (Figure 2,14) is expected to continue to grow rapidly over the next 15 years, Burnaby's (Figure 3,14) is expected to level off and West Vancouver's (Figure 4,14) is expected to decline. It is anticipated that net migration will make a massive contribution to growth in Langley, about eight times that of natural increase. By contrast, natural increase should coincide with or even overtake net migration as a factor contributing to growth in Burnaby. As illustrated by the curves in cell five (bottom left), Langley's population is expected to continue to be made up principally of people between the ages of 15 and 64, but to have a higher percentage of young people (0-14) than elderly (65+) (Figure 2,14). By contrast, the percentage of elderly in West Vancouver's population already exceeds that of young people and will continue to represent a growing proportion of that municipality's declining population (Figure 4,14).

The implications of this heterogeneity for health services planning are not insignificant. The service needs of the elderly will, clearly, differ from those of young families. The needs of the latter include, for example, obstetrics (maternity remains the leading single "cause" of hospitalization in all Canadian women[16]) and paediatric services. The hospital bed needs of the elderly will be quite different. To begin with, the risk of illness increases with age, leading to a proportionately higher rate of health care service utilization in the elderly.[17] Figures for British Columbia show that average lengths of hospital stay for those 70 and over (13.4 days) are more than three times longer than for those under age 14 (4.5 days).[18] Lengths of stay related to accidents in the over 74 age group average 46 days, more than four times that of the under 45 age group (under 10 days).[19] Special institutional care needs may be presented by some morbidities which affect the elderly to a greater extent than other age cohorts of the population. Severe dementia, for example, is present in about 5 percent of the 65 and over age group. Prevalence is believed to be as high as 20 percent in the 80 and over age group.[20]

Management of sub-regional differences is complicated further by the volatility of population growth, which is difficult to capture in population projections.[21] The 1990 forecasts, PEOPLE version 16, are based on data from the 1986 census, updated with reference to births and deaths for

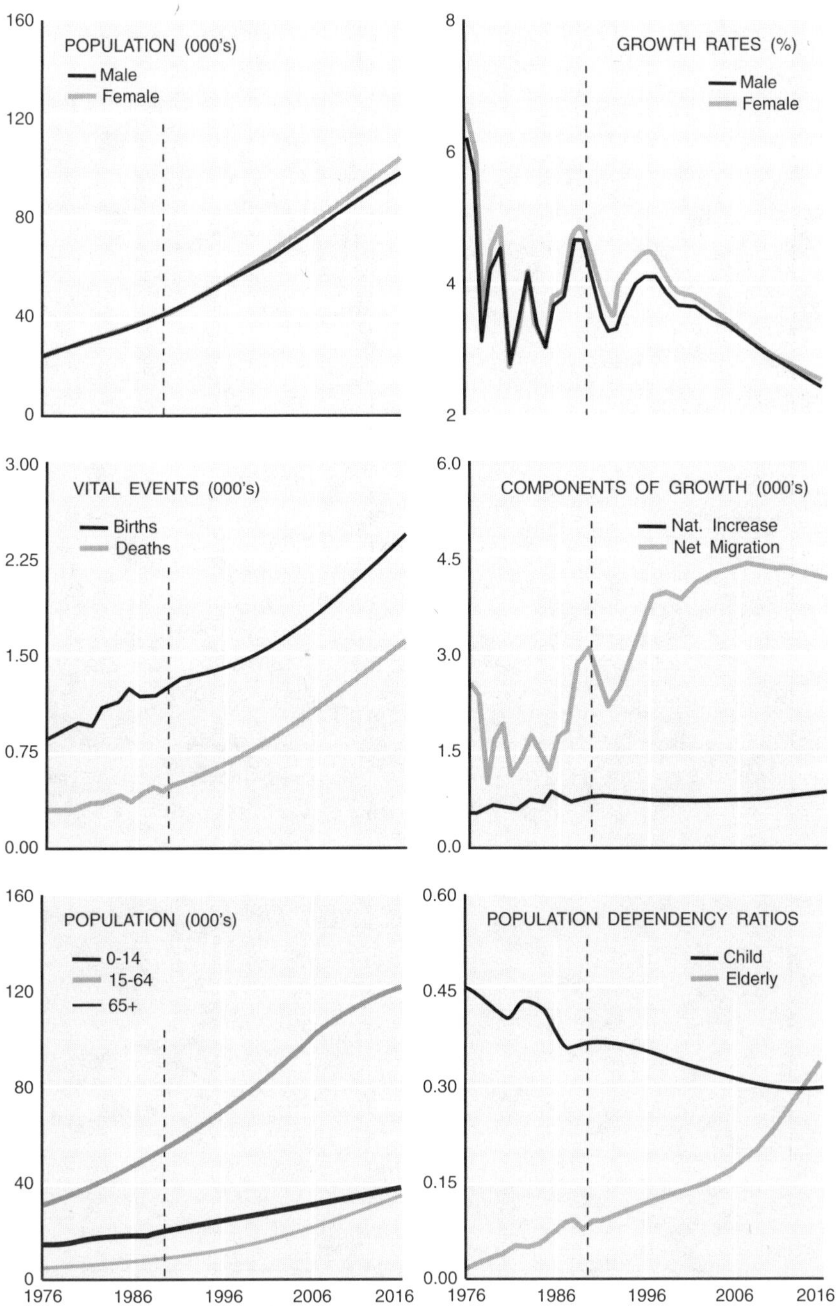

Figure 2,14 Population Profile: LHA 35 (Langley) 1976 to 2016

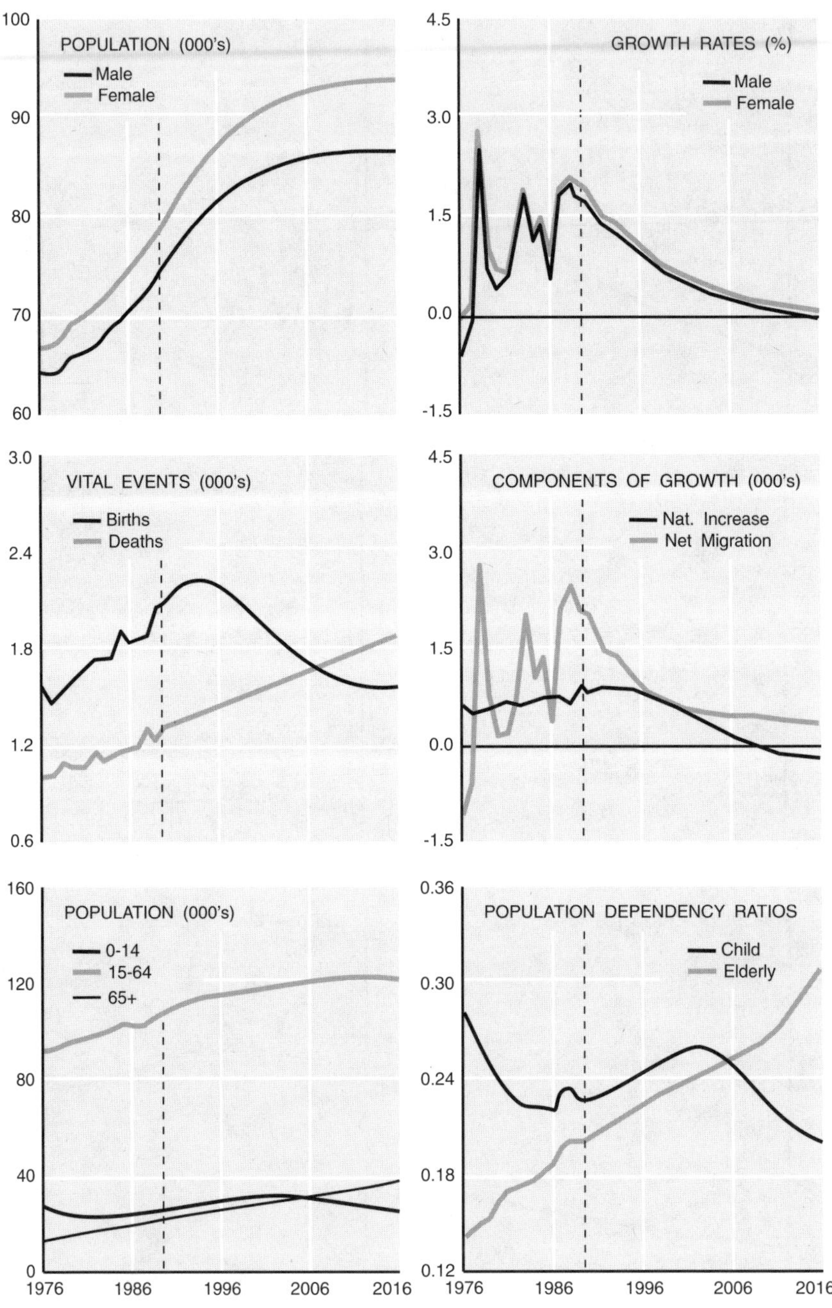

Figure 3,14 Population Profile: LHA 41 (Burnaby) 1976 to 2016

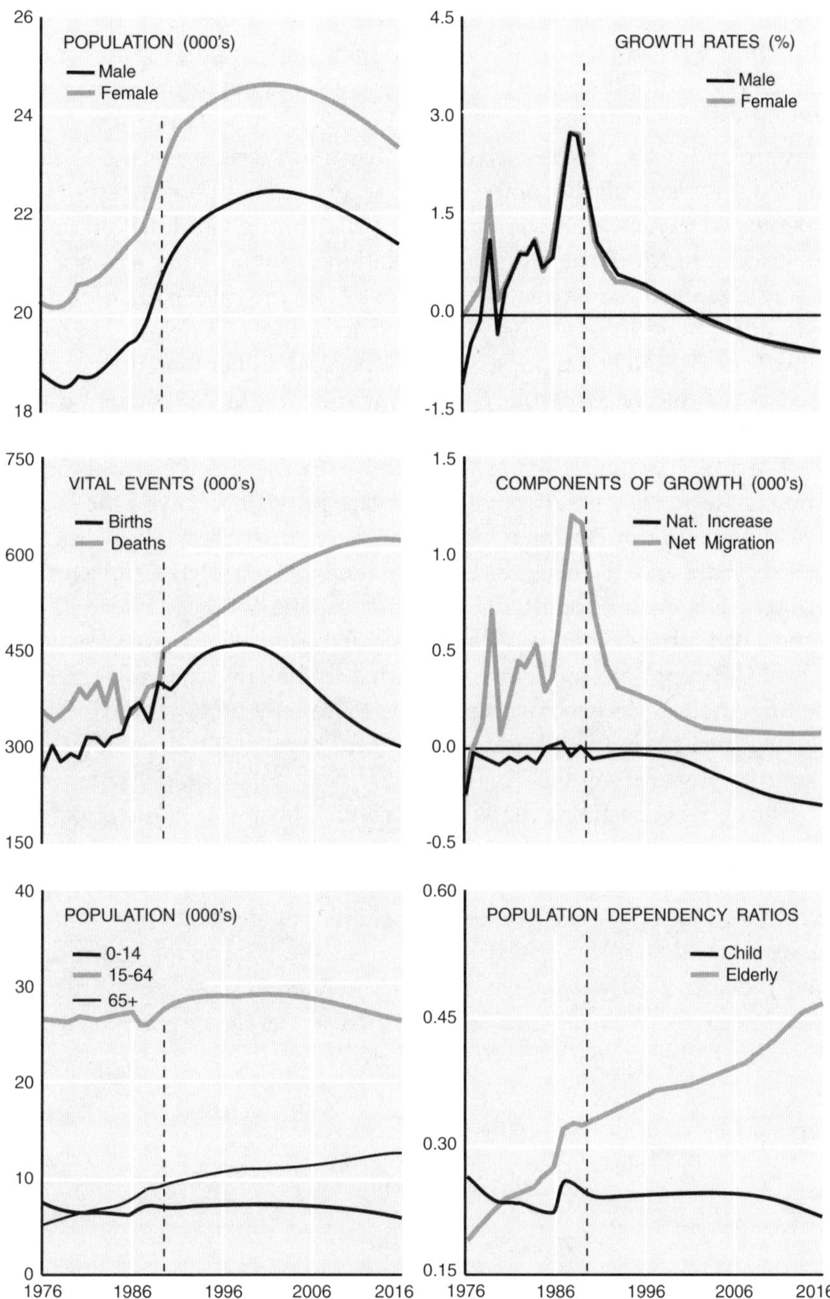

Figure 4,14 Population Profile: LHA 45 (West Vancouver) 1976 to 2016

1989, and "adjusted" according to professional assessments of "the way in which future growth will be affected by current and future developments".[22] Like all population forecasts, they represent "possible future scenarios of the size and characteristics of a defined population".[23] Long term projections are understood to be less reliable than short term projections. Their reliability is particularly compromised where large migration effects are experienced.[24] As was shown in Figures 1,14 through 4,14, net migration is a leading factor in the region's population growth. Thus, it is not surprising that even the two latest versions of the province's PEOPLE forecasts differ significantly. As illustrated in Table 1,14 below, the latest estimate of GVRHD total population is 6 percent higher than the previous forecast indicated it would be for 1990, while there is a 19 percent difference in the two forecasts of regional population for 2011.

The difficulty in predicting population growth can have serious implications in terms of cost-effective health care provision. Given the capital investment and time frame required for bed construction (on average, at least six years once the needs assessment has been completed and a project proposal has been submitted),[25] it is impossible to keep "ahead of the game". Bed numbers and types approved at the outset of a project may not be valid, appropriate, sufficient, or needed to the same critical degree, by the time the project is complete. While there may be some flexibility in locating and re-locating human resources, there is little or none with respect to physical capital.[26]

Figure 5,14 illustrates the degree to which hospitals in the region are clustered in Vancouver. In fact, 85 percent of the region's beds are concentrated in Vancouver, creating a total, unadjusted bed/population ratio five times that of Surrey/White Rock (LHA 36) (9.9 beds per thousand versus 1.9 beds per thousand).[27] Of course, the Vancouver bed complement includes a tertiary bed component, intended to serve the entire province as well as a referral component which serves the region. However,

Table 1,14 GVRHD Population Forecasts

Year	*People 15*	*People 16*	*Difference*	*%*
1990	1,397,152	1,487,790	90,638	6
2011	1,740,618	2,067,050	326,432	19

Adapted from: Central Statistics Bureau, Province of British Columbia, 1989; Planning and Statistics Division, 1990(b), *op. cit.*.

the clustering of these beds still creates physical or geographical *access* disparities within the region. The significance of accessibility is discussed in greater detail below.

Distribution and Utilization Issues[28]

The delineation of health care distribution and utilization *issues* is, to a certain extent, dependent on what social norms determine to be appropriate, fair and reasonable. At present, the GVRHD has no mechanism for gauging these. However, a number of basic values have been enshrined in the legislation which already governs the provision of health care in Canada, and the Province of British Columbia. Accessibility to health care is one of these. It was explicitly recognized as a key principle in the Medical Care Act of 1968 and, again, in the Canada Health Act of 1984.[29] Although these Acts, which saw the introduction of comprehensive public health insurance and the prohibition of extra billing and user fees, were designed mainly to address the financial barriers to access, the "reasonable access" sought may be seen also to include physical access. Indeed, if we are to believe Robert Evans,[30] "public concern for an appropriate geographic distribution of services [is] evidence of the sense of a social and public responsibility to assure access".

That many people have to travel to Vancouver for tertiary care (most of the provincial facilities are centred there), is not perceived to compromise the accessibility principle. It is "reasonable access" that is sought. It is not in the interests of quality assurance, nor is it economical or cost-effective, to scatter highly specialized, and often inter-dependent, services throughout the province. Indeed, one of the problems faced by the GVRHD is that a good many of the people coming into Vancouver do not require tertiary care and could be served by local providers in their own area. Vancouver facilities provide services for a "community" population from within the city, but they are not designed to provide primary level services to the entire region. Their high profile may work against them in drawing people from outside Vancouver who could secure care locally.

For example, a 1990 Ministry of Health study of emergency paediatric use in the area showed that the reputation of the British Columbia Children's Hospital (BCCH) was one of the factors leading to people utilizing the facility. The study also discovered that geographic proximity and use were correlated, with residents of Richmond and Burnaby, two adjoining municipalities, representing 47 percent of the non-Vancouver users of the BCCH. Patients from within the GVRHD constituted 91 percent of the facility's emergency utilization.[31]

It is worth noting that another reason parents brought their child to the BCCH was that the child's physician practised in Vancouver and/or had privileges at BCCH.[32] Although physician location clearly affects hospital use, neither the GVRHD nor any other government authority currently has the power to dictate physician practice location.[33]

The quality and cost-effectiveness of local services is dependent on a critical mass of users. If patients are drawn away from local hospitals, these facilities will not be able to develop and maintain the capacity to provide care, even to those who would prefer to use a service closer to home. While overutilization constitutes a problem, underutilization of hospital technology, widely construed, is neither economical nor safe.

PLANNING OBSTETRICAL BEDS FOR THE GVRHD

The GVRHD is currently considering the need for obstetrical beds in the region (Table 2,14). This exercise provides a good illustration of the difficulties associated with planning capital needs for the region. Many of the issues which arise are spatial in nature.

Demand for Obstetrical Beds

Demand for obstetrical beds can be seen to be considerably more predictable (within the short term) than demand for beds related to some randomly occurring morbidity, such as trauma. It is possible to isolate the affected population and calculate "latent demand" in terms of both potential volume and resource implication. The need for obstetrical beds is directly related to the size of the fertile female population (defined in British Columbia as those women between the ages of 15 and 44) and is calculated with reference to the projected number of births as follows:

$$\text{number of beds required} = \frac{\text{number of births x average length of stay}}{\text{365 days}}$$

The formula is deceptively simple. Its accuracy in predicting bed need, in the abstract, depends on the accuracy of the birth projections, the accuracy of the estimated length of stay, and the degree to which births are evenly spread out throughout the year (that is, not all births occurring at the same time and concentrating bed demand in one or two months of the year). We have already considered the unreliability of population projections under the conditions of volatile growth which are inherent in the GVRHD. Estimating average length of stay, too, is fraught with problems. Accepting Roemer's Law, that "a built bed is a filled bed", one may have reason to

Table 2,14 Vancouver Births by Month - 1986, 1987 and 1988

	Number of Births			
Month	1986	1987	1988	Mean
January	466	450	436	451
February	406	399	470	425
March	460	478	536	491
April	451	482	475	469
May	498	500	534	511
June453	495	498	482	
July	519	498	549	522
August	491	486	500	492
September	441	463	521	475
October	484	447	568	500
November	451	437	476	455
December	458	485	565	503
TOTAL	5578	5620	6128	5775
MEAN	465	468	511	481

Range: 399 to 568 Difference: 169 births

Bed Needs Implications:

Number of beds needed = (number of births x average length of stay)/number of days per month

Low: (399 x 4.1)/28 = 58 beds

High: (568 x 4.1)/31 = 75 beds

Difference: 17 beds

SOURCE: Vancouver Health Department Data presented in Evans, G.D., 1989.

question lengths of stay.[34] Finally, there is evidence of temporal variation in birth volume, although its impact is difficult to assess.[35]

Beyond these complications, bed need is dependent on a host of other less tangible factors such as consumer and care giver preferences for type or mode and location of care. The GVRHD is only beginning to devise mechanisms for identifying, let alone measuring and monitoring these. Its bed planning approach has, like most others', been one of extrapolating from the *status quo*.[36]

The Status Quo

There are nine hospitals providing obstetrical care in the GVRHD. Based on Ministry of Health approved operating capacity, these hospitals make available a total of 359 obstetrical beds, distributed across the region as illustrated in Figure 5,14.[37] This number differs from the hospitals' assessments of the number of obstetrical beds which are *actually* available (328 in November 1990), a number dependent on the resources available to operate them.[38] (It also differs from the "rated capacity" of the hospital, the number which, in theory, ultimately determines whether or not beds have to be physically *built*.) The most recent PSD birth forecasts for the region project a total of 21,709 births in 1991 and 22,295 by the end of the decade (1999).[39] The questions facing the GVRHD are (1) whether there are sufficient beds in the region to handle expected births, and (2) if not, where new beds should be built.

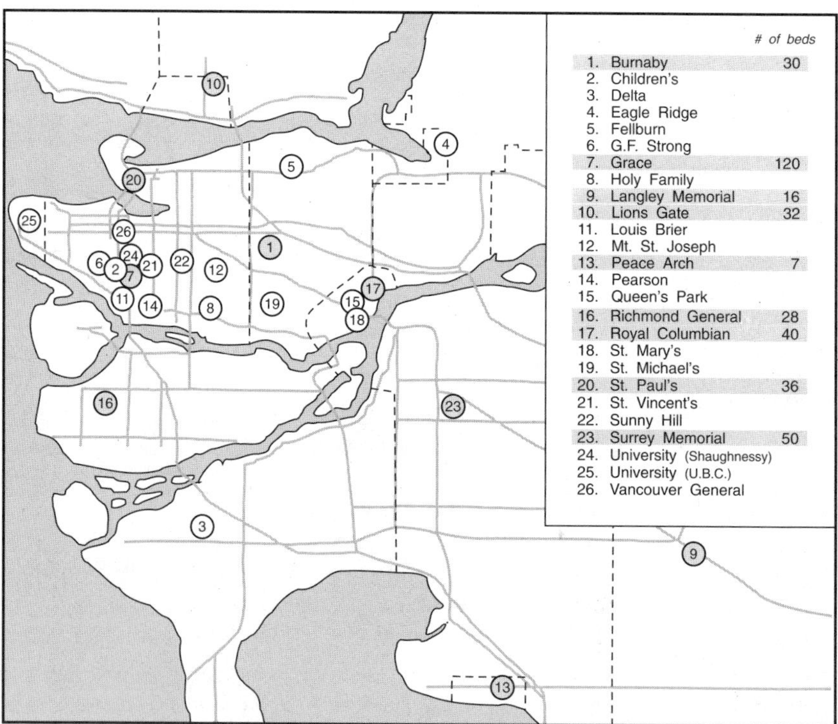

Figure 5,14 Map of GVRHD Hospitals Highlighting those with Obstetrical Beds

Reviewing Obstetrical Beds Requirements

In 1989 the GVRHD published an Obstetrical and Perinatal Services Report[40] which reviewed the need for obstetrical beds, and set out some protocol for their utilization. The report responded both to growing public concern regarding the over-utilization of Grace Hospital, a tertiary care maternity hospital located in Vancouver, and to other hospitals' demands for obstetrical department upgrading and expansion. According to the report, Grace Hospital handled over 8,000 births in 1987, although it was designed to accommodate only 6,000.[41] The hospital coped with this excessive demand by transferring patients quickly from delivery suites to postpartum modules, by housing some patients in less than ideal circumstances (for example, on stretchers, or in a "windowless daycare room"), and by reducing lengths of stay.[42] Based on 1988 PSD birth forecasts (PEOPLE version 14), the report concluded that careful management of Vancouver beds, renovation of other hospitals' facilities, including provision of birthing rooms, and construction of additional beds in the "South East Sector" of the region, would largely resolve the existing problems.

However, the 1989 PSD population forecast figures, PEOPLE 15, suggested that bed need would, in fact, be much greater than the previous year's forecasts had indicated. The difference between the two forecasts, PEOPLE 14 and PEOPLE 15, and the latest figures (PEOPLE 16) is illustrated in Figure 6,14. Within the region, the greatest difference in the two years' figures was with reference to the Vancouver population. Where Vancouver births were originally expected to peak by 1990 and decline rapidly thereafter, the PEOPLE 15 forecasts suggested that births would not peak before 1997 and would only decline gradually thereafter. The most recent population forecast indicates a peak in births a year later again, in 1998.

The difficulties for bed planning created by this inability to project births with reliability are obvious. The average construction project can take years to go from the planning stage through to construction. If the GVRHD agrees to construct new obstetrical beds, they may not be completed in time to avert the political damage done by "excessively overcrowded hospital wards". Alternately, birth volume may not materialize, the young families expected to settle in Vancouver's new housing developments may choose to live elsewhere in the region, and the GVRHD may be left with expensive and unneeded[43] hospital services, clustered in Vancouver.

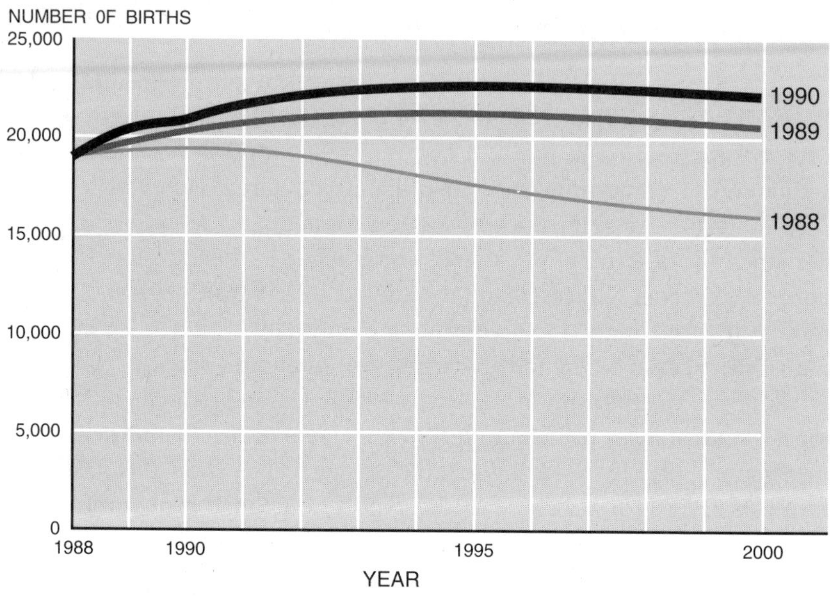

Figure 6,14 Comparison of Three Years of PSD Birth Projections for GVRHD

Underutilization of Vancouver's obstetrical beds, however, has not been a problem to date. In fact, it was overutilization, as noted above, which prompted the GVRHD to study obstetrical care in 1989. Family practitioners, one of several groups with an interest in maternity bed planning, posit that demand for Vancouver beds will be sustained. Their (simplified) scenario sees young "singles" settling initially in the downtown core close to their places of work but, as they decide to raise a family, moving to the periphery of the region, where housing is more readily available and affordable.[44] Having developed a relationship with a family practitioner in Vancouver, and continuing to work in reasonable proximity to their doctor's office, they continue to use these services when they move away from the core of the city.[45] It is reasonable, the family practitioners argue, for these individuals to expect their doctors to provide maternity care as required, including handling delivery. Practising within Vancouver, it is not surprising that these physicians have privileges at one of the city's two maternity facilities, St. Paul's and Grace hospitals, and that they should expect to be able to deliver there. Thus, maternity cases from outside the city are not always dealt with at their local hospitals but, instead, gravitate to Vancouver hospitals for care.

This is particularly problematic for Grace Hospital, which has tertiary care responsibilities for the entire province, and referral responsibilities for the region. However, both Vancouver hospitals will be overloaded if they attempt to deal with primary, low risk obstetrical cases from across the region. In addition to coping with a higher than expected volume of births from within their municipal bounds,[46] they face the prospect of imposing limits on utilization by people from communities outside the city. The question for GVRHD planners is what role can they play in shaping obstetrical bed utilization in the region.

Working together with the province and the hospitals, the GVRHD produced its Obstetrical and Perinatal Services Report in 1989. The report contained recommendations which sought to reduce the burden on Grace Hospital and to encourage use of local hospitals. However, the efforts undertaken to date have had only a very limited impact on utilization. The introduction of a priority access policy, favoring tertiary and high risk cases, followed by "community cases" from Vancouver, and a pre-registration system, have reduced Grace Hospital's occupancy rate from 96 percent in 1987/88 to 95 percent in 1988/89, still well outside the provincial standard of 85 percent.[47] Despite enhancement of maternity capabilities elsewhere in the GVRHD, including modernizing the maternity facilities and increasing the number of obstetrical beds at other hospitals (policies over which the GVRHD has some influence) and granting of reciprocal obstetrical privileges to physicians, Grace Hospital utilization by residents from other than LHA 39 (Vancouver) continues to be high. Figure 7,14 indicates the percentage of births delivered at Grace Hospital by LHA of origin. While utilization appears to exhibit a distance decay effect, high utilization by Burnaby (LHA 41) residents, who would essentially have to bypass their local 30-bed hospital en route to Grace, is more likely to relate to the image of Grace Hospital as a superior provider of maternity care.

Thus, while the GVRHD is able to make recommendations and circulate them widely, as they have done with the publication of the 1989 Obstetrical and Perinatal Services Report, they are dependent on the cooperation of the many other stakeholders involved in effecting even modest changes. Many of the factors influencing bed demand remain outside the control of the GVRHD and continue to compromise efforts to plan effectively.

THE ROLE OF SPATIAL RELATIONS

The issues discussed in this paper arise as a consequence of activity within and between politically constituted territories, or, as Moon[48] has referred to it, "formal space" (Figure 7,14). As he has pointed out, the creation of formal space, such as municipal or provincial jurisdictions:

> essentially defines for space a role as a container of activity. Within the defined spatial units, health care is provided by relevant authorities and consumed by locally resident users; the geographical area is simply a labelling or classifying device. For the authorities the spatial unit fulfils a delimiting role setting the bounds within which the legal responsibility for care provision falls.[49]

Unfortunately for those responsible for attending to the spaces so delimited, formally constituted boundaries are invisible. As one Ministry of Health official put it: "There is no fence between Burnaby and Vancouver". As a result, activity inevitably "spills over" formal boundaries. While this spillover may conform to a pattern, like that identified by a hospital as its catchment area, it is not always regular, creating intractable problems for the planners seeking to rationalize utilization.

There are several reasons why health care planning on a regional level could result in more effective planning and service rationalization. For example, the distinctiveness of each region can be recognized and addressed, specialized care need not be duplicated, and high technology may be shared. But, as Gosselin[50] has pointed out, power must accompany the will to plan at a regional level. Thus, of the three issues considered, the first, that of jurisdictional bounds and authority, is preeminent.

The GVRHD has no control over the policy levers with which to resolve population dynamics and service utilization issues, and has only limited influence with respect to service distribution.

The federal, provincial and municipal governments all have some power to affect population growth. The federal government directs immigration policy and, at the grossest level, can, like the province, introduce "pro-natal" policies (such as anti-abortion legislation on the negative side, or child care support programs on the positive side). Both the federal and provincial governments can introduce regional incentive or economic development programs to draw people to a particular part of the country or province. The municipalities have a finer level of control over population dynamics, as well as service distribution, through their land use

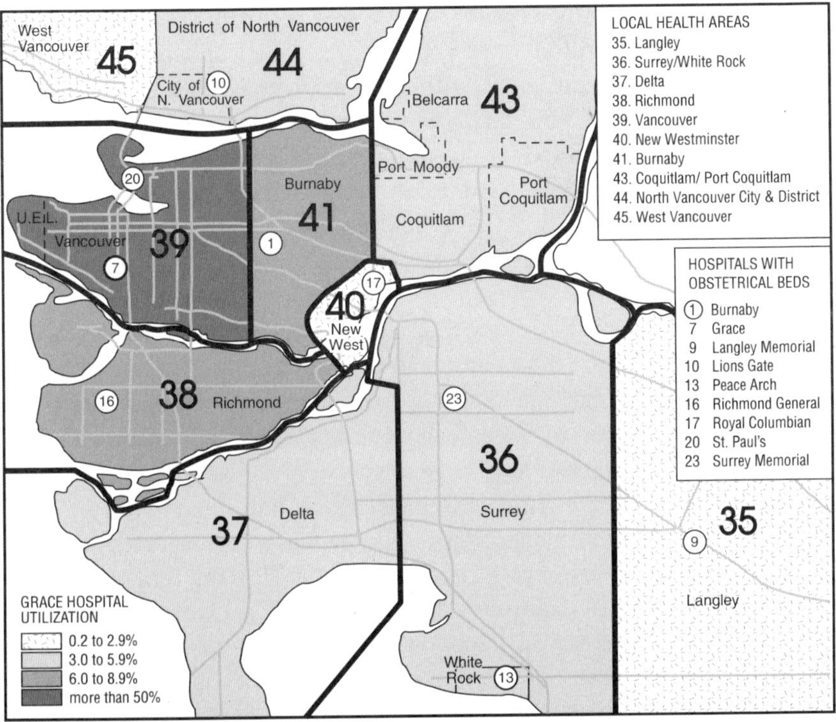

Figure 7,14 Percentage of Each LHA's Births Delivered at Grace Hospital.

classification powers. Their zoning decisions determine whether or not land may be used to build high density housing or hospitals.[51] The municipality can even shape the socio-demographics of a neighborhood by denying development permits for plans which do not include a subsidized housing component for low income renters.

With respect to service distribution and utilization, the provincial government has some ability to influence, if not direct, provider location through its provider remuneration programs. And, although the federal government has sought to restrict the use of user fees, the provincial government may have some latitude to shape utilization through the introduction of charges designed to inhibit use deemed inappropriate. Quebec has recently resorted to this policy lever, introducing an emergency department user fee in an effort to curb inappropriate use.[52]

Although nominally a player in decisions regarding facility construction or service distribution, the GVRHD has a history of being overlooked. It has fought a constant battle with provincial politicians who have announced the siting of new facilities without reference -- or even notice being given -- to the GVRHD. In theory, the GVRHD has the power to withhold its contribution to a capital project of which it does not approve. However, in practice, this power is seldom used.[53] None of the power to control population dynamics, or service utilization, resides with the regional government.

It can be argued, however, that although others may have a wider field of authority, each is to some extent constrained. The policy levers available to the province in shaping utilization or provider location, for example, are restricted by federal accessibility legislation and freedom of movement provisions in the Charter respectively. Similarly, municipalities are subject to overarching provincial legislation which effectively diminishes their power. For example, the province last year imposed an amendment to the Residential Tenancy Act, disallowing tenant discrimination of any kind (for example, adult only buildings) which obviated the municipalities' need to control of tenancy.[54] In a non-totalitarian state, the power to shape all the key variables is unlikely to be conferred to a single agency. Recognizing its inability to *impose* change, the GVRHD must seek to improve its ability to plan within existing constraints.

REFERENCES

[1] GVRHD. 1990. Report to the Royal Commission on Health Care and Costs, Vancouver: GVRHD, October,1990; Ministry of Municipal Affairs, Recreation and Culture. 1989. *Statistics Relating to Regional and Municipal Governments in British Columbia*. Victoria: Province of British Columbia, March 1989.

[2] Planning and Statistics Division. 1990. BC Stats, Victoria: Province of British Columbia, October, 1990(a).

[3] Ministry of Municipal Affairs, Recreation and Culture, *op. cit.*

[4] Walker, M.M., Manager Hospitals, Housing and Properties, GVRD. Personal communication, November 23, 1990.

[5] Health Manpower Research Unit. 1990. *Rollcall 89: A Status Report of Health Personnel in the Province of British Columbia*. Vancouver: University of British Columbia, March 1990.

[6] There are 27 recognized hospitals in the GVRHD. These include the Arthritis Foundation, which is a member of the GVRHD but which operates no beds. The remaining 26 facilities include 18 predominantly acute care hospitals, and 8 rehabilitation and extended care hospitals. Facilities located within the geographical bounds of the region, but providing only provincial or tertiary care, such as Riverview Hospital or the British Columbia Cancer Agency, are not members of the GVRHD.

[7] Planning and Statistics Division. 1990(a), *op. cit.*; GVRHD. 1990, *op. cit.*

[8] Bayne, L. 1990. Ministry of Health 1990/91 Budget, Hospital Advisory Committee Report, July 5, 1990.

[9] Stump, G. 1990. Five Year Capital Plan Policy, Board's Hospital Committee Report, May 22, 1990.

[10] Peat Marwick. 1989. *GVRHD Role Study - Part 1*. Vancouver: Peat Marwick, March 1989.

[11] Crichton, A., Professor Emerita, University of British Columbia, Department of Health Care and Epidemiology. Personal Communication, December 28, 1990.

[12] Taylor, M.G. 1986. The Canadian Health Care System 1974-1984, in *Medicare at Maturity: Achievements, Lessons and Challenges*. Calgary: University of Calgary Press, pp. 3-40; Barer, L. and Evans, R.G. 1986. Riding North on a South-bound Horse? Expenditures, Process, Utilization and Incomes in the Canadian Health Care System, in *Medicare at Maturity: Achievements, Lessons and Challenges*

[13] Crichton, A., Professor Emerita, UBC Department of Health Care and Epidemiology, Personal Communication, December 28, 1990.

[14] In the introduction it was noted that the GVRHD is made up of 18 municipalities. The remaining three members of the GVRHD are electoral areas A, B and C.

[15] Local Health Areas are sub-regional units used for planning and resource allocation purposes by the Ministry of Health. They correspond almost exactly to the boundaries of provincial School Districts and aggregate up to Health Units, jurisdictional units for public health services. The correspondence between LHAs and municipal boundaries is illustrated in Exhibit 1 of the Appendix.

[16] Canada. 1990. *Health Reports* vol. 1 no. 1.

[17] Barer, M.L., Evans, R.G., Hertzman, C. and Lomas, J. 1987. Aging and Health Care Utilization: New Evidence on Old Fallacies, *Social Science and Medicine*. 24(10), pp. 851-862; Barer, M.L., Pulcins, I.R., Evans, R.G., Hertzman, C. Lomas, J. and Anderson, G.M. 1988. Diagnosing Senescence: the Medicalization of B.C.'s Elderly. Vancouver: Health Policy Research Unit; Evans, R.G., Barer, M.L., Hertzman, C., Anderson, G.M., Pulcins, I.R. and Lomas, J. 1988. The Long Good-bye: The Great Transformation of the British Columbia Hospital System. Vancouver: University of British Columbia; Health and Welfare Canada. 1988. Elderly Persons With Psychiatric Disorders. Ottawa: Canada.

[18] Thomson, A.D., Medical Consultant, Hospital Programs Division, Ministry of Health, Personal communication, February 13, 1991.

[19] *Ibid.*

[20] Health and Welfare Canada, 1988, *op. cit.*.

[21] The Planning and Statistics Division distinguishes between "projections" and "forecasts", two terms which are often used interchangeably. As a "simulation based primarily on historical data", it would not, strictly speaking, be possible to capture volatility in a projection (see Planning and Statistics Division. 1990(b). *British Columbia Population Forecast: 1990-2016.*Victoria: Ministry of Finance and Corporate Relations, British Columbia, p. 29). As noted below, a "forecast" incorporates professional judgements regarding the future, including assessments of the possibility of unprecedented growth. The two terms are treated as roughly equivalent for the purposes of the discussion here.

[22] *Ibid.*

[23] Evans, G.D. 1989. *Projection of Births for the Vancouver Population 1986-2000.* Vancouver: City of Vancouver Health Department, p. 1.

[24] Curtis, S.E. and Taket, A.R. 1989. The Development of Geographical Information Systems for Locality Planning in Health Care. *Area.* 21(4), pp. 391-399.

[25] Walker, M.M., Manager Hospitals, Housing, and Properties, GVRD, Personal communication, January 18, 1991.

[26] This is not to suggest that there are no serious limits to effective centralized health care personnel planning as well. These are reviewed and discussed further; see Lomas, J. and Barer, M.L. 1986. And Who Shall Protect the Public Interest? The Legacy of Canadian Health Manpower Policy, in *Medicare at Maturity, op. cit.*, pp. 221-286.

[27] Ministry of Health. 1989. 1988/89 Acute Care Volume for Service Code 51 (Obst Del) by LHA and Hospital Used. Province of British Columbia, Internal document, May 7 1989; Planning and Statistics Division. 1990(b). *op. cit.*

[28] It is difficult to separate distribution from utilization. In a pure market, demand would dictate distribution - facilities/providers would not locate where a critical mass of patients/consumers (standing ready and able to pay for a commodity) existed. In the Canadian health care system where universal access is enshrined the market condition of economic viability must be weighed against an equity requirement (or "social" condition): all consumers, whether they use the service or not, pay equally into the health care system and, therefore, should be equally satisfied by the returns. Distribution may have to be managed in order to ensure, or at least improve, equity of (physical) access. While utilization or critical mass still drives distribution or service location to a great extent, the latter, too, influences the former. The degree to which availability drives utilization is discussed in depth elsewhere (see Evans, R.G. 1984. *Strained Mercy: The Economics of Canadian Health Care.* Toronto:Butterworths, esp. p. 85ff., p. 185ff).

[29] Taylor, G. 1986. The Canadian Health Care System 1974-1984, in *Medicare at Maturity, op. cit.*, pp. 3-40.

[30] Evans, R.G. 1984, *op. cit.*

[31] Ministry of Health and Children's Hospital. 1990. *Report of the Provincial Pediatric Hospital Utilization Study Steering Committee.* Victoria: Ministry of Health.

[32] See: Bayne, L. 1990. *Provincial Paediatric Hospital Utilization Study: Summary Report.* Hospital Advisory Committee Report.

[33] The Quebec government has had in place legislation to encourage physician practice location in specified regions of the province. Physicians practising within urban areas receive 80 percent of the standard fee-for-service rate, while those practising in designated areas of need are paid 120% of the standard. Recent, more prescriptive legislation, which requires newly licensed physicians to practice in "community settings", suggests that the earlier approach was not as effective as it was hoped it would be in inducing physicians to locate in outlying areas (see Globe and Mail, "Quebec set to reform health care", December 8, 1990, p. 1). For an exploration of B.C.'s unsuccessful attempts to manage physician practice location (through billing number restrictions) see Blomley, N.K.,

The Business of Mobility, *Canadian Geographer* (forthcoming). The Province has had some success with less "draconian" measures, including, for example, its Rural Training Program, which gives physicians the opportunity to train in rural settings with the hope that they will choose to establish a practice there when their training is complete (Hudson, R., Medical Consultant, Policy Planning and Legislation, Ministry of Health, Personal communication, January 25 1991).

[34] To what degree, for example, are lengths of stay influenced by demand? Are patients discharged earlier so as to make room for other, new patients? Or, are their stays being extended longer than necessary simply (a) because the bed they occupy is not needed, (b) because a more appropriate placement does not exist, or, (c) in the most cynical - but not unrealistic - scenario, because they can conveniently fill a bed which might, otherwise, be occupied by a more "expensive" patient?

[35] Vancouver Health Department figures, reproduced in Exhibit 2 in the Appendix, indicate that there is some variance in birth volume by month, ranging from a low of 399 to a high of 568 births for the 36 months covered. The implications for bed demand is a difference of 17 beds, more than 10 percent of the total Vancouver maternity bed complement. However, when it is considered that this effect is borne by two hospitals and that each has some latitude - at least to the level of a week - afforded by the ability to schedule some births (C-sections or inductions), the effect becomes less significant. Of course, the hospital's interests in spreading workload out over the week may conflict with the obstetrician's interest in preserving his or her weekend so that this flexibility may be more theoretical than real. Grace Hospital figures indicate a "seasonal variation" in birth volume, with lower admissions in the winter than in the summer months. The hospital concludes that a 5 percent bed planning adjustment is necessary (see Grace Hospital. 1990. Planning Database Update. Vancouver: Grace Hospital).

[36] The validity of this approach hinges on the assumptions (a) that there is a legitimate need, at some level, for hospital-based care (i.e."beds"), and (b) that the foreseeable future of health care service delivery is unlikely to represent a radical departure from the present. Implicit, here, is the assumption that government does not have a role in effecting substantive change. With reference to the GVRHD, the objective then becomes one of improving and/or containing existing programs and, by better detecting and monitoring trends, of moving the organization from a position of reacting to one of responding. For a discussion of the role of government and the limits to health policy-making in a plural democracy see Lomas, J. 1985. *First and Foremost in Community Health Centres: The Centre in Sault Ste. Marie and the CHC Alternative*. Toronto:Univeristy of Toronto Press; and Brown, L.D. 1983. *Politics and Health Care Organization: HMOs as Federal Policy*. Washington:Brookings Institution.

[37] Bayne, L. 1991. *Obstetrics and Perinatal Services*. Hospital Advisory Committee Report.

[38] *Ibid.*

[39] Planning and Statistics Division, 1990(b), *op. cit.*

[40] GVRHD. 1989. *Utilization of Obstetrical and Perinatal Services in the Greater Vancouver Region*. Vancouver: GVRHD.

[41] *Ibid.*

[42] *Ibid.*

[43] It should be pointed out that, although unneeded, these beds may still be used. As stated elsewhere: "capacity determines use and not the other way around". See Evans, R.G. 1986. Restraining Health Care: New Realities and Old Verities, in Allen, R.C. and Rosenbluth, G. (eds.), *Restraining the Economy*. Vancouver: New Star Books, p. 189.

[44] Collins, L., Head of Family Practice, Grace Hospital, Personal communication, November 14, 1990.

[45] Vreede-Brown, D.R. 1989. An Analysis of the Present Post Partum Bed Crisis from the Viewpoint of the Family Practitioners of Grace Hospital and a Proposal for Action. Presentation to the Department of Family Practice, Grace Hospital.

[46] The *actual* Vancouver births for 1988/89 were 6,281 compared with the PEOPLE 14 forecast of 5,975. Five thousand six hundred and thirteen (5,613) or 89% of these were handled at the two Vancouver hospitals. See Bayne, L., 1991, *op. cit.*; and Ministry of Health, 1989, *op. cit.*

[47] Grace Hospital 1990. Planning Database Update. Vancouver: Grace Hospital.

[48] Moon, G. 1990. Conceptions of Space and Community in British Health Policy. *Social Science and Medicine* 30(1), pp. 165-171, p. 165.

[49] *Ibid.*, pp. 165-171, p. 166.

[50] Gosselin, R. 1984. Decentralization/regionalization in Health Care: The Quebec Experience. *HCM Review* Winter, pp. 7-25.

[51] Walker, M.M., 1991, *op. cit.*

[52] Globe and Mail, *op. cit.*

[53] Indeed, it has only been used once in the history of the GVRHD when the Board refused to support the construction of UBC Hospital on the grounds that it would not be serving the local community.

[54] Walker, M.M., 1991, *op. cit.*

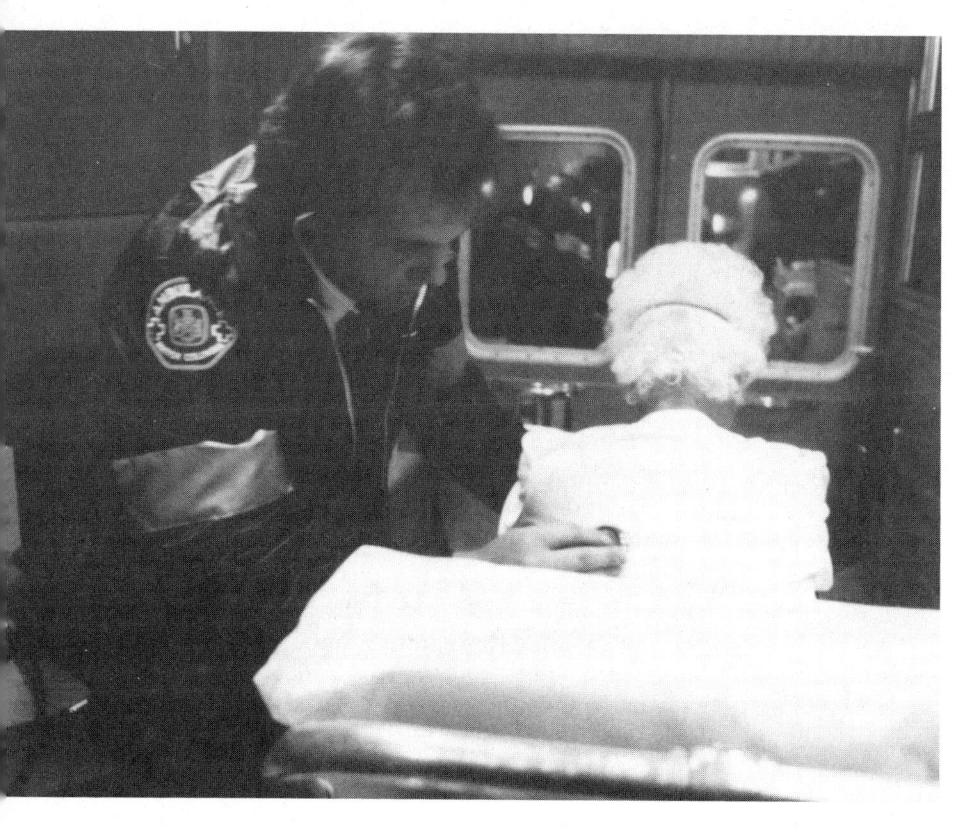

15

FACTORS RELATED TO THE ADOPTION OF MUNICIPAL BY-LAWS TO RESTRICT SMOKING:
An Analysis of Healthy Public Policy in Action

M. Hollander
L.T. Foster
G. Curtis
A. Galloway

Province of British Columbia, Ministry of Health and Ministry Responsible for Seniors

INTRODUCTION

The negative health effects of smoking, on smokers, have been well documented[1] and increasingly, today, unambiguous evidence is mounting on the health effects to non-smokers of second hand or environmental smoke.[2] Continued smoking, in light of the documented negative health effects, is of growing concern to all levels of government in Canada.[3] In 1982, Canada had the fourth highest annual per capita manufactured cigarette consumption among 110 countries.[4] Even though the percentage of regular cigarette smokers has decreased significantly from levels in the 1960s,[5] and total tobacco consumption in the 1980s fell by over 20 percent in Canada, the average annual consumption of tobacco was still 2.48 kilograms for each person 15 years or older in Canada in 1989.[6]

While this reduction is undoubtedly related to the increased cost of tobacco (from both taxes and manufacturers' price increases), other factors are important such as health promotion programs, increased social pressure from non-smokers, and the increasing recognition of adverse health effects. As well, the role of local governments has been particularly important in Canada. In the mid to late 1980s municipalities started to adopt local ordinances or by-laws that restricted smoking in public places and in work areas. These by-laws constitute an expression of what has come to be referred to as "healthy public policy".

BACKGROUND

The concept of healthy public policy is imbedded in an international movement to provide a new focus for, and emphasis on, public health and health promotion. Internationally, healthy public policy can be traced back to the 1977 World Health Assembly, the 1978 Alma Ata Declaration, the Health for All by the Year 2000 movement and the 1988 Conference on Healthy Public Policy in Adelaide, Australia. In Canada, this movement is reflected in the Ottawa Charter on Health Promotion and the *Achieving Health for All* document and has, more recently, been developed through the National Symposium on Health Promotion and Disease Prevention held in British Columbia in 1989.[7] While there is no generally agreed upon definition of healthy public policy, it is characterized by: a broad emphasis on a healthy lifestyle and a healthy society; a concern with ecology, the environment and social justice; a holistic approach to health; public participation in the health of one's community; and, an integrated and multisectoral approach to health and health care. Municipal by-laws to restrict smoking are reflective of a number of these aspects, especially public participation and the multisectoral approach. As a policy option, municipal by-laws also move the decision-making process to the local level so that people can more directly shape the nature of their own communities and changes can be readily made at a pace that the community finds acceptable.

Municipal by-laws reflect, and are generally supported by, public opinion. In 1985 some 53.6 percent of Canadians 15 years of age or older felt that it was very important for government to deal with the issue of smoking and, in 1986, 81.4 percent supported banning or restricting smoking in the workplace.[8] Many smokers themselves also supported restrictions on smoking. For example, prior to the adoption of the restricted smoking by-law in Vancouver, 89 percent of Vancouver smokers supported banning or restricting smoking in restaurants. The comparable figures for supporting restrictions in public transport, community centres and public buildings were 95 percent, 86 percent, and 78 percent respectively.[9] While there are few evaluations of by-laws in the published literature, existing references regarding municipal by-laws, or workplace restrictions, indicate that such initiatives have been well accepted by the public. There is also some evidence to indicate a relationship between by-laws and reductions in the prevalence and consumption of tobacco.[10]

Municipal by-laws to restrict smoking share some common characteristics. Almost all by-laws contain clauses related to signage stating that signs must be posted and be clearly visible. Proprietors may be required to enforce by-laws by bringing this signage to the attention of persons who are smoking and asking them to stop. There may also be requirements that proprietors or managers familiarize their staff with existing by-laws. Finally, it is also quite common for by-laws, while restricting smoking in general, to make provisions for designated smoking areas.

While British Columbia had the lowest percentage of regular cigarette smokers in the adult population of any province in Canada in 1986, 21.9 percent compared to a national average of 28.2 percent,[11] in 1988 it also had a disproportionately large percentage of municipalities with by-laws. A report entitled *Smoking By-Laws in Canada*[12] indicated that of 114 municipalities with by-laws in 1988, 33 or 28.9 percent were in British Columbia. This is, proportionately, some two and one half times as many as would be expected when compared to the percentage of the Canadian population residing in British Columbia.

Research Question

This is an exploratory study to analyze the pattern of adoption of no-smoking by-laws in British Columbia and to investigate factors that distinguish municipalities which adopt by-laws from those which do not. The results can provide useful information to those interested in the use of municipal by-laws as either a general instrument for healthy public policy or as a specific instrument to restrict smoking and reduce the associated health effects.

In particular, this study investigates three main areas of interest. First, what were the general trends in the adoption of no smoking by-laws by municipalities in British Columbia? Second, what relationship existed, if any, between relative "need" for a reduction in smoking, using morbidity rates of smoking-related diseases as a measure of "need", and the adoption of by-laws? Third, what demographic and socio-economic variables predisposed municipalities to adopt by-laws? The findings are then discussed in relation to additional steps taken by government to promote the adoption of municipal smoking by-laws.

METHOD

Unit of Analysis

The units of analysis of this study were the approximately 140[13] British Columbia municipalities in existence in 1986 governed by the *Municipal Act.* Section 692 of the Act states:

> Subject to the *Health Act*, the council may by by-law regulate persons, their premises and their activities, to further the care, protection, promotion and preservation of the health of the inhabitants of the municipality.

This section is clearly permissive and empowers municipalities to make by-laws in the area of public health, subject to the provisions of the *Health Act.* Such by-laws must be sent to the Minister of Health for approval. By-laws to restrict smoking enacted on or before December 31, 1989, were included in this study.

Twenty-six municipalities with a 1986 population of 1,000 or less were excluded from this analysis as they were not deemed to constitute an appropriate comparison group for municipalities that had adopted by-laws.[14] The statistical effect of this exclusion was to truncate the range of municipalities, thereby providing a more conservative estimate of the relationship between community size and the adoption of by-laws. Of the 110 remaining municipalities, six were deleted due to missing data or outliers on independent or dependent variables.

It was decided to categorize the remaining 104 municipalities into three groups as follows: non-adopters (61 municipalities); moderate by-law adopters (22 municipalities); and, comprehensive by-law adopters (21 municipalities). The latter two categories were based on the relative degree of smoking restrictions embodied in the by-law. (See Table 2,15 for details of the scoring system used to differentiate between these two categories.)

Morbidity Data

Two measures of morbidity were used: the number of smoking-related hospital days, and the number of smoking-related admissions per 1,000 population in the community. In order to obtain smoking-related measures of morbidity, smoking-attributable fractions (SAFs) developed by the United States Department of Health and Human Services[15] were utilized.

The eight groups of primary diagnoses, based on the International Classification of Diseases, version 9 (ICD-9) at the three-digit level, with the highest SAFs, were selected for inclusion in this study. These groups and their respective smoking-attributable fractions are presented in Table 1,15. Composite measures related to smoking-attributable days, and admissions, for the eight groupings of diseases were also developed.

Table 1,15 Groupings of Smoking-Related Diseases and Their Respective Smoking-Attributable Fractions (SAF)

Diverse Groupings	*ICD-9 Codes*	*SAF* Males	*SAF* Females
Neoplasms of the Lip, Oral Cavity and Pharynx	140-149	0.688	0.413
Neoplasms of the Esophagus	150	0.589	0.536
Neoplasms of the Larynx	161	0.806	0.413
Neoplasms of the Trachea, Lung and Bronchus	162	0.796	0.750
Aortic Aneurysm	441	0.624	0.468
Chronic Bronchitis and Emphysema	491-492	0.850	0.694
Chronic Airways Obstruction	496	0.850	0.694
Ulcers	531-534	0.479	0.445

Source: U.S. Department of Health and Human Services. Smoking-Attributable Mortality and Years of Potential Life Lost. *Center for Disease Control Morbidity and Mortality Weekly Report* 36(4), pp. 693-697.

There was considerable variation between SAFs by sex and, for the utilization rates of the eight groupings of diseases, by age. Thus it was decided to standardize morbidity rates for differences in the age and sex distributions of municipalities. Also, given the nature of these eight groupings, particularly with regard to smoking (their onset generally occurs later in life), persons under 35 years old were excluded from the analysis.

There was also considerable variation in the above-noted age and sex standardized, smoking-attributable morbidity rates for those 35 years of age or older, over time. This was due to the relatively small populations of most municipalities in which a variation of a few admissions, or the admission of a few long-stay patients, could significantly affect morbidity rates from year to year. Given that most by-laws to restrict smoking in British Columbia were passed between 1985 and 1988 and that morbidity measures, for this analysis, were desired, on average, for the period slightly before by-laws were passed, morbidity rates were averaged for the four year period from 1984 to 1987. These averaged, age- and sex-adjusted, smoking-attributable morbidity rates per 1,000 population for those 35 years of age or older provided a measure of the extent to which municipalities had been negatively impacted by the adverse effects of smoking around, or just before, the time that the majority of by-laws were passed.

Data for smoking-related hospital days and admissions were categorized by municipality of residence of the patients rather than municipality of the hospital that provided the service. This was done using a system that translated postal codes from the hospital abstracts of admissions and discharges to the municipalities of residence. Persons living outside municipal boundaries were not included in the study. Therefore the analysis was not subject to cross-boundary problems, as all population data, morbidity data, by-law data, and demographic and socio-economic data were keyed to the same geographic unit, the municipality of residence.

Demographic and Socio-Economic Data

With regard to demographic and socio-economic data, a set of relevant variables was derived from the 1986 Statistics Canada census data. These variables were selected because it was believed that they constituted appropriate operational definitions of theoretical constructs that were considered to be potentially related to the passage of municipal by-laws to restrict smoking. The constructs were urbanicity or population size, gender, social status, social conformity, social cohesion, social isolation and type of occupation.

TOTPOP Total population of the municipality (a measure of urbanicity).

POWNSD Percentage of persons who live in owned single family detached dwellings (a measure of social status and social conformity).

PUN Percentage of persons 15 years of age or older who are attending, or have attended, university (a measure of social status).

PONEPER Percentage of one person households (a measure of social isolation and an inverse measure of social cohesion).

PPAY700 Percentage of persons living in owned homes with mortgage payments of $700 or more per month (a measure of social status and social conformity).

TAVE	Average income per person (a measure of social status).
TUR	Unemployment rate (an inverse measure of social cohesion).
PFEMALES	Percentage of females (a measure of gender).
PTMAR	Percentage of married persons (a measure of social conformity and social cohesion and an inverse measure of social isolation).
PRSRCIND	Percentage of persons employed in resource industries (a measure of type of occupation). This category included: agricultural and related services; fishing and trapping; logging and forestry; and, mining, quarrying and oil.
PBLUE	Percentage of persons employed in blue collar work (a measure of type of occupation). This category included: manufacturing; construction; and, transportation and storage.
PWHITE	Percentage of persons employed in white collar occupations (a measure of type of occupation). This category included: communication and other utilities; wholesale trade; retail trade; finance and insurance; and, real estate operators and insurance agents.
PSERV	Percentage of persons employed in service industry occupations (a measure of type of occupation). This category included: business service; government service; educational service; health and social service; accommodation, food and beverage service; and, other service industries.

The variable, percentage of people 65 years of age or older, a measure of the relative age distribution of the residents of municipalities, was not included in the analysis as it was highly correlated with a number of other variables, particularly the percentage of females ($r = 0.81$).

Procedure

The chi-square statistic was used to test whether there were significant differences in morbidity rates between municipalities with no, moderate, or comprehensive by-laws to restrict smoking. A standard, or direct, discriminant function analysis was conducted to determine whether certain demographic or socio-economic variables could discriminate between municipalities with no, moderate, or severe by-laws. A stepwise discriminant function analysis using a composite measure of morbidity as a covariate was also conducted to determine whether demographic and socio-economic variables could still discriminate between municipalities after controlling for the effects of morbidity.

RESULTS

General Trends in By-Law Adoptions

In examining the adoption characteristics of by-laws to restrict smoking, several factors emerged as relavent. First, smoking by-laws were not adopted until 1984.[16] There was no major push to extensively introduce these by-laws until 1987, when 21 municipalities adopted non-smoking ordinances. By the end of the decade, a total of 43 communities had successfully introduced no smoking by-laws. While the growth in adoptions followed the classic 'S' shaped cumulative adoption curve noted by Everett Rogers,[17] only 31.6 percent of all municipalities eligible to adopt such by-laws had done so by the end of the decade. Despite this apparently low percentage, a closer examination revealed that those adopting ordinances tended to be larger communities, as the adopting municipalities represented an estimated 75.2 percent of the province's total population, and an estimated 87.5 percent of total municipal populations (Figure 1,15).[18]

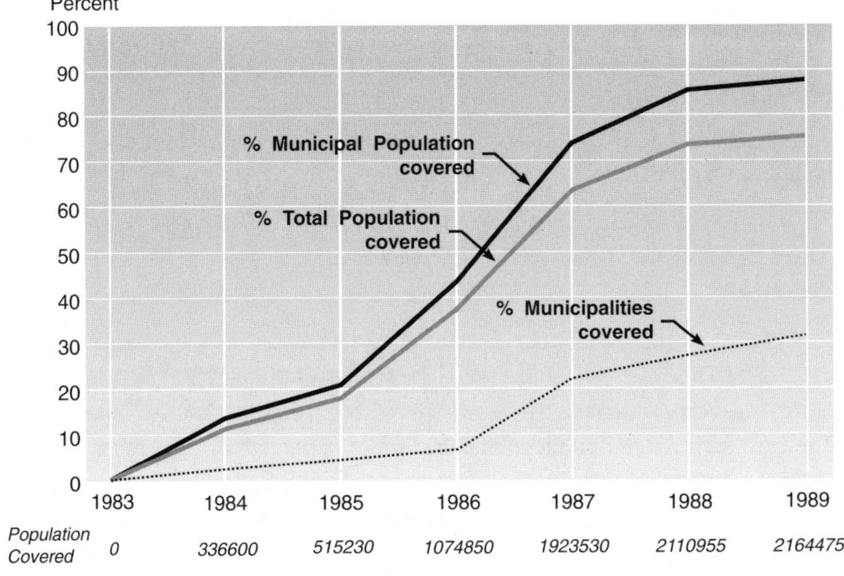

Figure 1,15 Extent of Municipal Smoking By-Laws (population based on 1986 Census)

Another characteristic that emerged was the location of the adopting municipalities. The provincial innovators and early adopters, to use Everett Rogers' terms, were all located in the south west part of the province. The subsequent diffusion path appeared to have travelled through the larger population centres (e.g. Vancouver, the largest and most influential municipality in the province) and then moved north and east to the larger centres in the interior (e.g. Kelowna and Prince George) followed by an "in filling" of contiguous communities, particularly in the lower mainland of the province (Figure 2,15). This pattern suggests that information about the use of by-laws moved through the larger communities in the province's urban system first, and then those communities acted as local cells of information in their own regions. Generally, smaller communities (less than 10,000 population) had not adopted by-laws by the end of 1989.

It is also interesting to note that while only three municipalities adopted by-laws in each of the years 1984 to 1986, 21 communities adopted by-laws in 1987. Given that the Vancouver by-law came into effect in June 1986 and there was a public advertising campaign concerning the by-law, it may be that the exposure of visitors to Expo 86, an international exposition on transportation and communications held in Vancouver, stimulated other municipalities to adopt by-laws. The impact of Expo 86 on by-law adoption would be an interesting question for further research.

A final characteristic that emerged was a tendency for communities to pass more restrictive by-laws over time. A criteria weighting system was developed based on 21 types of restrictions in which each was scored on a five point scale, with zero indicating no restrictions and four indicating no smoking allowed. The values for all criteria were added to give a total value for each community with the maximum value achievable being 84 (Table 2,15). Municipalities that passed revised by-laws subsequent to the passage of their original by-law, all passed increasingly restrictive by-laws (e.g. Capital Regional District and Prince George).

On average, the five areas with the most severe restrictions were: public transportation except school buses and taxis; school buses; elevators, escalators and stairways; service lines; and, public areas of retail stores and outlets. The five areas with, on average, the least severe restrictions were: common areas of shopping centres and malls; licensed premises; bowling alleys and billiard halls; workplaces; and nursing homes and extended care facilities.

As noted earlier, municipalities were broken down into three groups for purposes of analysis: those with no by-laws, those with moderate by-laws (criteria scores of 1-42) and those with comprehensive by-laws (criteria scores of 43-84). For purposes of categorizing municipalities into these

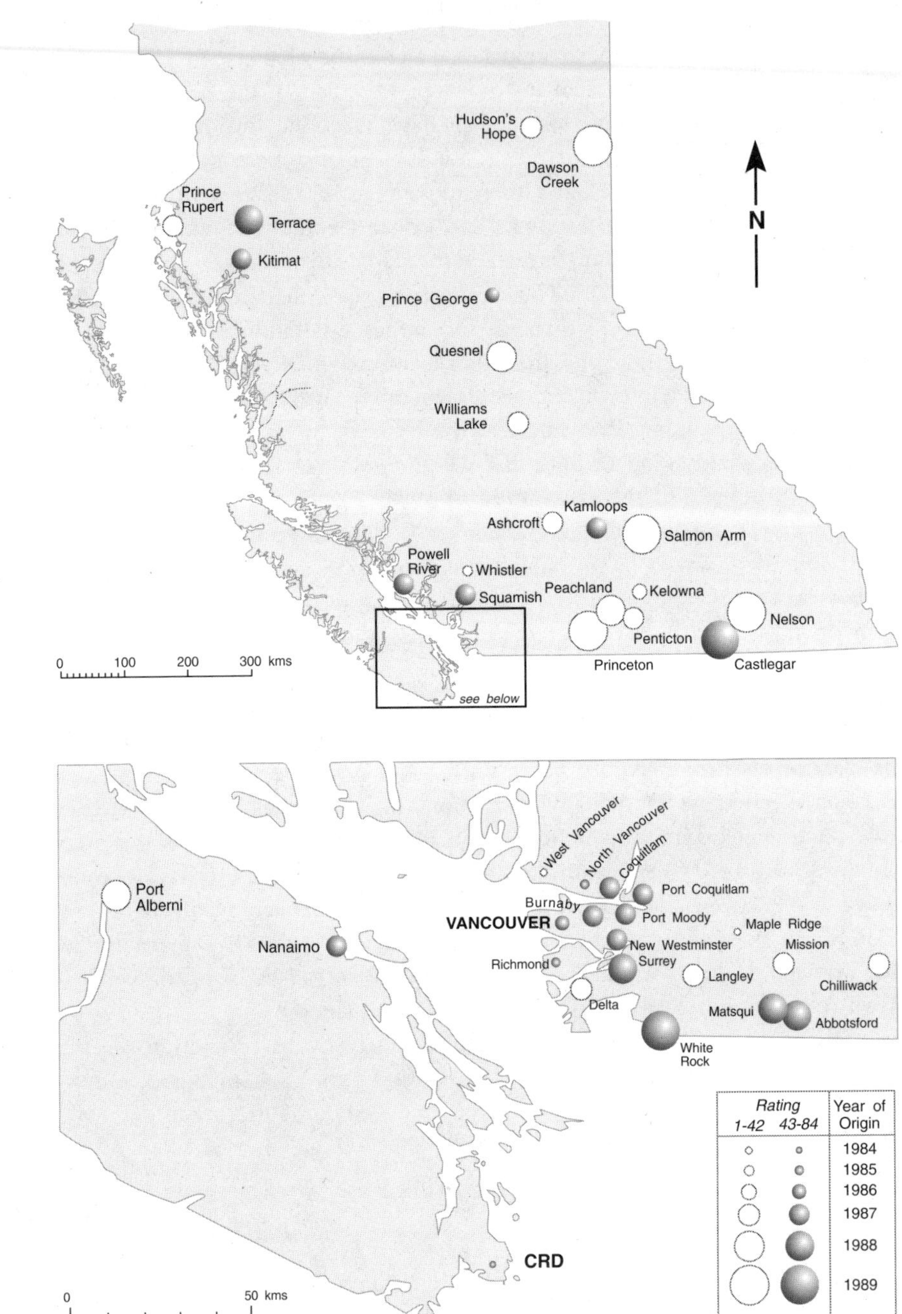

Figure 2,15 Distribution of Initial Smoking By-law Scores in B.C. 1984-89

three groups, and for the total and average scores in Table 2,15, the score of the first by-law adopted was used, for a given municipality, if more than one by-law was passed on or before December 31, 1989.

Municipal By-Laws and Morbidity

Although this study is exploratory, it was initially hypothesized that municipalities with higher rates of morbidity would be more likely to pass by-laws to restrict smoking. The logic behind this hypothesis was that as more people in a community were directly affected by having a friend or loved one hospitalized for illnesses attributed to smoking, they would take action to push for restrictions on the right to smoke, including the passage of municipal by-laws to restrict smoking.

Chi-square tests were conducted on the two aggregate measures of morbidity, average age- and sex-adjusted hospital days, and admissions, per 1,000 population 35 years of age or older for the period 1984 to 1987. It was found that the relationships between the rate of smoking-attributable hospital days and admissions, and the by-law adoption characteristics of municipalities, were significant (see Table 3,15).

The findings are significant but are in the opposite direction from what was expected. For example, Table 3,15 indicates that 36 (59.0 percent) of the municipalities with no by-laws, had high hospital admission rates compared to only one municipality with comprehensive by-laws. Conversely, 57.1 percent of municipalities with comprehensive by-laws had low admission rates compared to 14.8 percent for municipalities with no by-laws to restrict smoking.

Separate sub-analyses were conducted for the eight disease groupings comparing admission rates, and hospital days per 1,000 population by type of municipality. The scores for each of the eight disease groupings were collapsed into three categories: low, medium and high. It was found that the following disease categories, using the chi-square test, were significant at $p < 0.01$ for both analyses (i.e. for admissions and for days): ulcers; neoplasms of the lip, oral cavity and pharynx; neoplasms of the esophagus; neoplasms of the trachea, lung and bronchus; and, neoplasms of the larynx (here the relationship was the opposite of the aggregate measure of admission rates; i.e. municipalities with no by-laws had the lowest morbidity rates).

These findings raise possible alternative hypotheses for future research. It may be that municipalities with a greater overall concern for health status are more likely to have fewer smokers and, as a consequence,

Table 2,15 By-Law Scores for Municipalities

Area to be Restricted	*Abbots-Ford 1988*	*Ashcroft 1987*	*Burnaby 1987*	*CRD 1984*	*CRD 1986*	*Castlegar 1989*	*Chilli-wack 1987*	*Coquitlam 1987*	*Dawson Creek 1989*	*Delta 1987*	*Hudson's Hope 1987*	*Kamloops 1987*
Restaurant	3	0	1	1	1	1	1	2A	2	1	3H	3
Public Areas of Government Offices	4D	2E	0	4D	4	4D	1E	0	4D	0	0	0
Public Areas of Financial Institutions	4D	0	0	4D	4	4D	1	0	0	0	0	0
Reception Areas	4	0	2	4	2	3	1	2	2	0	4	3
Service Lines	4	0	4	4	4	4	1	4	0	4	0	4
Elevators/Escalators/Stairways	4	0	4	4	4	4	0	4	4	4F	0	4
Public Washrooms	4	0	4	0	4	4	0	4	0	0	0	4
Health Care Offices/Facilities	3	0	1	1	1	3	1	2	2J	2	1G	3H
Nursing Homes/Ext. Care Facilities	0	0	1	1	1	3	1	2	2	2	0	3H
Public Transportation except School Buses/Taxis	4	0	4	4K	4	4	0	4	4K	4K	4K	4
School Buses	4	0	4	4	4	4	0	4	4	4	4	4
Taxis	4	0	4	4	4	4	0	4	4	0	4	4
Places of Public Assembly for Entertainment Education, Worship, Business	3	0L	2	2	2	3	1	2	2	2	3H	3
Places of Public Assembly for Recreation	3	0L	2	2	2	3	1	2	2	2.	3H	3
Bowling Alleys/Billiard Halls	1	0	1	0	2	1	1	1	2M	0	1	1
Licensed Premises	1	0	0	0	0	1	1	0	0	0	1	1
Public Area of Retail Stores Outlets	3	0	4	4	4	4	1	4	1	0	1	4
Common Area of Shopping Centres/Malls	0	0	0	0	2	0	1	0	0	0	0	0
Personal Service Establishments	3	0	1	2P	2P	3	1	2P	2P	0	1	3
Educational Institutions	3	0	2	0	0	3	1	2	2	0	0	3
Workplaces	3	0	3	1	1	1	1	3	0	0	1	3
TOTAL SCORE	**62**	**2**	**44**	**46**	**52**	**61**	**16**	**48**	**39**	**25**	**31**	**57**

Table 2,15 continued

Area to be Restricted	*Kelowna 1986*	*Kitimat 1987*	*Langley Township 1987*	*Maple Ridge 1984*	*Matsqui 1988*	*Mission 1987*	*Nanaimo 1987*	*Nelson 1989*	*New Westminster 1987*	*N. Van (City) 1987*	*N. Van. (Dist.) 1985*	*N. Van. (Dist.) 1989*
Restaurant	1	1	2C	1	2	1	3	0	2A	3	1	2
Public Areas of Government Offices	4	4	1	4D	4	0	4D	4E	0	4D	4D	4D
Public Areas of Financial Institutions	0	4	1	4D	4	0	4D	0	0	4D	4D	4D
Reception Areas	2	2	4	0	4	4	3	0	2	3	2	2
Service Lines	4	4	4	4	4	4	4	0	4	4	4	4
Elevators/Escalators/Stairways	4	4	4	4F	4	4F	4	0	4	4	4	4
Public Washrooms	0	4	0	0	4	0	4	0	4	4	0	4
Health Care Offices/Facilities	1	1	1	2	1I	2	3H	0	2	3H	2	2
Nursing Homes/Ext. Care Facilities	0	0	1	2	1	0	3H	0	2	3H	2	2
Public Transportation except School Buses/Taxis	4	4K	4K	4K	4	4K	4	0	4	4	4K	4
School Buses	4	4	4	4	4	4	4	0	4	4	4	4
Taxis	0	4	0	0	4	0	4	0	4	4	4	4
Places of Public Assembly for Entertainment Education, Worship, Business	2	2	2N	2	2	2	3	4L	2	3	2	2
Places of Public Assembly for Recreation	2	2	2N	2	2	2	3	4L	2	3	2	2
Bowling Alleys/Billiard Halls	0	0	1	0	0	0	1	0	0	1	2	0
Licensed Premises	0	0	1	0	0	0	1	0	0	1	0	0
Public Area of Retail Stores Outlets	4	4	4	2O	4	2O	3H	0	4	4	4	4
Common Area of Shopping Centres/Malls	0	0	0	0	1	0	3	0	0	0	0	0
Personal Service Establishments	2P	1	0	0	1	0	3H	0	2P	3	4P	2
Educational Institutions	2	1	0	0	2	0	3	0	2	3	2	2
Workplaces	0	1	1R	0	1	0	3H	4L	3R	3	0	1
TOTAL SCORE	**36**	**47**	**37**	**35**	**53**	**29**	**67**	**16**	**47**	**65**	**51**	**53**

Table 2,15 continued

Area to be Restricted	*Peach-land 1988*	*Penticton 1987*	*Port Alberni 1988*	*Port Coquitlam 1987*	*Port Moody 1987*	*Powell River 1987*	*Princeton 1989*	*Prince George 1986*	*Prince George 1989*	*Prince Rupert 1987*	*Quesnel 1988*	*Richmond 1985*
Restaurant	0	1	1	2	2A	3	1	0	1	0	1	2
Public Areas of Government Offices	2E	0	1	4D	0	4D	4D	0	3	0	3	4D
Public Areas of Financial Institutions	0	0	0	4D	0	4D	4D	0	3	0	1	4D
Reception Areas	0	2	0	4	2	3	2	4	4	3	4	2
Service Lines	0	4	4	4	4	4	4	0	4	0	0	4
Elevators/Escalators/Stairways	0	4	4F	4	4	4	4	0	4	0	0	4
Public Washrooms	0	0	0	0	4	4	0	0	4	0	0	0
Health Care Offices/Facilities	0	1	1	1	2	3	1	0	1	0	1	1
Nursing Homes/Ext. Care Facilities	0	1	1	1	2	3	0	0	0	0	0	0
Public Transportation except School Buses/Taxis	0	4K	4K	4K	4	4	4K	4K	4K	3K	4K	4K
School Buses	0	4	4	4	4	4	4	4	4	3	4	4
Taxis	0	0	0	4	4	4	0	4	4	4	4	4
Places of Public Assembly for Entertainment Education, Worship, Business	0L	2	1	2	2	3	2	3	3	3	3	2
Places of Public Assembly for Recreation	0L	2	1	2	2	3	2	3	3	3	3	2
Bowling Alleys/Billiard Halls	0	0	0	0	1	1	0	0	1	0	1	0
Licensed Premises	0	0	0	0	0	1	0	0	1	0	1	0
Public Area of Retail Stores Outlets	1	4	0	4	4	4	4	1	1	1	1	4
Common Area of Shopping Centres/Malls	0	0	0	0	0	0	0	0	0	0	0	0
Personal Service Establishments	0	2P	0	2P	2P	3	2	0	1	0	1	2P
Educational Institutions	0	2	1	2	2	3	0	3	3	0	3	2
Workplaces	0	0	0	0	3	0	0	1	1	1	1	0
TOTAL SCORE	**3**	**33**	**23**	**48**	**48**	**62**	**38**	**27**	**50**	**21**	**36**	**45**

Table 2,15 continued

Area to be Restricted	*Richmond 1988*	*Salmon Arm 1989*	*Squamish 1987*	*Squamish 1989*	*Surrey 1988*	*Terrace 1988*	*Vancouver 1986*	*Vernon 1987*	*West Vancouver 1984*	*Whistler 1985*	*Whistler 1986*	*White Rock 1989*
Restaurant	2	1	1	3	3	1	3	1	2A	0	2	2C
Public Areas of Government Offices	4D	0	1D	4D	0	4	3	4D	4D	4	4	4
Public Areas of Financial Institutions	4D	0	0	4D	0	4	3	4D	4D	4	4	0
Reception Areas	2	0	4	4	4	2	3	2	0	0	0	4
Service Lines	4	0	4	4	4	4	4	0	4	0	0	4
Elevators/Escalators/Stairways	4	0	4	0	4	4	4F	0	4F	4F	4F	4
Public Washrooms	0	0	4	4	4	4	4	0	0	0	0	4
Health Care Offices/Facilities	1	0	1	3H	3	1	3	1	2J	0	0	2
Nursing Homes/Ext. Care Facilities	0	0	1	3H	3	1	3	1	2J	0	0	2
Public Transportation except School Buses/Taxis	4K	0	4	4	4	4K	4	0	4K	0	0	4
School Buses	4	0	4	4	4	4	4	0	4	0	0	4
Taxis	4	0	4	4	4	4	4	0	0	0	2	4
Places of Public Assembly for Entertainment Education, Worship, Business	2	4L	2	3	3	2	3	2	2	2	2	2
Places of Public Assembly for Recreation	2	4L	2	3	3	2	3	2	2	2	2	2
Bowling Alleys/Billiard Halls	0	0	1	1	1	0	1	0	0	0	0	2
Licensed Premises	0	0	1	0	1	0	1	0	0	0	0	2C
Public Area of Retail Stores Outlets	4	1O	1T	3H	4	4	4	4	0	4	4	4
Common Area of Shopping Centres/Malls	2	0	0	0	4	0	3	0	0	0	0	0
Personal Service Establishments	2P	0	1	3H	3	1	3	2P	0	0	0	2P
Educational Institutions	2	0	3	3	3	1	3	2	0	2	2	2
Workplaces	0	0	1	3H	3	1	3	0	0	0	0	3
TOTAL SCORE	**47**	**10**	**44**	**60**	**62**	**48**	**66**	**25**	**34**	**22**	**26**	**57**

Table 2,15 continued

Area to be Restricted	*Williams Lake 1987*	***Total**** ***Score***	***Average**** ***Score***
Restaurant	1	63	1.47
Public Areas of Government Offices	0	98	2.28
Public Areas of Financial Institutes	0	74	1.72
Reception Areas	4	100	2.33
Service Lines	0	121	2.81
Elevators/Escalators/Stairways	0	128	2.98
Public Washrooms	0	68	1.58
Health Care Offices/Facilities	1	62	1.44
Nursing Homes/Ext. Care Facilities	1	51	1.19
Public Transportation except School Buses/Taxis	4K	143	3.33
School Buses	4	143	3.33
Taxis	4	108	2.51
Places of Public Assembly for Entertainment Education, Worship, Business	3H	97	2.26
Places of Public Assembly for Recreation	3H	97	2.26
Bowling Alleys/Billiard Halls	1	23	0.53
Licensed Premises	1	16	0.37
Public Area of Retail Stores Outlets	0	115	2.67
Common Area of Shopping Centres/Malls	0	12	0.28
Personal Service Establishments	1	61	1.42
Educational Institutions	3H	68	1.58
Workplaces	1	50	1.16
TOTAL SCORE	**32**	**1698**	**39.49**

*Scores based on initial by-laws only.

RATING SCALE FOR TABLE 2,15

The following point system is used to rate the by-laws.

Points for Public Areas and Work Places:

0 - no restrictions on smoking covered by the law
1 - provision for designated area but proprietor/person in authority may designate unlimited space for smoking area
2 - proprietor/person in authority must not allow smoking except in designated area(s) and this is defined as less than the total area
3 - same as above but also the designated area for smoking must be separately ventilated and must be an area to which nonsmokers do not need access
4 - no smoking allowed in the indoor area

Provisions for Enforcement Area Also Rated

A - restaurants of more than 30 seats must have some nonsmoking seating
B - restaurants of 30 seats or less
C - restaurants or licensed premises of 40 seats or more must have some nonsmoking seating
D - public areas of government or financial institutions and offices, only counters smoke free
E - public areas of government offices - applies only to municipal buildings
F - applies to elevators and escalators only
G - hospital only
H - entire area can be smoking area if no ventilated space made available
I - health care offices smoke-free
J - applies only to specific areas within institution
K - public transportation - applies to buses only
L - applies to municipal buildings only
M - seating area only
N - limited definition of "places of public assembly"
O - applies to feed stores only
P - personal service establishments with more than 10 seats
Q - education institutions - rated according to inclusion under places of public assembly connected with education
R - refers to workplaces in offices only
S - fines according to offence act

Table 3,15 The Relationship Between Morbidity Rates and Municipal By-Laws to Restrict Smoking

HOSPITAL DAYS

		Average Annual Hospital Days Per 1,000 Population		
By-Laws		Low (0-74)	Medium (75-149)	High (150 or more)
No By-Laws	N	13	23	25
	%	21.3	37.7	41.0
Moderate By-Laws	N	8	11	3
	%	36.4	50.0	13.6
Comprehensive By-Laws	N	11	10	0
	%	52.4	47.6	0.0

$X^2(4, \underline{N}=104) = 17.51, \underline{p} < .002$

HOSPITAL ADMISSION RATES

		Average Annual Admission Rates Per 1,000 Population		
By-Laws		Low (0-5)	Medium (6-9)	High (10 or more)
No By-Laws	N	9	16	36
	%	14.8	26.2	59.0
Moderate By-Laws	N	8	9	5
	%	36.4	40.9	22.7
Comprehensive By-Laws	N	12	8	1
	%	57.1	38.1	4.8

$X^2(4, \underline{N}=104) = 25.76, \underline{p} < .0001$

reduced morbidity. Residents of these municipalities may also be more likely to push for by-laws to restrict smoking as an expression of their overall desire to ensure good health. Alternatively, given that the largest communities are more likely to have the most restrictive by-laws (of the 10 largest municipalities in British Columbia, eight had comprehensive by-laws), it may be that the anonymity of urban life allows people to push for the adoption of by-laws. People in smaller communities may hesitate to do so as such action would pit them against known friends and neighbors. In other words, the desire for community peace and cohesion may override the desire to insist on individual rights to a smoke-free environment.

Municipal By-Laws and Socio-Economic Variables

As noted in the Method section, 13 demographic and socio-economic variables were selected as independent variables to be included in a standard, or direct, discriminant function analysis to predict membership in the three groups of municipalities. All communities with a population over 1,000 were included. There were no missing data and all but two of the independent variables were considered to be normally distributed. These two variables, total population of the municipality and the percentage of persons working in resource industries, were positively skewed and were subjected to log transformations to normalize their distributions.

An inspection of the data revealed that, to the extent the variables were skewed, they were positively skewed. A one-tailed cut-off of 3.72 standard deviations, $p \leq 0.0001$ was selected to inspect univariate outliers for each of the three groups of municipalities.[19] One municipality had an outlier on the variable percentage of persons with mortgage payments of $700 or more per month and another on the variable percentage of persons 15 years of age or older who had attended, or were attending, university. These two communities were excluded from the analysis. The 13 variables were also analysed to identify the existence of multivariate outliers using Mahalanobis distance, but none were found.

While there were no problems of multicollinearity or singularity, there was some intercorrelation between predictor variables. The highest such correlation was 0.637 between the variables percentage of married persons and percentage of persons who live in owned single family detached dwellings.

There were no identified curvilinear relationships between predictor variables. The data were also deemed acceptable regarding the homogeneity of variance-covariance matrices. Finally, the groupings of municipalities with moderate and with comprehensive by-laws had 22 and 21 cases each, respectively. This was considered adequate as a sample size.

A similar analysis was conducted for the morbidity variables. It was found that each of the distributions for the eight disease groupings was skewed. Given these distributions, and the high degree of fluctuation from year to year, it was decided to use the two aggregate measures of the average age- and sex-adjusted hospital days, and admissions, per 1,000 population for those 35 years of age or older for the period 1984 to 1987. For these variables, one community was a univariate outlier using the one-tailed 3.72 standard deviations cut-off criterion and therefore was excluded from the analysis.

Finally, three municipalities which had missing data respecting the aggregate morbidity measures were excluded from the analysis. After the exclusion of all six municipalities, it was found that the two aggregate measures of morbidity were adequately normally distributed so that transformation would not be necessary when they were included in the discriminant analysis. A final check for multivariate outliers was conducted with the combined morbidity, demographic and socio-economic variables but none were found.

After deleting six municipalities and performing log transformations on TOTPOP and PRSRCIND, it was determined that the data to be used for the discriminant analyses conformed to the assumptions of multivariate analysis related to cases per group, missing data, outliers, normality, linearity, singularity, multicollinearity and the homogeneity of variance-covariance matrices. The total adjusted sample was comprised of 104 municipalities, 61 with no by-laws, 22 with moderate by-laws and 21 with comprehensive by-laws.

A direct discriminant function analysis was conducted in which demographic and socio-economic variables were used to predict group membership between municipalities with no, moderate or comprehensive by-laws to restrict smoking. Table 4,15 presents the pooled within-group correlation matrix of the predictor variables.

Two discriminant functions were calculated for the three groups, with a combined $x^2(26, N=104) = 102.83$, $p < 0.0001$. After removal of the first function, the second function was found not to be significant $x^2(12) = 10.86$, $p = 0.541$. The two discriminant functions accounted for 93.1 percent and 6.9 percent of the between-groups variance, respectively. The percentage of grouped cases correctly classified was 77.9 percent. As expected, an inspection of the canonical discriminant functions evaluated at group means (i.e. group centroids) revealed that the first discriminant function maximally separated the three groups.

The loading matrix of correlations between predictors and discriminant functions showed that the best predictor for distinguishing between the three types of municipalities was the log of total population (LOG-POP), and the next best predictor was the log of the percentage of persons

Table 4,15 Pooled Within-Group Correlation Matrix of Predictor Variables for Municipalities with No, Moderate, or Comprehensive By-Laws to Restrict Smoking

	POWNSD	PUN	PONEPER	PPAY700	TAVE	TUR	LOGRSRC	PBLUE	PWHITE	PSERV	PFEMALES	PTMAR	LOGPOP
POWNSD	1.00000												
PUN	-0.02499	1.00000											
PONEPER	-0.34267	-0.00571	1.00000										
PPAY700	-0.17046	0.43044	-0.46189	1.00000									
TAVE	-0.05922	0.39935	-0.52414	0.50336	1.00000								
TUR	-0.00550	-0.34102	0.20679	-0.44217	-0.50345	1.00000							
LOGRSRC	-0.05932	-0.34323	-0.13208	-0.13347	0.06595	0.08395	1.00000						
PBLUE	0.06039	-0.24327	-0.22477	-0.06325	0.18418	-0.01345	-0.35309	1.00000					
PWHITE	0.10988	0.13004	0.24853	0.00042	-0.36149	0.14687	-0.39905	-0.32202	1.00000				
PSERV	-0.05758	0.46340	0.30788	0.01273	-0.36298	0.02048	-0.36954	-0.49168	0.22614	1.00000			
PFEMALES	0.14154	0.00426	0.59474	-0.34180	-0.54485	0.18986	-0.26886	-0.27873	0.62853	0.32658	1.00000		
PTMAR	0.63704	0.04412	-0.29918	-0.15807	0.01756	-0.10631	-0.01118	-0.12525	0.11472	0.07489	0.15166	1.00000	
LOGPOP	0.02577	0.20452	0.00072	0.26482	0.05379	-0.14063	-0.20892	-0.08982	0.36019	0.09814	0.26269	-0.00936	1.00000

working in resource industries (LOGRSRC). The variable LOGPOP was positively related to the three groups, while the variable LOGRSRC was inversely related to the three groups. The means for LOGPOP and LOGRSRC for municipalities with no, moderate, and comprehensive by-laws were: 3.52 and 1.15; 4.13 and 1.00; and 4.59 and 0.68, respectively. Other predictors with loadings of 0.3 or higher were: the percentage of persons 15 years of age or older who have attended, or were attending, university; the percentage of persons in white collar industries; and, the percentage of persons who live in homes with mortgages of $700 or more per month.

Four other comparisons were also performed between municipalities with no by-laws versus those with either moderate or comprehensive by-laws; no by-laws versus moderate by-laws; no by-laws versus comprehensive by-laws; and, moderate by-laws versus comprehensive by-laws. The latter comparison was not significant. The results of the significant discriminant function analyses are presented in Table 5,15 and show that the same set of five variables consistently discriminated between all combinations of municipalities with statistically significant discriminant functions. Population, and the percentage of persons working in resource industries (in log form), were the best discriminators between the different by-law adoption categories of municipalities. Other variables that discriminated between municipalities were: the percentage of persons 15 years of age of older who had attended, or were attending university, the percentage of persons in white collar occupations, and the percentage of people living in homes with mortgages of $700 or more per month. In comparing the two extreme groups of municipalities, those with no by-laws and those with comprehensive by-laws, the discriminant function was able to correctly classify 92.7 percent of the cases and accounted for 69.4 percent of the variance.

In examining Table 5,15, one can see that the major construct that discriminates between municipalities is urbanicity. The total population (in log form) of the municipality was the major discriminating variable. The percentage of persons working in resource industries is typically higher in small towns. However, it may also be that small resource-based communities are more likely to be populated by "rugged individualists" who oppose government interventions such as by-laws to restrict smoking.

It should be noted that measures of social cohesion and/or conformity such as the unemployment rate, the percentage of people living in owned, single family dwellings, and the percentage of married persons were not significant discriminators. Social isolation, as measured by the percentage of one-person families, was also not a significant discriminator.

Table 5,15 Results of Standard Discriminant Function Analyses of Demographic and Socio-Economic Variables

Predictor Variables	*Correlations of Predictor Variables With Discriminant Functions*			
	No By-Laws Versus Moderate Versus Comprehensive By-Laws	No By-Laws Versus Moderate and Comprehensive By-Laws Combined	No By-Laws Versus Moderate By-Laws	No By-Laws Versus Comprehensive By-Laws
LOGPOP	0.819	0.899	0.793	0.815
LOGRSRC	-0.516	-0.497	-0.274	-0.491
PUN	0.380	0.435	0.341	0.429
PWHITE	0.372	0.449	0.382	0.328
PPAY700	0.342	0.372	0.249	0.364
Related Statistics				
Canonical R	0.788	0.721	0.657	0.833
Chi-Square	102.83	70.02	42.13	87.20
N	104	104	83	82
df	26	13	13	13
p	0.0000	0.0000	0.0001	0.0000
Percentage of Cases Correctly Classified	77.88	85.58	84.34	92.68

A stepwise discriminant analysis using the 13 demographic and socio-economic variables as predictors of membership in the differing municipal by-law adoption categories, with hospital admission rate (RALL: the average age- and sex-adjusted admissions per 1,000 population, 35 years of age or older for the period 1984 to 1987) as a covariate, was also conducted. In order to determine the ability of the demographic and socio-economic variables to predict group membership while controlling for the effects of morbidity, RALL was entered into the equation first and the other predictor variables were then entered one by one in a stepwise manner.

In comparing the results of stepwise discriminant analyses using the demographic and socio-economic predictor variables only, and the same analysis with RALL as a covariate, it was found that there was little difference in the outcome except that RALL did correlate with the first discriminant function (r = -0.347).

This finding indicates that the ability of the demographic and socio-economic variables to predict group membership is not significantly affected by, or is independent of, morbidity rates for smoking-related diseases.

DISCUSSION

What can be learned from this preliminary analysis about the use of municipal by-laws to restrict smoking and about the use of municipal by-laws, in general, as an instrument of healthy public policy?

Given that only 31.6 percent of British Columbia's municipalities had adopted by-laws by the end of 1989; that many of the by-laws were relatively weak; that, based on smoking-related morbidity, those with the highest need for by-laws, especially resource industry towns, had not adopted them; and, that there had been a reduction in the growth of new municipalities adopting by-laws, it might be concluded that this approach has not been successful as a measure to restrict smoking in public areas or as a general instrument of healthy public policy. However, such conclusions would appear to be premature.

In the six years from 1984 to 1989, an estimated 75.2 percent of the province's population was covered by some type of no-smoking restriction ordinance. Larger communities were more likely to adopt by-laws[20] and such by-laws were more likely to be restrictive in nature. Furthermore, 6 of the 43 municipalities with by-laws had revised them, by the end of 1989, to make them more restrictive, a trend that is likely to continue.

Indeed, given that there has been only moderate direct involvement by senior levels of government in the adoption process (other than to give final approval to a by-law), the permissive process has worked quite well. Communities have adopted restrictions at their own pace and two natural diffusion processes appear to have helped the adoption process: diffusion among major, large communities in the urban system, followed by diffusion to neighboring, often adjoining communities.

Nevertheless, it is apparent that two items still need addressing. First, municipalities that have not yet adopted restrictions need to be encouraged to do so. Second, municipalities with weak by-laws need to be encouraged to make them more restrictive. The second issue will be easier to achieve than the first. Those municipalities that have adopted by-laws have already made the initial "psychological" breakthrough of adopting restrictions. These municipalities are generally larger, relatively more diverse economically and culturally, have higher average levels of education. An increase in the severity of smoking restrictions will occur, and has occurred, naturally over time as information on the actions of other neighboring and trend setting communities becomes available to them. Furthermore, due to their size, larger communities may have a greater likelihood of having champions who will push for innovations and restrictions in the name of the public good.

Getting non-adopters to introduce restrictions will be more difficult. Analysis indicates that the non-adopters are typically small: only 8 of the 96 communities with 1986 populations below 10,000 have adopted smoking restrictions, and only one has a comprehensive by-law. In addition, such communities are often one industry resource towns, and relatively isolated interior and northern communities. As suggested earlier, people in smaller communities may hesitate to push for by-laws that affect individual behavior, as such action could pit them against friends and neighbors. Furthermore, a detailed analysis of such communities would likely show that smoking-related diseases were of lesser importance than other factors, particularly mental health issues (including suicide), alcohol abuse, and occupational and travel related accidents, on the overall health of the community. However, this can only be verified by further research and analysis.

To get a better understanding of a community's decision to adopt smoking restrictions, it would be useful to undertake some in-depth case studies of the decision-making process, especially in those communities that are small, resource based, relatively isolated and have comprehensive anti-smoking by-laws (e.g. Castlegar). Such a study would undoubtedly shed light on how barriers can be overcome and would no doubt provide worthwhile information on the role of individual "champions" in such a process.

Nevertheless, in the absence of this more detailed information, several "cajoling" tactics have been used to try to get non-adopters to introduce no-smoking restrictions. In 1990, the British Columbia government banned smoking in almost all its provincial buildings. Such buildings are widely scattered throughout the province and most non-adopting municipalities have provincial government offices within their boundaries. It is anticipated that there will be some "example" effect in those communities.

Second, in conjunction with the 1990 annual meeting of the Union of British Columbia Municipalities (an organization that represents all local governments in the province), the Minister of Health made a strong statement on the need for no-smoking by-laws and the need for local politicians to act as "champions" for such a cause within their own communities.[21]

Third, following this statement, the Minister of Health wrote to the mayors of all non-adopting municipalities with information on the need for no-smoking restrictions and provided examples of by-laws that have been used elsewhere in the province. In addition, the assistance of the local Medical Health Officer, a provincial government employee, was offered to each mayor to help put such a by-law through local council.

Fourth, during the week of January 21 to 27, 1991 (National Non-Smoking Week in Canada), the Ministry released an information backgrounder on the health effects of smoking and environmental tobacco smoke. Copies of this backgrounder were sent to each municipality. Finally, the 1989 Annual Report of the Division of Vital Statistics[22] was released in March 1991, along with a news bulletin that stressed the relationship between the leading causes of death (heart disease, cancer and strokes) and smoking. Again each municipality received a copy of the report and news release.

It is hoped that these "non-intrusive" approaches will encourage municipalities without bylaws to take action, at a pace that is perhaps faster than that at which they may otherwise have proceeded, and will encourage municipalities with existing by-laws to strengthen them.

CONCLUSION

While it is evident that more research is required to fully understand and assess the relative success of municipal by-laws in restricting smoking and ultimately improving community health, it is also clear that such instruments allow for adoption of healthy public policies at a pace consistent with community desires. This is particularly important when major behaviorial changes, such as restricting smoking, are involved. It seems that this approach works best for larger, more diverse communities and well connected urban systems. The diffusion of new ideas, and policies, such as no smoking by-laws, occurs efficiently among larger communities and their immediate neighbors. However, the barriers of remoteness may slow this process down in sparsely populated areas like some parts of British Columbia, and some external assistance may be required to overcome these barriers. Other studies of this nature need to be undertaken to see if these conclusions apply to the diffusion of healthy public policies in general.

ACKNOWLEDGEMENT

The authors would like to acknowledge the contributions of Mrs. Angela Van Raamsdonk for undertaking the computer programming required to manipulate and integrate the several different data bases used for this study, and Mrs. Sandy Ho for her patient preparation of this manuscript through its several drafts.

REFERENCES

[1] U.S. Department of Health and Human Services. 1987. Smoking-attributable mortality and years of potential life lost. *Center for Disease Control Morbidity and Mortality Weekly Report* 36(42), pp. 693-697; U.S. Department of Health and Human Services. 1982. *The Health Consequences of Smoking - Cancer: a Report of the Surgeon General.* Rockville, MD.: U.S. Public Health Service; U.S. Department of Health and Human Services. 1985. *The Health Consequences of Smoking - Cancer and Chronic Lung Disease in the Workplace: a Report of the Surgeon General.* Rockville, MD.: U.S. Public Health Service; U.S. Department of Health and Human Services. 1983. *The Health Consequences of Smoking - Cardiovascular Disease: a Report of the Surgeon General.* Rockville, MD.: U.S. Public Health Service; U.S. Department of Health and Human Services. 1984. *The Health Consequences of Smoking - Chronic Obstructive Lung Disease: a Report of the Surgeon General.* Rockville, MD.: U.S. Public Health Service.

[2] Altman, L.K. 1990. The Evidence Mounts on Passive Smoking. *The New York Times*, May 29, 1990, pp. B5, B7; Glantz, S.A. 1990. Achieving a Smokefree Society. *Circulation* 76(4), pp. 746-752.

[3] British Columbia Ministry of Health. 1989. *British Columbia Smoking Reduction Status Report.* Victoria: Unpublished Manuscript; Health and Welfare Canada. 1989. *Active Health Report.* Ottawa: Health and Welfare Canada.

[4] Masironi, R. 1986. World Trends in Smoking, in Skirrow, J. (ed.), *Strategies for a Smoke-Free World.* Edmonton: The Alberta Alcohol and Drug Abuse Commission, pp. 46-47.

[5] Health and Welfare Canada. 1988. *Canada's Health Promotion Survey: Technical Report.* Ottawa: Health and Welfare Canada, p. 25; Health and Welfare Canada. 1988. *Smoking Behaviour of Canadians - 1986.* Ottawa: Health and Welfare Canada, p. 2.

[6] Kaiserman, M.J. 1990. Trends in Canadian Tobacco Consumption, 1980-89, *Canadian Medical Association Journal* 143(9), pp. 905-906.

[7] Epp, J. 1986. *Achieving Health for All: A Framework for Health Promotion.* Ottawa: Health and Welfare Canada; Wolczuk, P., McDowell, J. and Ainslie, K. (eds.) 1989. *Proceedings, National Symposium on Health Promotion and Disease Prevention.* Victoria, B.C.

[8] Health and Welfare Canada, *Canada's Health Promotion Survey: Technical Report. op.cit.*, p. 212; Health and Welfare Canada, *The Smoking Behavior of Canadians - 1986, op.cit.*, p. 36.

[9] Rowlands, G.D. 1987. *The Strategic Planning and Marketing Processes Behind the First Workplace Smoking By-Law in Canada.* Unpublished Manuscript, p. 5.

[10] Millar, W.J. 1988. Evaluation of the Impact of Smoking Restrictions in a Government Work Setting, *Canadian Journal of Public Health* 79, pp. 379-382; Petersen, L.R., Helgerson, S.D., Gibbons, C.M., Calhoun, C.R., Ciacco, K.H. and Pitchford, K.C. 1988. Employee Smoking Behavior Changes and Attitudes Following a Restrictive Policy on Worksite Smoking in a Large Company, *Public Health Reports* 103(2), pp. 115-120; Stachnik, T. and Stoffelmayr, B. 1983. Worksite Smoking Cessation Programs: A Potential for National Impact, *American Journal of Public Health* 73, pp. 1395-1396; Stanwick, R.S., Thompson, M.P., Swerhone, P.M., Stevenson, L.A. and Fish, D.G. 1988. The Response of Winnipeg Retail Shops and Restaurants to a By-law Regulating Smoking in Public Places, *Canadian Journal of Public Health* 79, pp. 226-230.

[11] Health and Welfare Canada, *The Smoking Behaviour of Canadians - 1986, op.cit.*, p. 26.

[12] Calgary Health Service, Health Education Division. 1988. *Smoking By-Laws in Canada.* Calgary: Calgary Health Services, pp. 2-5.

[13] One of these communities was a District (North Vancouver), one was a Township (Langley), and one was a Regional District (Capital Regional District). There were 144 communities in British Columbia in 1986. Nine of these were aggregated into the Capital Regional District to make a base of 136 communities for analysis. For purposes of this study, the term "municipality" will be used for all such communities for ease of discussion.

[14] The smallest municipality to adopt such a by-law had a population of 1,160 in 1986.

[15] U.S. Department of Health and Human Services. Smoking-attributable Mortality and Years of Potential Life Lost. *Centre for Disease Control Morbidity and Mortality Weekly Report, op.cit.*, pp. 693-697.

[16] The Municipality of Burnaby passed a by-law in 1968 that was primarily a fire prevention measure. Given the difference between that by-law and the by-laws passed in the 1980s, the 1968 Burnaby by-law was excluded from the analysis. Burnaby passed a by-law to restrict smoking in 1987.

[17] Rogers, E.M. 1962. *Diffusion of Innovations.* New York: The Free Press, 1962.

[18] Parts of the province do not have a formal local government administration and are therefore not covered by the *Municipal Act.*

[19] For a discussion of screening data prior to multivariate analysis, see Tabachnick, B.C. and Fidell, L.S. 1989. *Using Multivariate Statistics.* New York: Harper and Row.

[20] Eight of the 10 largest municipalities have comprehensive by-laws and only 5 out of 40 municipalities with a population of 10,000 or more did not have an approved by-law by the end of 1989.

[21] The speech was made by the Honourable John Jansen, Minister of Health, at the Annual General Meeting of the Associated Boards of Health of British Columbia on September 17, 1990. This meeting was held in conjunction with the Annual Meeting of the Union of British Columbia Municipalities (September 18-21, 1990). Generally, the same municipal representatives attend both functions.

[22] Danderfer, R.J. and Foster, L.T. (eds.) 1990. *1989 Annual Report of Vital Statistics.* Victoria: British Columbia Ministry of Health.

16 PERCEPTIONS AND REALITIES: Medical and Surgical Procedure Variation - A Literature Review

Samuel Sheps
Susan Scrivens
Jennifer Gait

Department of Health Care and Epidemiology and the Centre for Health Services and Policy Research University of British Columbia

INTRODUCTION

Although variety is said to be the spice of life, geographic variations in health services procedures pose major conceptual and policy problems. At both levels, variations in patterns of health care utilization between geographic areas raise concern because of the fundamental assumption that individuals should have equal access to health care. In this context, procedure or practice variations create the dilemma that either access to, or availability of, health care services is geographically uneven or that morbidity patterns are variable. Both possible violations of the assumption of equity pose political problems.

While at a policy level, health care utilization variations are a source of real concern regarding the organization and cost of health services, at the conceptual and methodological level, the issue of the magnitude of utilization variations and the rigor and consistency with which explananations for such variations are undertaken pose equally substantial problems. For if there is no convincing evidence to explain observed variations, assuming these observations to be correct, the solutions remain obscure. Moreover, it remains unclear if variations are a good thing or a bad thing given that normative criteria for the frequency of health care utilization in general, and procedure variations specifically, do not exist.

An assumption is made, particularly in the "small area analysis" literature, that the population groups being studied are sufficiently similar in terms of morbidity that other strucutural and functional factors must be responsible for any variations observed. However, to date there have been limited data to support such an assumption due to the lack of a concerted effort to support rigorous epidemiological research. Until such information is available, the ongoing discussion of variations in health care utilization will remain largely speculative.

Despite the lack of a clear indication that variations exist, and notwithstanding the limited effort to explain variations, there is no question that a literature has developed arguing that utilization variations do exist, that they are of sufficient magnitude to stimulate concern, and that there is an urgent need for a clearer understanding of the factors responsible. This perception of significant and meaningful variation, reported often in editorial and other non-empirically based statements, probably represents a certain degree of truth, but the exact nature and magnitude of that truth is unclear. Moreover, despite the identification that variation exists, there remains uncertainty regarding the factors producing it, whether these factors are consistent across procedures, and whether the studies documenting variations or assessing associated (causal) factors are sufficiently rigorous to provide useful information for planners and policy makers.

An in-depth literature review was undertaken with the overall goal of assessing the current state of knowledge of procedure variations in medical and surgical practice with particular reference to research attempting to assess the impact of specific variables on the variation observed. Our explicit objectives were to review the recent literature using *a priori* criteria in order to: 1) identify papers using sound methodology; 2) determine the magnitude of variation found to exist and whether there is consistency in these estimates across studies; 3) determine, in general and for specific procedures, which factors seem consistently related to variations (if any); and 4) identify directions for future research.

Background

Prior to discussing the methodology and results of our own study, it is useful to provide an overview of the health care utilization literature. This is conveniently done by considering two recent papers.

McPherson et al.[1] provide a broad overview of both international variations in surgical procedure rates and relevant methodological issues. The magnitude of the variations observed (Japan generally exhibits much lower rates than the United States or Canada) clearly beg explanantion and McPherson et al. provide a useful review of factors that may be responsible. Even though their discussion is largely speculative (based on the fundamental limitations of the literature they are reviewing), they do point to critical areas for future research.

McPherson et al. correctly identify functional factors such as varying day surgery rates, differential procedure coding and completeness, comparability of computer file definitions and formats, and protocols regarding whether primary, secondary or tertiary diagnoses are counted as potential sources of artifactual differences. Perhaps more importantly, McPherson et al. consider structural differences such as true differences in morbidity, and the age and sex structure of the population. These factors are critical since a common asssumption in the variation literature is that these health status and demographic variables are sufficiently similar across study populations as to be discounted. They also stress the importance of *clinical uncertainty* as a source of variation and distinguish this factor from the more frequently cited, but no more intensively studied, variables of prevailing local custom and what has been termed "practice style". Finally they discuss the critical issue of basing rates on an accurate estimate of the population at risk. Although McPherson et al. believe this to be a source of artifact, we believe this to be a more fundamental methodological problem, which is pervasive in the literature.

The major limitations of the McPherson et al. review, as indeed with the literature generally, are that in attempting to explain international differences one is immediately confronted with the problem of the ecological fallacy. This issue is compounded by a lack of normative expectation, thus, although differences are observed, it is not clear what to make of them: i.e. are they "real". Moreover, McPherson et al. fail to consider a host of methodological problems across studies (e.g. inadequate definition of the independent variables or incomplete description of analytic techniques) which may cumulatively be responsible for at least some of the differences in observed rates. Finally, broad geographic comparisons have not considered culturally based differences in: 1) determinants of health; 2) patterns of utilization and criteria for normative utilization; or 3) even clinical uncertainty.

Paul-Shaheen et al.[2] provide at once a more comprehensive discussion of small area analysis and a more limited review since it is restricted to the North American literature. Their complex paper cannot be easily summarized, but several general points emerge which are of value to discuss at this point. The first is that they set forth a framework for categorizing papers which we found conceptually useful, but have adapted. Second, the striking feature of the data they present is the relatively small magnitude of the observed variation. Across the 59 "core" papers they discuss in detail, the magnitude of variation was generally on the order of one to three fold, rarely exceeding five to six fold. Of particular interest is that Paul-Shaheen et al. limited their review to papers which attempted, in a formal way, to asssess the impact of specific factors and found that, generally speaking, none of the usual variables studied was correlated with rate variations with a coefficient of greater than 0.7 ($r^2 = 0.49$). They concluded that variables such as bed or physician supply, morbidity, socio-economic status, etc., individually do not account for significant variation and that a combination of practitioner and community variables provides a better explanatory model. However, differing combinations of variables accounted for huge differences in the variation of utilization: 2 to 73 percent of the variation of discharge rates; 14 to 76 percent of the variation in length of stay; and 3 to 89 percent of the variation of patient day rates. This magnitude of variation explained by independent variables strongly suggests that methodological differences rather than true relationships are reflected in these data.

Paul-Shaheen et al. provide an excellent discussion of future research approaches and issues. Among these are the need: 1) to assess the impact of the unit of analysis under study, 2) to create normative estimates to allow judgments of under- or over-utilization, 3) to provide greater detail in defining factors that may influence rate variations, 4) to have better estimates of underlying mortality and morbidity patterns, and 5) to be aware of policy changes (such as a move from inpatient to outpatient service provision for a particular procedure) which may produce observed variations. Specific methodological considerations include the standardization of rates by age and sex, an attempt to document the validity of underlying asssumptions regarding the similiarity of the populations being compared, and the use of standard measures across studies such as procedure specific rates with appropriate at-risk denominators.

DATA SOURCES AND METHODS

Sources used to identify papers describing procedure variations or examining factors associated with variation included: 1) the 1985 and 1987 bibliographies prepared by the Copenhagen Collaborating Centre;[3]

2) the recent study by Paul-Sheehan et al.[4] which focussed on North American literature; and (3) and several Medline searches using different key word combinations: geographic variation; small area analysis; small area variation; regional variation; utilization rate; and physician practice patterns. We sought papers published (or abstracted) in English since 1975 (until May 1990), although a few papers published prior to 1975 were included if they were cited frequently.

All papers identified were reviewed and categorized as follows, using the categories adapted from Paul-Sheehan et al.:

I. *Utilization of Medical and Surgical Procedures*

   A. Studies which presented primary data on variations and examined factors associated with variations.

   B. Studies which presented primary data on variations and performed statistical assessment on the rate variations but performed no analysis on causal factors

   C. Studies which presented primary data without any analysis

   D. Studies which presented secondary data or editorials, letters, and commentaries, etc.

II. *Utilization of Health Services*

   All papers in this group lacked any primary data on medical or surgical procedures. These papers focused on physician or hospital use, or discussed analytic approaches to the assessment of procedure variations.

All papers categorized as IA were reviewed in detail and for each paper the following information was abstracted: 1) focus of study; 2) procedure(s) studied; 3) relevant medical or surgical speciality; 4) level of comparison (i.e. national, regional, provincial); 5) study population; 6) unit of study definition (i.e. city, state, province, etc.); 7) data sources; 8) use measures; 9) statistical calculations; 10) independent variables analysed; 11) results; 12) conclusions; 13) limitations recognized by the authors; and 14) shortcomings (i.e. problems not recognized by the authors but noted by us).

Prior to reviewing the 51 core papers, the two reviewers independently assessed a sample of papers to ensure consistency of abstraction.

RESULTS

General Comments

Over 360 papers from 56 journals were reviewed and categorized. Of these, 51 papers met the criteria of Category IA and were reviewed in depth. In general, most of these papers which we would define as having reasonably good methodology were published since 1980. Of these 51 papers, only 17 utilized Canadian data and of the 106 papers presenting primary data of some sort, only 28 utilized Canadian data (Table 1,16).

What follows will provide an overview of the findings regarding 6 of the 14 characteristics abstracted for each paper.

Table 1,16 Breakdown of Variation Literature

	Canadian	*American*	*International Comparisons*	*Specific Country*	*Total*
Utilization of Procedures					
Data presented and analyzed re Independent Variables	17	23	4	7	51
Data presented and analyzed in Utilization Trends	5	13	5	4	27
Data presented on Trends - no analysis	6	15	4	3	28
				Subtotal	106
Specific Procedures Not Assessed					
Editorial, Thinkpiece, Letter, Policy					62
Models (Analytic Approaches)					48
Health Services					38
Physician Services					25
Hospital Services					58
Incidence of Disease					27
				TOTAL	364

Procedures studied: The papers were roughly divided between those examining variations in a single procedure and those examining several or numerous procedures. In the mid 1980s, a few high technology procedures were examined (e.g. coronary artery bypass), but overwhelmingly the literature has evaluated variations in common procedures. Recently, appropriateness defined by panels of physicians has been added as a variable in the assessment of variations, thus the distinction between area variation and quality assurance as study objectives is becoming blurred.

Levels of comparison: Almost half of the papers focus on within state or province variations. More recently there has been an increase in the number of papers using physicians as the unit of analysis.

Area definitions: Canadian studies tended to use well defined geographic boundaries (i.e. counties), while U.S. studies used hospital service areas or Standard Metropolitan Statistical Areas.

Data sources: North American and U.K. studies used government or administrative databases and calculated population based rates of utilization, while European studies tended to focus on individual hospital data or surveys.

Independent variables: The primary focus for most papers assessing relationships between procedure variations and other factors was supply factors (number of physicians or beds) and economic factors (measured either by insurance coverage or socio-economic status), since these are generally the easiest to measure. Physician characteristics such as specialty, and place or time since graduation, were examined in only a few studies.

A major problem was the lack of rigorous analysis, either because appropriate methods were not applied, or because the data were sufficiently limited to make such analysis impossible. A second problem was the limited assessment of patient demand factors (either defined as need on the basis of morbidity patterns across communities, or expressed demand as measured by requests for procedures). This limitation has arisen primarily because the physician is generally perceived as the "gate keeper" or initiator of service demand, therefore it has been of greater interest to determine if physician supply characteristics influence utilization patterns.

Results

All studies found variations in utilization rates and, although this variation was somewhat greater for elective than non-elective procedures, the magnitude of the variation was, generally speaking, modest, on the order of one to three fold. Interestingly, variation within states or provinces was generally higher than that between states or provinces. While this may reflect the instability of small geographic area rates (especially if the data are for single years), inter-provincial or inter-state comparisons may obscure real variations since one is examining average rates. There has only recently been an adequate consideration of these statistical issues in the literature. International comparisons tend to show wide variations. The relationship between procedure variation and physician or bed supply was unclear; papers presented conflicting results both across procedures and for the same procedure in different jurisdictions. In general, however, no one factor could be said to account for a substantial proportion of variation in procedures and most studies concluded that several factors were operative.

An important observation emerging from this review is that procedures must be examined independently. The range of variation and extent to which various independent factors influence utilization patterns varies from procedure to procedure. In addition, the relationship between underlying patterns of morbidity and procedure utilization rates has been inadequately addressed.

Table 2,16 provides some estimates of rates from recently published papers for those procedures for which we could identify at least 10 papers meeting the criteria for Category IA and which presented data on individual procedures. An exception was made for coronary artery bypass which, although examined in only three papers, is of considerable interest to policy makers because of its increasing frequency, high cost and political impact in terms of claims regarding long waiting times.

Shortcomings

A host of methodological issues may contribute to the highly inconsistent nature of the results presented in Table 2,16, and to the general impression that dramatic rate variations exist. These problems are of two types; limitations which are problems noted by the authors and shortcomings which are problems we observed.

The most frequently encountered problems were: (1) failure to describe the methods of analysis in sufficient detail so that the nature of the analysis being undertaken was clear; and (2) failure to specify whether patients were counted more than once, particularly in papers describing rate variation for more than one procedure. Surprisingly, the third most common problem was a failure to test for statistical significance when differences in rates were observed. This is particularly critical since, as will be discussed in more detail below, small area analysis may involve small numbers of procedures within each area studied, thus producing unstable rates, and the problem of adequate power to detect a significant differences in rates. This latter problem is highlighted by the fact that a small number of papers utilized very small study populations. Other problems worth commenting on include the failure to utilize a denominator for rate calculations relevant to the objectives of the study, failure to include a comment (if not data) on all the independent variables stated to be included in the analysis, and failure to define clearly the geographic unit of analysis. Many of these problems are easy to overcome and every effort should be made to do so in future research.

DISCUSSION

Several impressions emerge from this review. First, although it appears large, the literature contains relatively few "good" papers, and thus the available information on geographic variations for specific procedures is sparse. Moreover, there are few recent "good" papers, thus it is not clear if the rate variations reported for the 1970s and early 1980s still exist; particularly in the United States after the introduction of prospective payment in the mid-1980s. Data on the effect of reductions in Medicaid fees for surgical procedures in Massachusetts, for example, would suggest that prospective payment may have had a dramatic effect, reducing some rate variations (e.g. tonsillectomy) and increasing others (e.g. disc surgery/spinal fusion).[5] Moreover, the direction of change varied geographically.

In addition, given the considerable inconsistency of results across papers, even for those which focus on a single procedure, it is apparent that no clear pattern of factors emerges which would guide efforts to reduce substantially the variations observed. This finding is not surprising with respect to patient characteristics since these were assessed relatively infrequently. However, it is surprising with regard to supply factors (beds, number of physicians, etc.) since these have generally been the

Table 2,16 Variation in Rates of Selected Procedures

Year	*Hernia Repair*	*Hemorrhoidectomy*	*Hysterectomy*	*Cholecystectomy*	*Cesarean*
1965	17.8 - 42.5 Regions in State Lewis (46)	11.4 - 34.6 Regions in State Lewis (46)		12.1 - 42.3 Regions in State Lewis (46)	
1965-1966					
1967-1968			21.3 - 47.2 International Vayda (92)	Vayda (92)	14.7 - 17◊ International
1969-1971	38 - 52 Regions in State Gittlesohn & Wennberg (34)		30 - 60 Regions in State Gittlesohn & Wennberg (34)	18 - 53 Regions in State Gittlesohn & Wennberg (34)	
1968 - 1972	**13.5 - 24.6** **Province** **Vayda et al. (97)**	**5 - 10.9** **Province** **Vayda et al. (97)**	**43.9 - 70.9** **Province** **Vayda et al. (97)**	**18.4 - 40.9** **Province** **Vayda et al. (97)**	
1971-1972	**31 - 85**** **Other** **Roos & Roos (72)**			**36 - 99**** **Other** **Roos & Roos (72)**	
1973	27.8 - 37.9 Regions in State Detmer & Tyson (26)	3 - 19 Regions in State Wennberg & Gittlesohn (106)	20.4 - 33.6 Regions in State Detmer & Tyson (26)	21 - 29 Regions in State Detmer & Tyson (26)	

	35 - 60 Regions in State Wennberg & Gittlesohn (106)		39 - 93 Regions in State Wennberg & Gittlesohn (106)	27 - 55 Female Regions in State Wennberg & Gittlesohn (106)	
1974			**41 - 202** **Regions in Province** **Stockwell &** **Vayda (89)**	**20.6 - 101.7** **Regions in Province** **Stockwell &** **Vayda (89)**	
1975	19 - 35* Other Wennberg et al. (101)		25 - 90* Other Wennberg et al. (101)	14 - 34 Other Wennberg et al. (101)	
1970-1975			78 - 114◊ Other Walker & Jick (99)		
1976				15.3 - 26 Regions in Nation Fowkes (31)	
1970-1976	14.2 - 24.3 International Vayda, et al. (96)		23.5 - 50.1 International Vayda et al. (96)	6.3 - 33.2 International Vayda et al. (96)	5.4 - 8.3* International Vayda et al. (96)

*	Not age standardized	+	Insurance groups	◊	Per 10,000 women at risk
**	65+ or 66+ years old	++	Occupational groups	**Bold**	**Canadian Study**
***	25+ years old	+++	Mean annual rate		
****	<15 years old	++++	Medicaid patients		

Table 2,16 continued

Year	*Hernia Repair*	*Hemorrhoidectomy*	*Hysterectomy*	*Cholecystectomy*	*Cesarean*
1974-1976			**47 - 128*** Regions in Province Roos (68)**		
1977			**33 - 68 Province Mindell et al. (55)**	**41.5 - 52.2 Regions in Province Cageorge et al. (15)** **21 - 30 Provinces Mindell et al. (55)**	**10.4 - 14.8 Provinces Wadhera & Nair (98)**
1973-1977			9.19 - 26.9 Regions in Province Vayda et al. (94)	24.3 - 51.4 Regions in Province Vayda (94)	3.6 - 13.8* Regions in Province Vayda et al. (94)
1974, 1977			34.5 - 198*** Regions in Province Cohen (20)		
1978	**14.5 - 18.9 Regions in Province Halliday & Le Riche (37**	**4.7 - 9.6 Region in Province Halliday & Le Riche (37)**	**42.2 - 72.8 Regions in Province Halliday & Le Riche (37)** 27.1 - 47.3 International Savage (81)	**21.4 - 31.7 Regions in Province Halliday & Le Riche (37)**	**37.2 - 66.1 Regions in Province Halliday & Le Riche (37)** 13.9 - 17.6 Regions in Nation Placek & Taffel (64)

1975-1978		2.6 - 3.2* Other Shwartz et al. (84)	1.51 - 2.94* Other Shwartz et al. (84)	0.91 - 1.76* Other Shwartz et al. (84)	
1980		6 - 8 Regions in State Griffith et al. (36)	43 - 46 Regions in State Griffith et al. (36)	19 - 24 Region in State Griffith et al. (36)	
1978-1980					8.6 - 13.3 Other Williams (110)
1979-1980					9.97 - 17.08 Other Evans (30)
1981	38 - 53** Other Chassin et al. (17)	13 - 31*** Other Chassin et al. (17)		34 - 52* Other Chassin et al. (17)	17.1 - 20.0 Regions in Nation Placek (65)
1979-1981	45.8 - 65.9 Other Schact (82)	13.7 - 24.5 Other Schact (82)	81.4 - 151.6 Other Schact (82)	34.3 - 58.6 Other Schact & Pemberton (82)	

*	Not age standardized	+	Insurance groups	◊	Per 10,000 women at risk
**	65+ or 66+ years old	++	Occupational groups	**Bold**	**Canadian Study**
***	25+ years old	+++	Mean annual rate		
****	<15 years old	++++	Medicaid patients		

Table 2,16 continued

Year	*Hernia Repair*	*Hemorrhoidectomy*	*Hysterectomy*	*Cholecystectomy*	*Cesarean*
1982					**17.1 - 20.2** **Regions in Province** **Anderson & Lomas (4)**
1977-1982					16.1 - 21.31 Other Haynes De Regt et al. (38)
1985					13.5 - 30.4 Other Acker et al. (1)
1983-1985			45.6 - 114.3 Other Coulter et al. (25)		
1986					15.6 - 29.1 Other Stafford (86)

Table 2,16 continued

Year	*Appendectomy*	*Prostatectomy*	*T & A*	*Cataract*	*CABS*
1965	14.6 - 61.8 Regions in State Lewis (46)				
1965-1966	21.7 - 22 male 18 - 22.3 female International Bunker (13)		32.2 - 63.7 male 32.1 - 64.1 female International Bunker (13)	47.2 - 65.3 male 69.1 - 82.5 female International Bunker (13)	
1966-1967	26.1 - 118.8+ 29.9 - 94.6++ Other Lichtner & Pflanz (47)				
1967-1968		72 - 183 International Vayda (92)			
1968		**14 - 26** **Province** **Mindell et al. (55)**	**53 - 120** **Province** **Mindell et al. (55)**		

* Not age standardized
** 65+ or 66+ years old
*** 25+ years old
**** <15 years old
+ Insurance groups
++ Occupational groups
+++ Mean annual rate
++++ Medicaid patients
◊ Per 10,000 women at risk
Bold **Canadian Study**

Table 2,16 continued

Year	*Appendectomy*	*Prostatectomy*	*T & A*	*Cataract*	*CABS*
1947-1971	8.1 - 26.5 Other Yoshida & Yoshida (113)				
1969-1971	14 - 31 Regions in State Gittlesohn & Wennberg (34)	15 - 32 Regions in State Gittlesohn & Wennberg (34)	23 - 122 Regions in State Gittlesohn & Wennberg (34)		
1968-1971	**17.3 - 30.3** **Provinces** **Vayda et al. (97)**		**42.7 - 97.9** **Provinces** **Vayda et al. (97)**	**4.36 - 9.56** **Provinces** **Vayda et al. (97)**	
1971-1972		***125 - 282**** **Other** **Roos & Roos (72)**		**31 - 131**** **Other** **Roos & Roos (73)**	
1969-1973			102 - 126.6**** Other Wennberg et al. (102)		
1973	11.9 - 25.6 Regions in State Detmer & Tyson (25) 11 - 22 Regions in State Wennberg & Gittlesohn (106)	18 - 40 Regions in State Wennberg & Gittlesohn (106)	24.4 - 39.9 Regions in State Detmer & Tyson (25) 23 - 122 Regions in State Wenberg & Gittlesohn (106)		

1972-1974			**80.8 - 163.6** **Regions in Province** **Roos et al. (74)**	
1974	**11.8 - 56.6** **Regions in Province** **Stockwell &** **Vayda (89)**		**23.2 - 191.1** **Regions in Province** **Stockwell &** **Vayda (89)**	
1974-1975				73 - 142& of the national rate Other Sanderson (78)
1975	10 - 28* Other Wennberg et al. (101)	13 - 42* Other Wennberg et al. (101)	11 - 61* Other Wennberg et al. (101)	
1973-1977	**11.6 - 42.8+++** **Regions in Province** **Vayda et al. (94)**	**4.8 - 36+++** **Regions in Province** **Vayda et al. (94)**	**21.0 - 89.2 +++** **Regions in Province** **Vayda et al. (94)**	
1975-1978			4.18 - 11.1+++ Regions in State Shwartz et al. (84)	

*	Not age standardized	+	Insurance groups	◊	Per 10,000 women at risk
**	65+ or 66+ years old	++	Occupational groups	**Bold**	**Canadian Study**
***	25+ years old	+++	Mean annual rate		
****	<15 years old	++++	Medicaid patients		

Table 2,16 continued

Year	*Appendectomy*	*Prostatectomy*	*T & A*	*Cataract*	*CABS*
1977		**16 - 25** **Provinces** **Mindell et al. (55)**	**37 - 68** **Provinces** **Mindell et al. (55)**		
1978	**16.3 - 29.3** **Regions in Province** **Halliday &** **La Riche (37)**	**17.4 - 25.7** **Regions in Province** **Halliday &** **La Riche (37)**	**32 - 57.4** **Regions in Province** **Halliday &** **La Riche (37)**	**6.5 - 14.4** **Regions in Province** **Halliday &** **La Riche (37)**	
1977-1979				15 - 26 male 22 - 43 female Bernth-Peterson & Bach (9)	
1979-1981	50 - 98.3 Other Schact & Pemberton (82)	38.7 - 61.1 Other Schact & Pemberton (82)	52.8 - 108.5 Other Schact & Pemberton (82)	29.0 - 65.5 Other Schact & Pemberton (82)	
1980	12 - 17 Regions in State Griffith et al. (36)	20 - 24 Regions in State Griffith et al. (36)	24 - 34 Regions in State Griffith et al. (36)		
1981	2 - 5** Other Chassin et al. (17)	57 - 97** Other Chassin et al. (17)	267 - 643** Other Chassin et al. (17)	120 - 180** Other Chassin et al. (17)	7 - 13** Other Chassin et al. (17)

1978-1984			**2.63 - 5.61+++** **Regions in Province** **Roos & Cageorge (70)**
1979-1985		**30 - 130**** **76 - 161**** **Other** **Roos et al. (75)**	**4.28 - 7.78** **Regions in Province** **Anderson &** **Lomas (6)**

*	Not age standardized	+	Insurance groups	◊	Per 10,000 women at risk
**	65+ or 66+ years old	++	Occupational groups	**Bold**	**Canadian Study**
***	25+ years old	+++	Mean annual rate		
****	<15 years old	++++	Medicaid patients		

This table and noted references is based on Table 4 in Report No. HPRU 90:24D, by the same authors and published by the Health Policy Research Unit of the University of British Columbia in December 1990.

focus of most papers assessing the effect of independent variables on rate variations. For example, although bed supply has often been assumed to be a major contributor to procedure rate variations, the maximum number of times this variable was examined for any one procedure was 14, and of these studies it was found significant in only 4. The inconsistency we found agrees with the observations of Paul-Sheehan et al. who concluded that results are inconsistent and conflicting, and that a combination of community and provider variables provided a better explanation of the observed variation than either type of variable when analyzed separately. We found data from the U.S. to be primarily concerned with insurance issues and thus of somewhat limited relevance to Canada.

Turning to a consideration of some general conceptual and methodological issues, few researchers have addressed the possibility that the observed variations in procedure rates may result from the non-comparability of data sources (an assumption often made in small area analysis that has not been tested,[6] the clarity with which the jurisdictions are actually defined,[7] the effect of the statistical analyses undertaken, or even the impact of random fluctuations of rates and their magnitude relative to the observed rates.[8] The issue of comparability arises most dramatically in international comparisons where a host of differing disease definitions, reporting procedures, and data sources may be involved. For example, in the United Kingdom primary procedures are reported, whereas in the United States primary, secondary and tertiary procedures are reported. The degree to which this differential reporting has resulted in higher observed rates in the United States has not been adequately addressed. Even in jurisdictions which may report more than one procedure, the protocols for which procedure to report as primary may vary, which will in turn give rise to differential rates. Sauter and Hughes[9] found, for example, differences in reporting protocols to have created rate variations on the order of six percent for the 10 procedures they describe, with about half of these procedures decreasing in frequency. Despite the fairly obvious potential for differences in methodology to influence reported rates of utilization, it is surprising how little attention is paid to the implication of this finding.

With regard to small area analysis, a popular topic in the late 1970s and early 1980s, the difficulties of small populations, noted above, raises questions regarding the stability of rates (and thus rate variations), the problem of power, and the fact that extreme variations may result from comparing high and low rate jurisdictions without adequate consideration of the overall distribution of rates and the degree to which the high and low rate jurisdictions are representative of the data. A similar problem is noted in specific subpopulations, like the elderly, where relatively small numbers may make rates unstable. Data from Roos et al.,[10] which consistently reveal quite high rates and rate variations across procedures, may in part reflect both the fact that they studied the 65 year plus population and the fact that some of the health regions in Manitoba have small populations overall. Morbidity is another factor which creates rate differentials and, although this factor has been assessed in a number of studies, we would agree with Joffe[11] (who commented on this phenomenon some time ago) that insufficient attention has been paid to its effect.

The issue of the appropriate denominator is critical, and we noted above that this was not an uncommon shortcoming in the papers we reviewed. A rate in epidemiology is defined as the number of events occurring in a population at risk for the event. Most studies use general population estimates as the denominator, and while this will not distort rate estimates for large population too severely, it will in small populations. The lack of major distortion arises because, relative to the population generally, the number of individuals who have procedures is small. However, for those procedures in which an organ is removed (e.g. hysterectomy), obviously the population at risk varies with the rate of the procedure since a woman cannot have two hysterectomies. Using general populations as denominators may, in fact, underestimate the rate variations because the denominator is actually inappropriately large.

Where rates and rate variations for jurisdictions are compared, it is important to undertake some form of age standardization since it is clear from the literature that the rate for some procedures is highly related to age (i.e. cataract surgery). The need for this epidemiological methodology is widely recognized, yet many of the early studies by Wennberg[12] fail to age standardize. This may account for the relatively high rate variations he observed which created the impression that significant rate variations existed for which there was no obvious explanation.

It cannot be assumed that large rate variations are in fact statistically significant. All variations should be tested. As noted above, rate comparisons across small areas may be unstable both across jurisdictions and over time. Diehr et al.[13] have recently observed, using computer simulations, that relatively large statistically significant rate variations can occur by chance.

The issue of the *ecologic fallacy* is important in this field. The ecologic fallacy arises when inferences based on data from one level of analysis are applied to a different unit of analysis. For example, considering appendectomy rates and socioeconomic status (SES), if one observes that high rates occur in *jurisdictions* of low SES and that low rates occur in *jurisdictions* of high SES, and although there may be a statistically significant negative association between these two variables, causal inferences are questionable because one does not know if it is the *individuals* with low SES who have a higher frequency of appendectomy; the unit of analysis is not the individual. Wilson,[14] for example, found a high correlation between median community income and overall community mortality rates and general surgery rates for patients over 65 years of age in Michigan, yet it remains unclear whether this correlation, particularly with regard to income, would be observed at the level of the individual. Thus, analyses seeking to associate various factors with rate variations must always be cognizant of this problem. Unfortunately, this is rarely the case.

Finally, as noted at the outset, McPherson et al.[15] raised the issue of clinical uncertainty as a factor in producing rate variations. They did not define this concept very precisely, thus it is not immediately apparent what is meant by it. However, clinical uncertainty has an intuitive explanatory appeal because no physician can be absolutely certain about any diagnosis, or the efficacy of any procedure in any patient (i.e. outcome). The data from Young et al.[16] on the differences between family practitioners and cardiologists in recommending coronary arteriography may reflect such differences in uncertainty. A related concept is practice style. This concept was first invoked by Wennberg and Gittelsohn[17] to explain small area variations in their early studies. In a sense it is, like clinical uncertainty, an explanation of exclusion. Moreover, Stano[18] found, in a study of physicians services in Michigan, that rate variations differences were not only dependent on practice style differences, but that these related to the number of physicians a patient saw, rather than intensity of service.

Given these conceptual and methodological issues, how then is one to judge the degree to which procedure rate variations are an important phenomenon worthy of attention and, if possible, action? In addition, how is one to interpret the magnitude of rate variations? Is a two fold variation (if valid) of sufficient importance to warrant action, or should one concentrate on rate variations of greater magnitude? For the 10 procedures we have reviewed in depth, it is clear that there is considerable variability in rate variations, some of which is immediately explainable (at least on a speculative basis since precise or relevant data are not available) and some of which remains obscure.

It is claimed for nearly all the procedures examined in the literature that the United States has higher rates than the United Kingdom (with Canada having rates somewhere in between), that rates for discretionary (elective) procedures are higher than rates for non-discretionary procedures, and that these variations arise primarily because of differences in supply variables. The main problem with this inference is that supply variables have been most often studied, but have not often been studied along with other variables, particularly differences in morbidity. Moreover, even given the relatively high frequency with which supply variables have been assessed, it is clear from Table 2,16 that there is a lack of consistency in the frequency with which these variables have been found to be significant and, even when found to be statistically significant, whether they really explain sufficient amounts of the observed variation to warrant attention.

In terms of patient characteristics, age seems to be a fairly consistent determinant of rates in that most studies examining surgical rates among the elderly (those 65 years of age or more) find both higher rates and wider rate variations than observed for general populations. Unfortunately, single studies rarely, if ever, looked at procedure rate variations in younger and older age groups using the same methodology, thus methodological issues may be in part responsible for the perception that older patients have higher rates. Nevertheless, since this population is at greater risk of untoward reactions to surgery, is growing in absolute numbers and in percent of the population, tends to stay in hospital longer, and may represent a group in whom clinical uncertainty is high, the general observation that this group has higher rates deserves attention.

Given the very large number of possible explanatory variables, and rate variations on the order of two to three fold, *we remain to be convinced that procedure rate variations have been clearly demonstrated to be a major problem* requiring urgent attention, or that policy relevant solutions are clearly available. However, we must also state that a concerted effort to continue to document and analyse such geographic variation is a valid policy objective. Moreover, we believe that there is an urgent need to assess the relationship between procedure rates, and morbidity and health status at the community level. This can either be done directly, which would require good morbidity estimates (often unavailable), or indirectly by assessing the appropriateness of procedures in individual patients. The indirect approach, as exemplified in the work of Winslow et al.[19] and Leape et al.,[20] provides, in our view, a better index of the meaning of rate variations since it affords not only a method for addressing the issue of clinical uncertainty or practice style, but allows for the assessment of supply and patient variables on appropriateness rather than simply overall rates.

It is interesting to note that these two studies come to different conclusions. Winslow et al. examined coronary artery bypass surgery in three hospitals for the years 1979, 1980 and 1982 and found that 44 percent of bypass procedures were either equivocally or clearly inappropriate, when evaluated against a comprehensive list of indications developed by a national panel of experts. In addition, the proportion of inappropriate procedures across the hospitals studied varied from 22 percent to 63 percent, but did not vary with patient age. Winslow et al. concluded that "eliminating the performance of inappropriate procedures *may* [emphasis added] lead to reduction in health care expenditures or to improved patient outcomes". Leape et al., using a similar methodology in a paper that was published two years later, examined procedure variations for coronary angiography, coronary endarterectomy, and upper gastrointestinal endoscopy among Medicare beneficiaries in 1981. The proportions of inappropriate procedures varied widely by procedure (coronary angiography 8 to 75 percent, coronary endarterectomy 0 to 67 percent, and endoscopy 0 to 25 percent). Twenty-eight percent of the variance in coronary angiography rates across counties was accounted for by inappropriate use, but inappropriateness was stated not to be significantly related to rate variations in the other two procedures studied.

Given these mixed results (derived from differing patient populations), there is an urgent need to continue the development of appropriateness criteria.

Another outstanding issue in the rate variations literature concerns whether low or high rates are "good" or "bad" (an important question in and of itself) and whether this interpretation of rates varies across populations. For example, Wennberg et al.[21] recently noted in his study of medical and surgical variations in New Haven and Boston, that medical procedures had higher rates of variation than surgical procedures among patients 65 years of age and older. Thus, the issue of low rates representing underutilization and high rates representing overutilization may be more germane to medical than surgical procedures, especially among the elderly in whom we have already noted a fairly consistent pattern of higher rates and greater rate variations.

CONCLUSION

It is critical that research focus on individual procedures. Many papers assess variations in many procedures and it is not always clear in the analyses which variables are associated with rate variations for which procedure. Moreover, in papers reporting variations for more than one procedure, the actual relationships between variables and procedure variations may be mixed, leading to complexity in reporting and confusion in interpretation. Assessing single procedures both simplifies the analysis and clarifies the conclusions.

Such a research agenda is a departure from the supply focus of much of the research to date which has failed to yield any consistent patterns of supply effect (either of physician or beds). Although Roemer's Law ("a built bed is a filled bed") may be valid, and despite the intuitive appeal of the relationship between number of physicians (or surgeons) and rate of procedures, these effects seem to be neither large nor consistent; thus other explanations must be sought. Moreover, it is important from a policy perspective that assessment of procedure variations account for underlying estimates of need in addition to supply side factors since altering supply, which may be relatively easy to do, might in fact be inappropriate even though supply factors may be associated to some degree with the observed variations. For example, consideration should be given to estimating the variation in rates of procedures that *should* have been done, but were not. Since most commentators observe that rate variations *per se* tell us nothing about appropriateness,[22] variation in legitimate unmet need should be an integral part of future research: altering supply factors without concern for the critically important issue of appropriateness may well do more harm than good.

Finally, a host of geographically related variables need consideration in future research. For example, proximity to services, defined as an issue of access (i.e., resulting either from governmental, or other payer, restrictions in the number or distribution of relevant resources, or perception of quality), and cultural effects on utilization pattern may affect procedure variation rates but have been neglected in the research we have reviewed.

In summary, given the substantial Canadian contribution to the procedure variations literature, it is our view that Canada is in an excellent position to continue to build a significant research agenda in this area. Although we have noted methodological limitations and shortcomings in both the international and Canadian literature to date, most of the these problems are relatively easy to overcome. We believe that the Canadian health care system is an ideal setting for the continued development of procedure variation methodology, particularly the assessment of appropriateness. The principles of comprehensiveness, universality, portability, and accessibility enshrined in the Canada Health Act create the perfect laboratory for the evaluation of those structural and functional factors which may influence rate variations, other than individual ability to pay, and thus which maintain the focus on the main policy issues underlying our collective interest in an efficient and effective health care system.

REFERENCES

[1] McPherson, K., Epstein, A. and Jones, L. 1981. Regional Variations in the Use of Common Surgical Procedures: Within and Between England and Wales, Canada, and the United States of America. *Soc. Sci. Med.* 15A, pp. 273-288.

[2] Paul-Shaheen, P., Clark, J. and Williams, D. 1987. Small Area Analysis: A Review and Analysis of the North American Literature. *Journal of Health Politics, Policy and Law* 12(4), pp. 741-807.

[3] Copenhagen Collaborating Centre for the Study of Regional Variations in Health Care. 1985. *Bibliography on Regional Variations in Health Care.* Geneva: World Health Organization; and, Copenhagen Collaborating Centre for the Study of Regional Variations in Health Care. *Number 2: Bibliography on Regional Variations in Health Care.* Geneva: World Health Organization.

[4] Paul-Shaleen et al., *op. cit.*

[5] Shwartz, M., Martin, S.G., Cooper, D.D., Ljung, G.M., Whalen, B.J. and Blackburn, J. 1981. The Effect of a Thirty Percent Reduction in Physician Fees on Medicaid Surgery in Massachussets. *American Journal of Public Health* 71(4), pp. 370-375.

[6] Sauter, V. and Hughes, E. 1983. Surgical Utilization Statistics: Some Methodologic Considerations. *Medical Care* 11(3), pp. 370-377.

[7] Diehr, P., Cain, K., Connell, F. and Volinn, E. 1990. What is Too Much Variation? The Null Hypothesis in Small-Area Analysis. *Health Services Research* 24(6), pp. 740-771. See also Paul-Shaheen, *op. cit.*

[8] Diehr, P. 1984. Small Area Statistics: Large Statistical Problems. *American Journal of Public Health* 74(4), pp. 313-314. See also Diehr, *ibid.*

[9] Sauter and Hughes, *op. cit.*

[10] Roos, N.P. and Roos, L.L. 1981. High and Low Surgical Rates: Risk Factors for Area Residents. *American Journal of Public Health* 71, pp. 591-600.

[11] Joffe, J. 1979. Mobility Adjustments for Small Area Utilization Studies. *Inquiry* 16, pp. 350-355.

[12] Wennberg, J.E. and Gittelsohn, A. 1975. Health Care Delivery in Maine I: Patterns of Use of Common Surgical Procedures. *Journal of the Maine Medical Association* 66(5), pp. 123-130, 149; and, Wennberg, J.E. 1979. Factors Governing Utilization of Hospital Services. *Hospital Practice* 14, pp. 115-121.

[13] Diehr et al., *op. cit.*

[14] Wilson, P.A. 1981. Hospital Use by the Aging Population. *Inquiry* 18, pp. 332-44.

[15] McPherson et al., *op. cit.*

[16] Young, M.J., Fried, L.S., Eisenberg, J., Hershey, J. and Williams, S. 1987. Do Cardiologists Have Higher Thresholds for Recommending Coronary Arteriography than Family Physicians? *Health Services Research* 22(5), pp. 623-635.

[17] Wennberg and Gittelsohn, *op. cit.*

[18] Stano, M. 1986. A Further Analysis of the "Variations in Practice Style" Phenomenon. *Inquiry* 23, pp. 176-182.

[19] Winslow, C.M., Kosecoff, J.B., Chassin, M., Kanouse, D.E. and Brook, R.H. 1988. The Appropriateness of Performing Coronary Artery Bypass Surgery. *Journal of the American Medical Association* 260(4), pp. 505-509.

[20] Leape, L.L., Park, R.E., Solomon, D.H. et al. 1990. Does Inappropriate Use Explains-Area of Variations in the Use of Health Care Services? *Journal of the American Medical Association* 263, pp. 669-672.

[21] Wennberg, J.E., Freeman, J.L., Shelton, R.M. and Bubolz, T.A. 1989. Hospital Use and Mortality Among Medicare Beneficiaries in Boston and New Haven. *New England Journal of Medicine* 321(17), pp. 1169-73.

[22] See, for example, McPherson et al., *op. cit.*, and Paul-Shaheen et al., *op. cit.*

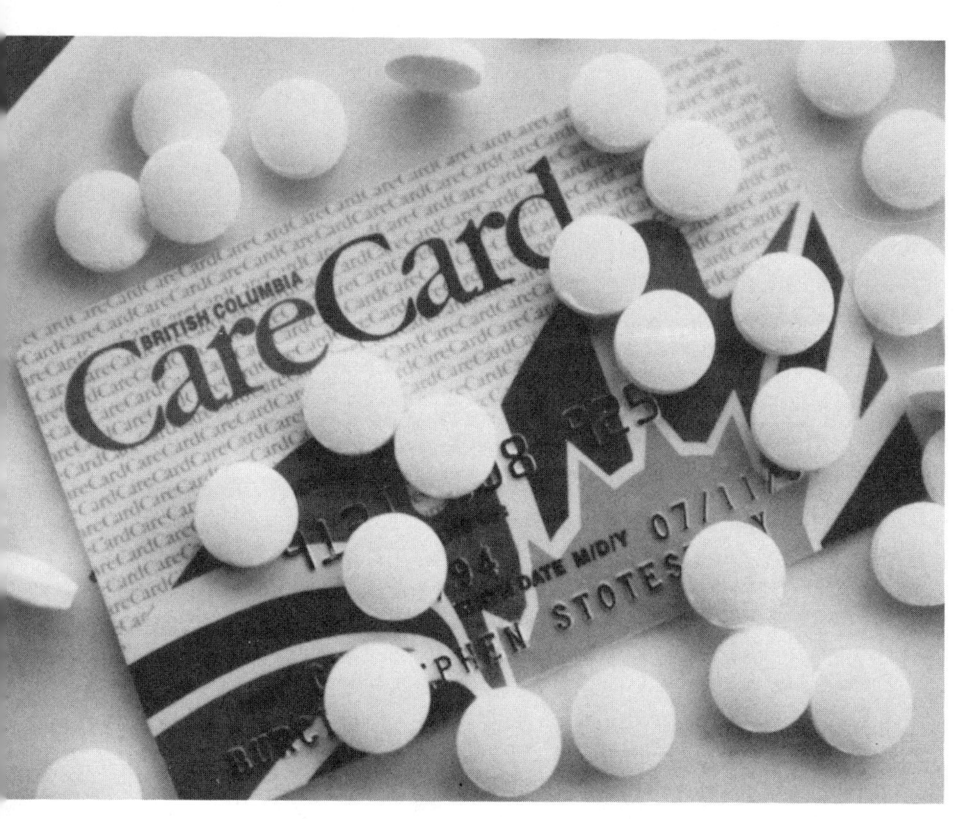
BRITISH COLUMBIA
CareCard

17 THE CANADIAN AND AMERICAN HEALTH CARE SYSTEMS FROM A VALUES PERSPECTIVE

Jonathan D. Mayer

Geography, Medicine (Infectious Diseases), Family Medicine, and Health Services
University of Washington

INTRODUCTION

To the average citizen of the United States, the Canadian health care system may appear to be appealing indeed. Recent survey research indicates that nearly two thirds of the respondents to a systematic survey indicated a preference for the Canadian health care system over that found in the United States when asked the question:

> In the Canadian system of national health service, the government pays most of the cost of health care out of taxes and the government sets all fees charged by doctors and hospitals. Under the Canadian system -- which costs the taxpayer less than the American system... -- people can choose their own doctors and hospitals. On balance, would you prefer the Canadian system or the system we have in the United States?[1]

The research was done jointly by the Harvard School of Public Health and by Louis Harris and Associates in 1988. It was validated in 1990, and included over 2000 adults; the error at the 95 percent confidence level was approximately plus or minus 4 percent. While the respondents may have been unaware of some aspects of the Canadian system that have proven to be problematic, there is impressive support in the United States for a system similar to that of Canada, at a time when there are major barriers to access to health care in the United States for a significant proportion of its citizens -- nearly 40 million lack health insurance -- and at a time when the national cost of health care is approaching 12 percent of the Gross National Product. In a related survey of 10 developed countries, the greatest dissatisfaction with the health system was in the United States, and the least dissatisfaction was in Canada.[2] Indeed, it appears that the Canadian health care system distinguishes itself as the most popular social program in Canada.

This chapter assumes that the reader is familiar with the basic structure of the American and Canadian health care systems -- the U.S. system being an amalgam of fee-for-service and government programs for the elderly and the poor, and the Canadian system being a series of joint federal-provincial programs of universal entitlement designed to provide easy access by all Canadians to inpatient and outpatient care, as well as preventive and domiciliary services. Given this understanding of the basic differences in structure, how can the evolution of two very different systems in two similar although subtly different countries be explained?

It is the author's contention that the answers to this question lie deeply within the social fabric and tapestry of ideologies found within the two countries. The health care systems provide not only health care, but also provide views into the deeper structure of the two societies. Viewed in this way, health care systems are socially constructed and culturally interpreted sets of institutions. Thus, this chapter will not discuss differences in the micro-costing of services, the applicability of DRGs (diagnostic related groups) to Canada, or other topics which are hot issues in the health services research community in both countries. Rather, it will seek to explain the differences in the two systems as cultural artifacts. Finally, this chapter will explore why the Canadian system seems so desirable by an overwhelming majority of the U.S. citizenry.

The reasons that the two countries have dissimilar health care systems are threefold. First, the systems are reflections of different value systems concerning the role of government in service provision; second, they reflect historically differing political climates; and third, they reflect the historically different power of political interest groups in the two countries as the systems evolved.

VALUES AND THE HEALTH CARE SYSTEMS

As social institutions, health care systems can be expected to be influenced by any of the same factors which influence other social institutions. Some of these are ideology, political power and its distribution, social organization, and historical tradition. It is in this framework that a partial understanding may be gained concerning the differences between the two health care systems.

There are numerous definitions of individual and social values. This is not the place to delve into these differences. A working definition of values developed by this author is: *"those individually or socially held beliefs which generate moral judgment, and give meaning to individual actions or socially constructed institutions"*. It should be noted that economists have attempted to define values in measurable terms based upon behavior oriented towards utility maximization, and that sociologists and social psychologists have attempted to measure values based on beliefs and behavior which transcend the simple maximization of utility. It is notable, however, that there are few value-oriented discussions of health care systems. Indeed, in the morass of esoterica dealing with some of the fundamental fundable issues in health services research, there is a vacuum of such studies.

Several aspects of the value systems of Canada and the United States are different. These differences are unseen by the traveller, and even by the health care analyst who seeks to explain the operational details of the health care systems. Political sociologist Seymour Martin Lipset frames his analysis of values and institutions in Canada and the United States as follows:

> Americans do not know but Canadians cannot forget that two nations, not one, came out of the American Revolution. The United States is the country of the revolution, Canada of the counterrevolution. These very different formative events set indelible marks on the two nations. One celebrates the overthrow of an oppressive state, the triumph of the people, a successful effort to create a type of government never seen before. The other commemorates a defeat and a long struggle to preserve a historical source of legitimacy: government's deriving its title-to-rule from a monarchy linked to a church establishment. Government power is feared in the south; uninhibited popular sovereignty has been a concern to the north.[3]

This statement has many implications. The anti-statist ideology of the United States has resonated through many election campaigns, as embodied in George Wallace's cry for "state's rights" and the associated independence from federal influence. Populism also is derived from the historical origins of the United States in its emphasis on individualism and its distrust of government. Indeed, the vehement opposition of many against any tax increases can be seen as deriving not only from economic motives, but as a symbolic resistance against government incursion into the daily lives of citizens. The issue of potential tax increases is frequently used as justification against "socialized" medicine such as that found in Canada. While "socialized" has become a buzzword in the lexicon of those who oppose greater government involvement in public service provision, it has also developed the political and ideological overtones of resistance to *becoming socialist*, a step on the road to Communism, which is seen by many as the logical consequence of "socialized medicine". Communism has been so feared by most of the American public because it is seen as threatening the basic value structure of society, particularly individualism. It is for this reason -- to avoid the ideological closemindedness against that which is "socialized" -- that the term "national health insurance" seems much more accurate in the long run, and certainly more acceptable to those who are ideologically committed to the present system.

Lipset further suggests that the history of the United States has led to a state of "rule derived from the people", with its emphasis on individualism, while the history of Canada led to a desire to build free institutions within the more paternalistic framework of a monarchical state. While the United States, according to Lipset, is "unique among developed nations in defining its raison d'etre ideologically", the respected American historian Richard Hofstadter notes that "It has been our fate as a nation not to have ideologies but to be one".[4]

Canada, on the other hand, is not ideologically defined, but maintains its roots in a shared sense of the past and of identification with the monarchy. Canadian historian Kenneth McNaught argues that this emphasis on monarchy leads to "an abiding respect for continuity", and "continuity as the basis of legitimacy".[5] Lipset suggests that the clear accent on ideology, and on the ideology of individualism in the United States, is exactly akin to Winston Churchill's characterization of a country defined by a sense of history, and one defined by ideology. Lipset's interpretation of this is that "In Europe and Canada, nationality is related to community; one cannot become un-English, or un-Swedish. Being an American, however, is an ideological commitment. It is not a matter of birth. Those who reject

American values are un-American".[6] There are two relevant implications of this statement. The first is that American radicalism has tended more towards anarchy than toward collectivism, and the second is that anti-statism, as the political expression of individualism, is part of the American ideology.

Canada, on the other hand, is more open to collectivism and to a collective identity. The often quoted statement of Canadian historian William Stahl embodies much of the contrast between the Canadian ideology and the American ideology: while the "Fathers of Confederation spoke of peace, order, and good government", the Founding Fathers of the United States spoke of "life, liberty, and the pursuit of happiness".[7] This is deeply embodied in Robert N. Bellah et al.'s *Habits of the Heart: Individualism and Commitment in American Life* -- a profound and searching exploration of the role and depths of individualism in the United States.[8]

Put in these terms, it is not so difficult to understand why, in the United States, a chaotic system of public and private medicine exists, with any suggestion of government financing being interpreted by many as "socialized medicine". A collective and monarchical identity actually allows more freedom in institution-building than does the dogged cum ragged individualism of the United States. New institutions may be proposed, debated, and built without their becoming seen as threats to the prevailing ideology. In the United States, radically new institutional forms can be seen as threats to the American ideology, and thus to the state itself. U.S. medicine therefore remains structurally similar over the years, in spite of crying needs for both accessibility and cost containment. In Canada, however, a radically new health care system could be allowed to develop without any fundamental threat to the Canadian identity, but rather as a responsibility of the collective identity -- to provide access to health care as an essential component of a humane state, and as part of the social contract. This is not the place to trace the philosophical and ethical justifications for health care as a right -- this has been done admirably, based upon Rawls' theory of justice, by Norman Daniels in *Just Health Care*[9] and in other places. Suffice it to note that the argument that health care is a right is steeped in the notion that somehow health care is "special" as a social service, and that proof of this argument is emotionally easy but analytically complex.

The whole point of this analysis is that there can be more freedom in institution-building in a system which does not conform to an ideological imperative, such as that of individualism, but rather in a state that is not defined by several maxims. With the anti-statist tradition in the United

States, it becomes more obvious why, historically, there has been so much resistance to centrally-financed health care: it would be contrary to one of the major tenets of the American ideology and would be taken not only as a symbolic thereof, but would also threaten some of the interest groups which have developed in health care as a result of this particular value system. Three of the most influential are the American Medical Association, the American Hospital Association, and the Health Insurance Association of America, although all have recently acknowledged the need for improved access to health care for all citizens of the United States.

Improved access for the uninsured does not necessarily entail fundamental structural change in the ways in which health care is organized and provided, but could be accomplished in a fee-for-service system (as the Canadian and other examples have proven) without centrally financing *all* health care. Rather, some guarantee could be provided only for those who lacked insurance, or did not qualify for some preexisting government program such as Medicaid and Medicare, and this could be accomplished through some sort of means testing. Indeed, it is exactly this position that many influential parties, themselves artifacts of the U.S. value system, have taken. One of the most powerful professional associations, the American College of Physicians, representing specialists in internal medicine, has taken the position that insurance coverage must be expanded for the underinsured and for the uninsured, but within the existing framework of health care financing, i.e. they have advocated patchwork systemic change without comprehensive change and centralization:

> A nationwide program is needed to assure access to health care for all Americans, and we recommend that such a program be adopted as a policy goal for the nation. The College believes that health insurance coverage for all persons is needed to minimize financial barriers and assure access to appropriate health services...
>
> A comprehensive and coordinated program to assure access on a nationwide basis is essential. In the near term, given the urgency of the need, it should build on the strengths of existing health care financing mechanisms. In the longer term, careful consideration of new and innovative alternatives, including some form of nationwide financing mechanism, will be necessary.[10]

Some of the more radical physicians in the United States have recently called for a national insurance plan,[11] but there is no evidence that this group represents the majority of practitioners; moreover, it is easy to exaggerate the influence of practitioners or groups of practitioners within the context of all of the parties and interest groups which influence contemporary health care in the United States.

The Canadian health care system is itself the product of numerous actions and bills at the provincial and then the federal level. Under the *Medical Care Act*, the four criteria which had to be met to qualify for federal cost-sharing are very revealing about values in Canada. These four are universal access and coverage for health care broadly taken; inter-provincial portability of benefits; comprehensiveness of coverage for all necessary services; and a federally administered non-profit program. The *Canada Health Act* of 1984 prevented "extra billing" by physicians, over and above provincially negotiated rates.

Universality is itself a significant statement to serve as the basis of a national health system. It means that other differences aside, there is a social contract between the citizens of a nation and the provision of medical care services -- services that are somehow "special", according to bioethical analysis. It means that there should be no exclusion of social groups or sub-populations, and that all citizens should have access to the health care system. In this respect, Canada is not unique, but shares this feature in common with most industrialized nations.

This goal was quite apparent in the disputes over "extra billing" whereby some physicians were charging their patients in excess of the provincially negotiated rates. This practice is no longer legal and, as John K. Iglehart, an astute observer of health policy has noted, its exclusion is rooted in the "strong belief that all citizens should have equal access to medical care, regardless of ability to pay. In essence, Canadian policy says that simply because people can afford to pay, they should not be able to purchase care that is better or more readily available to the less well off".[12] Writing on the same topic, Robert G. Evans, a noted authority on health care policy in general and on the Canadian system in particular, argues that the Canadian population has a deep suspicion of any elitist or class-based system in general. He notes that there are few private schools in Canada, for example. In short, as Iglehart also notes, in the words of Evans, "equality before the health care system" is as important a value as equality before the law.[13] In this sense, the Canadian system is even more egalitarian than the environment surrounding the government-run National Health Service in the United Kingdom, because in that system nearly 10 percent of the population has private insurance which may be spent in the private sector of the medical system, either in NHS facilities using NHS staff, or in completely private facilities using staff which has no connection to the NHS.

As a goal, there should be no geographical discrimination in access so that, for example, rural residents should not be at any significant disadvantage vis-a-vis urban residents. In point of fact, this is an unattainable goal, so there is an often quoted understanding that rural residents have the "right to be at risk"; i.e., the right to place themselves in a relatively disadvantageous location with respect to access to the health care system. This concept is very egalitarian as far as it goes, but it does contain some implicit elements of the bioethical concept of "informed consent", in which the physician provides the patient with all relevant information concerning a treatment or procedure, and the patient then makes an informed choice among treatment options. The right to be at risk assumes that all geographically remote populations have considered the pros and cons of where they are living, and have made an informed decision that they would rather live in a rural area than in an urban area, in spite of the greater health threats that such a decision may imply, given the relative lack of accessibility to health care services in rural areas -- especially tertiary services. As a goal, though, equality bespeaks the equitable aims of the Canadian health care system, the legislation that created the current system, and the people who support this system. In not meeting the goals of spatial equality, again Canada is not alone. No medical care system has or can provide total equality of access because of low population density of demand in rural areas. This low density in turn leads to an uneven distribution of facilities because of the economically mediated need to allocate scarce resources.

The portability of benefits from province to province is a realistic recognition of the fact that Canadian society is highly mobile, and that even if an individual moves from one province to another, there should be transferability of benefits. In a sense, this may be seen as an attempt to ensure relative uniformity of benefits from one province to another, although it theoretically does allow for some inter-provincial variation of benefits which is realized in fact.

Comprehensiveness of coverage is another egalitarian concept, though one that has proven to be troublesome in implementation, particularly for tertiary, specialized referral services. The implication of this provision is that all medically necessary services should be covered. This leads to two problems, one of them theoretical, and the other operational. The theoretical problem is in the definition of which services are medically necessary. The State of Oregon has done this as part of its rationing scheme in the allocation of medical resources and has not included, for example, cardiac transplantation as a medically necessary service. The reason for this is that it is a capital-intensive and costly set of services, a set which is thought to be less vital than prenatal care or childhood immunization. Just what is necessary and what is not necessary? Under what conditions? How is necessity defined? What about those procedures for which there is no clear consensus in terms of their efficacy, either in prolonging life or in terms of improving the quality of life? These continue to be problems in the Canadian system as well as health care systems of other industrialized countries.

The operational problem with the provision for the comprehensiveness of services has been operational, and this is the single most common controversial point about the Canadian health care system, both within Canada, and in the United States, when the Canadian model is considered as an alternative to the U.S. model. The problem is one of queueing. When resources are constrained, as they are in every medical care system in the world, supply of medical care services is limited at some point. Just as was found in Oregon, these constraints are felt most poignantly at the highly specialized centers which perform costly procedures. Because of economic constraints, those needing costly procedures such as cardiac and other transplantation may need to wait some time because of the queueing problem. Indeed, in 1990, the Ministry of Health in British Columbia contracted with two medical centers in Seattle to provide open-heart surgery on some of the patients in the province awaiting the procedures, since the waiting time had grown to be intolerably long. In sum, it is much easier to provide in practice for the comprehensiveness of coverage when referring to primary care and procedures which are not costly, than it is to provide comprehensiveness of service for very costly procedures whenever there is a budgetary constraint, which there always is in every country. Nonetheless, the intent of this provision speaks for itself.

The provision of the *Medical Care Act* that provides for a federally financed, nonprofit medical care system is also testimony to the egalitarian basis of this ironically monarchical society. The historical tradition of collectivism in Canada contrasts blatantly with the prevailing ideology in the United States, although the fact that the aforementioned survey research shows that most American citizens would prefer the Canadian system to the U.S. system blurs this contrast. However, there is no evidence that the now popular preference for a Canadian-style system in the United States is out of social concern and a commitment for egalitarianism for health care. The survey question cited at the beginning of this chapter describes the bare outlines of the Canadian system but makes no mention either of the resource allocation problems, or of the fact that a major point of emphasis in Canada is on egalitarianism. This author suggests that the resource allocation problems and relatively slow diffusion of high technology would not appeal to the citizenry of the United States, and that the egalitarianism of the Canadian system would be regarded as being irrelevant -- another testimony to the fact that there is more room for innovation in a society with a strong collective identity versus one which is defined by the very ideologies of individualism and the ensuing dominance of the private entrepreneur in every sector of society, including health care. Indeed, the way in which the original question was worded almost certainly determined its answer, for who could object to a system which provides high quality health care at a lower cost than in the United States, and especially at a tax rate lower than that of the United States?

This chapter is an analysis of some of the implications for human values of the differences between the health care systems found in the United States and Canada, and indeed is based upon the predicate notion that health care systems, as social institutions, cannot be anything other than reflections of the value systems of their respective societies. This is what Navarro, in discarding the value-oriented approach, argues is the "popular choice" explanation for why some countries have national health programs while the United States lacks such a program. Ultimately, Navarro argues, this approach suggests that the answer to his question is that the citizens of the United States do not want a national program.[14] Yet, as Navarro is quick to highlight, "The majority of the U.S. population supports a national health program and has supported it for quite awhile".[15] Navarro, after careful analysis, rejects the popular choice explanation because it is "limited, when not faulty". Further, Navarro argues persuasively that it is a misconception among the popular choice theorists that values shape institutions, rather than the other way around:

> To claim that popular values explain reality is to beg the question of how these values appear and are reproduced in these societies. In other words, the first assumption...is that there is a free market of ideas, all competing on an equal footing for the eyes and minds of our population... Values are produced and reproduced within highly controlled political environments by the promotion of values favorable to the powerful and by repression of values perceived to be threatening to them.[16]

Navarro's analysis adds to the exploration of values in relation to health care in the sense that he delves more deeply into values and explores the question of how they are generated. Indeed, as Navarro suggests, values do not exist autonomously but are created by those with a vested interest in those values. A more conventional Marxist interpretation might hold that actual institutions are created by those groups vested with power, whereas Navarro is suggesting a more profound penetration of the popular mind, whereby it is not only the institutions which are created by the "ruling class", but indeed that the mode of thought and the underlying sentiments are also generated by those groups in power. It is ultimately satisfying to question what gives rise to values, for it is not enough merely to state that health care systems are creatures of the dominant social values, for what creates and nurtures those values in the first place? How is power legitimized in society, and what are the relationships to dominant social institutions? In the end, what does give adequate explanation to the question of how institutions arise and are perpetuated, and undergo modification? Navarro argues:

> ...to understand the causes for this diversity [in the organization of health care] we have to shift our analytical paradigm: 1) we must look at the forest (the powers in society) rather than the trees (the different visible actors within the institutions of medicine); and 2) we must realize that power is not only distributed according to region, race, gender, or professional interest groups, but also and primarily according to class.[17]

This chapter is not a critique of the Marxist paradigm; however, aren't the powers and the forces that generate, maintain, and change institutions more diverse than class alone? Navarro would argue that all else is a product of class, and bases his ultimate argument of why the United States has no national health program while other developed nations do, within a class approach. In particular, Navarro justifies his argument that the reason that there is a national health insurance system in Canada was that the New Democratic Party, in reality a socialist party, was influential and instrumental in the establishment of the premier insurance program in the Saskatchewan Medical Care Insurance Plan. According to Navarro, Canada has had a strong and influential labor party and strong labor unions, and "The different types of funding and organization of health services are explained primarily by the degree to which the differing class aims in the health sector...have been achieved through the realization of class power settings".[18]

This chapter began with the argument that health care systems are consequences of, and reflect, the underlying social values of the societies in which they are constructed. It ends by acknowledging that this is a valuable question, but also by recognizing that the question may be too limited, for Navarro is quite correct in his saying, in effect, "fine, health care systems reflect the values of the societies in which they occur. But what gives rise to those values in the first place?" This author rejects the notion that it is only the class interests of society that shape social institutions. Future research on the social context of medicine and health care must come to grips with these questions, while at the same time realizing that health care systems should be viewed only in context: social context, and the organizational context which argues that health care systems are social institutions in the same way as are many other social institutions, though with powers that can touch upon the beginning of life to the end of the process of death.

REFERENCES

[1] Blendon, R.J. et al. 1988. Satisfactions with Health Systems in Ten Nations, *Health Affairs*, p. 186.

[2] *Ibid.*, p. 188.

[3] Lipset, S.M. 1990. *Continental Divide: The Values and the Institutions of the United States and Canada*. New York: Routledge, Chapman and Hall, Inc., p. 1.

[4] Quoted in Lipset, p. 19.

[5] McKnaught, K. 1984. Approaches to the Study of Canadian History, *Annual Review of Canadian Studies* 5, p. 89.

[6] Lipset, p. 19.

[7] Quoted in Lipset, p. 43.

[8] Bellah, R. et al. 1985. *Habits of the Heart: Individualism and Commitment in American Life*. Berkeley: University of California Press.

[9] Daniels, N. 1985. *Just Health Care*. Cambridge: Cambridge University Press.

[10] American College of Physicians. 1990. Access to Health Care, *Annals of Internal Medicine* 112, p. 643.

[11] Himmelstein, D.U. and Woodhandler, S. 1989. A National Health Program for the United States: A Physicians' Proposal, *New England Journal of Medicine* 320, pp. 102-108.

[12] Iglehart, J.K. 1990. Canada's Health Care System Faces its Problems, *New England Journal of Medicine* 322, p. 562.

[13] Evans, R.G. 1988. 'We'll Take Care of it for You' Health Care in the Canadian Community, *Daedalus* 117, pp. 155-189.

[14] Navarro, V. 1989. Why Some Countries Have National Health Insurance, Others Have National Health Services, and the United States Has Neither, *Social Science and Medicine* 28, pp. 887-898.

[15] Navarro, *op. cit.*, p. 888.

[16] *bid.*

[17] *Ibid.*, p. 890.

[18] *Ibid.*, p. 892